HVAC TESTING, ADJUSTING, AND BALANCING MANUAL

Third Edition

Sponsored by National Environmental Balancing Bureau

John Gladstone, B.S., M.A.
W. David Bevirt, P.E., CMS

Boston, Massachusetts   Burr Ridge, Illinois
Dubuque, Iowa   Madison, Wisconsin   New York, New York
San Francisco, California   St. Louis, Missouri

Library of Congress Cataloging-in-Publication Data

Gladstone, John.
    HVAC testing, adjusting, and balancing manual / John Gladstone,
W. David Bevirt.—3rd ed.
        p.   cm.
    Rev. ed. of: Air conditioning testing, adjusting, balancing / John
Gladstone. 2nd ed. c1981.
    Includes index.
    ISBN 0-07-024184-8
    1. Air conditioning—Equipment and supplies—Testing—
Handbooks, Mechanical Engineering
manuals, etc.   I. Bevirt, W. David.   II. Gladstone, John. Air
conditioning testing, adjusting, balancing.   III. Title.
TH7687.7.G58   1996
697.9'3'0287—dc20   96-18012
                                                              CIP

Sponsored by National Environmental Balancing Bureau

McGraw-Hill

*A Division of The **McGraw·Hill** Companies*

        3 4 5 6 7 8 9   BKM BKM   9 0 9 8 7 6 5 4 3 2 1 0

ISBN 0-07-024184-8

*The sponsoring editor for this book was Robert Esposito, the editing
supervisor was Pro Image and the production supervisor was Donald
F. Schmidt. It was set in Century Schoolbook by Pro Image.*

McGraw-Hill books are available at special quantity discounts to use
as premiums and sales promotions, or for use in corporate training
programs. For more information, please write to the Director of
Special Sales, McGraw-Hill, 11 West 19th Street, New York, NY
10011. Or contact your local bookstore.

Contents

Foreword

In the past, when environmental or heating, ventilating, and air conditioning (HVAC) systems were less complex and when system designers, owners, and occupants of indoor environments were less demanding, system balancing was considered to be a simple adjustment of some dampers in the ductwork and perhaps a few valves in the piping. Systematic measurements with test instruments were scarcely, if ever, undertaken.

Later, some serious attempts were made to balance HVAC system airflow with little or no thought to the hydronic flow. In recent years HVAC installations have become sophisticated systems of complex environmental control, not only for human comfort in schools, hospitals, auditoriums, offices, and hotels, but for a variety of industrial processes as well. Testing, adjusting, and balancing (TAB) personnel can no longer be just instrument readers; they must have a comprehensive knowledge of system function fundamentals, theory of fluid flow, heat transfer, psychrometrics, and state-of-the-art control systems.

Today's TAB specialists must have an exhaustive understanding of indoor air quality (IAQ) and be certified to perform sound and vibration measurements as well as cleanroom performance testing. Patently, system designers and design engineers must be responsible to provide the proper equipment and accessibility to that equipment, so that HVAC systems can be adequately tested, adjusted, and balanced to meet the expectations of today's owners and users who have become better educated and critical consumers.

Unfortunately, published work in the general TAB field is still somewhat meager, and the educational community has not met the challenge of training large numbers of TAB personnel to coincide with new energy conservation needs and the demands of an expanding technology in environmental controls. Design engineers, although largely recognizing the need for the TAB function, have not made adequate advances with regard to system designs more congruent with the test and balance technician and the field mechanic tasks.

Before 1965, when a small group of individuals from TAB firms, including John Gladstone, banded together to form the Associated Air Balance Council (AABC) and write the first industry standard for TAB, few engineers or contractors had ever heard of the term "TAB." In 1970, the American Society of Heating, Refrigerating, and Air-Conditioning Engineers (ASHRAE) published a "Testing, Adjusting, Balancing" Chapter for the first time in their Handbook series. The following year the National Environmental Balancing Bureau (NEBB) was founded. Today NEBB is the foremost and largest national and international organization producing an ongoing systematized body of information on TAB and related subjects.

In 1974 Gladstone wrote the first independently published book on this subject, *Air Conditioning Testing/Adjusting/Balancing: A Field Practice Manual* (Van Nostrand Reinhold, New York), and in 1984, NEBB published W. David Bevirt's *Environmental Systems Technology*, a 768-page textbook that received wide acclaim across the nation. In this present collaborative effort, W. David Bevirt and John Gladstone offer a practical *handbook* for contractors and engineers, TAB specialists, teachers and students, plant engineers and managers, and building executives who have a special interest in making well-designed building systems work the way they were intended to work. All equations and examples with solutions are presented in this handbook in both U.S. and metric units.

References

Figures, tables, and other reference data and materials used in this handbook without specific notation as to source generally have been developed by the National Environmental Balancing Bureau (NEBB) or jointly developed by NEBB and the American Society of Heating, Refrigerating and Air-Conditioning Engineers (ASHRAE) and/or the Sheet Metal and Air Conditioning Contractors National Association (SMACNA) over many years of mutual cooperation. Some of the others have been developed by one of the above three national organizations and traditionally interchanged under blanket copyright agreements and then updated and revised as needed, until the original source may no longer exist or be recognizable. Some others of the unidentified figures and tables were copied from the second edition of *Air Conditioning Testing/Adjusting/Balancing: A Field Practice Manual* by John Gladstone.

Where identified, the figures or tables have been reprinted with permission of the noted source for use in this handbook. No artwork, figures, or tables have been developed for exclusive use in this handbook only, as all have been or are being used in various editions of the many other NEBB publications.

The authors and NEBB wish to thank the following organizations for allowing the use of the noted materials in this publication:

Air Movement and Control Association, Inc. (AMCA)

Alnor Instrument Company

American Society of Heating, Refrigerating and Air-Conditioning Engineers, Inc. (ASHRAE)

Amtrol Inc.

Carrier Corporation

Cole-Parmer Instrument Company

Dwyer Instruments Inc.

Fluke Corporation

Gates Rubber Company

Greenheck Fan Corporation

Heating/Piping/Air Conditioning (HPAC)

Sheet Metal and Air Conditioning Contractors National Association (SMACNA)

Shortridge Instruments Inc.

Stanford Research Institute

Introduction

TAB Background

Since the late 1960s, heating, ventilating, and air conditioning (HVAC) system designers have realized that even "well-designed" HVAC systems require testing, adjusting, and balancing (TAB) to make them function properly. As the need for qualified TAB supervisors and technicians developed, international, national, and local certification programs for organizations of TAB firms were established during that period of time. Procedural standards for TAB work were published, training programs and study courses were established, the passing of written and practical examinations were required for TAB personnel, and proof of sound business practices became a certification requirement for TAB firms.

The National Environmental Balancing Bureau (NEBB) is a nonprofit organization cosponsored by the Mechanical Contractors' Association of America (MCAA) and the Sheet Metal and Air Conditioning Contractors' National Association (SMACNA). The function of NEBB is to establish and direct a management-oriented national and international program to upgrade and maintain uniform standards for the testing, adjusting, and balancing of environmental systems, for the performance testing of cleanrooms and clean air devices, for building systems commissioning, and for the measuring of sound and vibration in environmental systems. NEBB has over 600 certified firms throughout the world, with over 900 qualified supervisors.

The Certification Programs

The purpose of NEBB certification programs is to offer tangible proof of competent firms qualified in the proper methods, skills, and procedures for:

1. The testing, adjusting, and balancing (TAB) of environmental systems
2. The performance testing of cleanrooms and clean air devices
3. Building systems commissioning
4. The measuring of sound and vibration (S&V) in environmental systems

Objectives of the NEBB TAB Program

The purpose will be accomplished by meeting the following objectives:

- To establish industry standards, procedures, and specifications of the testing, adjusting, and balancing (TAB) of environmental systems
- To set minimum education standards and other requirements for the qualification of supervisor personnel employed by firms who perform this work
- To establish an educational program of instruction to train supervisory personnel in the proper methods and procedures in the testing, adjusting, and balancing of environmental systems; to accredit schools established by local NEBB chapters; and to publish the necessary procedural standards, books, manuals, and study courses
- To certify for the performance and supervision of testing, adjusting, and balancing of environmental systems those firms who meet the requirements for certification as established by NEBB, who comply with the objectives of NEBB, and who employ supervisor personnel who have met the qualifications established by NEBB
- To promote the concept of total responsibility for the testing, adjusting, and balancing of environmental systems

Other NEBB Programs

1. Cleanroom performance testing
The NEBB *Cleanroom Performance Testing Certification Program:*
   a. Establishes and promotes industry procedural standard and specifications for cleanroom performance testing.
   b. Sets minimum educational and experience requirements for the qualification of cleanroom performance testing supervisory personnel.
   c. Develops educational programs, procedural standards manuals, and study courses for cleanroom performance testing.
   d. Certifies firms that meet established requirements of the program.

2. Building systems commissioning

Building systems commissioning is an authorization to act in a prescribed manner to ready all building mechanical and electrical systems for active service. Essentially, building systems commissioning is the process of providing a building owner with a building that is complete, in compliance with the plans and specifications, and operationally and functionally ready to be taken over by the owner. This includes:

   a. Verifying the operation of system components under various conditions

   b. Verifying interactions between systems and subsystems

   c. Documenting system performance in reference to design criteria

   d. Instructing operators how to operate the building systems and equipment

The work is performed according to the NEBB *Procedural Standards for Building Systems Commissioning* manual by technicians coordinating, witnessing, verifying, and reporting on the work of specialist contractors and manufacturer's representatives under the supervision of the NEBB Certified System Commissioning Agent.

3. Sound and vibration measurement

The NEBB *Sound and Vibration Measurement Certification Program:*

   a. Establishes and promotes industry procedural standards and specifications for measuring sound and vibration (S&V) in environmental systems.

   b. Sets minimum education and experience requirements for the qualification of S&V supervisor personnel.

   c. Develops educational programs, texts, procedural standards manuals, and study courses for measuring sound and vibration.

   d. Certifies firms that meet established requirements of the program.

Standard TAB Procedures

1.1 Introduction

Well-performed testing, adjusting, and balancing (TAB) work is essential to the proper performance of building HVAC systems and the resultant indoor air quality. Chapter 34 of the ASHRAE *1995 HVAC Applications Handbook* gives the following definition of TAB:

> System testing, adjusting, and balancing is the process of checking and adjusting all the environmental systems in a building to produce the design objectives. This process includes (1) balancing air and water distribution systems, (2) adjusting the total system to provide design quantities, (3) electrical measurement, (4) establishing quantitative performance of all equipment, (5) verifying automatic controls, and (6) sound and vibration measurement. These procedures are accomplished by (1) checking installations for conformity to design; (2) measuring and establishing the fluid quantities of the system, as required to meet design specifications; and (3) recording and reporting the results.

The following ASHRAE definitions also are used by NEBB firms:

Test. To determine quantitative performance of equipment.

Adjust. To regulate the specified fluid flow rate and air patterns at the terminal equipment (e.g., reduce fan speed, adjust a damper).

Balance. To proportion flows within the distribution system (submains, branches, and terminals) according to specified design quantities.

Procedure. An approach to and execution of a sequence of work operations to yield repeatable results.

Report forms. Test data sheets arranged for collecting test data in logical order for submission and review. The data sheets should also form the permanent record to be used as the basis for any future testing, adjusting, and balancing.

Terminal. A point where the controlled medium (fluid or energy) enters or leaves the distribution system. In air systems, these may be variable air or constant volume boxes, registers, grilles, diffusers, louvers, and hoods. In water systems, these may be heat transfer coils, fan coil units, convectors, finned-tube radiation or radiant panels.

1.2 Preliminary Procedures

1. Review contract documents and plans for all of the HVAC systems.
2. Review approved shop drawings and equipment submittals.
3. Prepare system schematics.
4. Insert preliminary data on test report forms.
5. Review electrical characteristics of equipment and assure that safety controls are operating and that all motor starters have the proper heater coils or overload protection.
6. Complete *Systems Ready To Balance* checklists and verify that balancing devices have been installed.
7. Confirm that all HVAC and temperature control systems have been tested, strainers cleaned, systems flushed, etc., and are ready to balance. Clean or temporary air filters should be in place as specified.
8. Confirm that all building components, such as ceiling plenums that affect system balance, are in place and sealed, and that all windows, doors, etc., are installed and closed.
9. Confirm that all TAB instruments are in good order.

1.3 Air System Preliminary Procedures

1. Set all volume control dampers and variable air volume (VAV) boxes to the full open position, unless *system diversity* requires balancing in zones.
2. Set outside air dampers to the *minimum* position.
3. Verify correct fan rotations and speeds.
4. Check fan drives and adjust fans to design conditions or slightly above, on all systems.
5. Check motor amperages and voltages; make necessary adjustments.

1.4   Hydronic System Preliminary Procedures

1. Set all balancing devices to full open position.

2. Set mixing valves and control valves to full coil flow; close coil by-pass valves.

3. Verify correct pump rotations and proper drive alignments.

4. Measure and record pump motor amperages and voltages.

5. Confirm that systems are filled and expansion (compression) tanks correctly connected and charged; check basin levels of cooling towers.

6. Confirm that all automatic air vents are operational and system and all coils properly vented.

7. Confirm operation of boiler(s), chiller(s), and cooling towers. Set flows and temperatures to approximate design conditions.

8. Proceed with air systems TAB work.

1.5   HVAC Air System TAB Procedures

1. All related HVAC and exhaust air systems should be operating.

2. Determine whether any other HVAC or exhaust air systems could affect the system ready to be balanced.

3. Make Pitot tube traverses on all main supply and major branch ducts where possible to determine the air distribution.

4. Adjust balancing dampers of each major branch duct that is high on airflow. A minimum of one branch duct balancing damper shall remain fully open.

5. Measure and record the airflow of each terminal device in the system without adjusting any terminal outlet. Flow measuring hoods are the preferred airflow-measuring device.

6. The total airflow for the terminal outlets should be close to the Pitot tube traverse air measurement of that branch, and the main duct traverse air measurement should be within 10% of the total of all terminal outlet air measurements.

7. Check for excessive duct leakage if total terminal outlet air measurements are less than 90% of the main duct traverse air measurement.

8. Adjust the terminals that are highest on airflow to about 10% under design airflow.

9. Next, adjust each terminal outlet throughout the zone or system to design airflow and record measurements and make any necessary branch damper adjustments.

10. An additional adjusting pass throughout the system may be necessary. Make final adjustments to the fan drives where required. Record all data.

11. Adjust terminal device vanes to minimize drafts and for proper air distribution.

12. Measure and record system static pressures.

13. Measure and record all required outdoor air, return air, mixed air, and supply air drybulb and wetbulb temperatures. Measure and record all plenum static pressures.

14. Measure and record all coil entering air and leaving air drybulb and wetbulb temperatures. Measure and record all coil pressure differentials.

15. Measure and record final fan motor full load amperages and voltages.

16. *Proportional balancing* procedures may be found in the NEBB *Procedural Standards for Testing, Adjusting, Balancing of Environmental Systems*. Many TAB technicians may find these procedures more accurate and easier to use. Others may find them more complicated.

1.6   Hydronic System TAB Procedures

1. Continually check system and vent air from high points and circuits with lower flows during hydronic balancing. Periodically check and clean strainers.

2. Using "pump shutoff head," verify each pump head, operating curve and impeller size.

3. Adjust pumps to design flow and record data.

4. Adjust boilers and/or chillers to design flows and temperatures and record data.

5. If flow-measuring devices are used, record flow data throughout the systems before making adjustments.

6. Measure and record pressure drops through all coils and units. Compare with submittal data for high and low flows.

7. Adjust high flows to near design.

8. Adjust pump flows to design and check pressures, amperages, and voltages.

9. Set bypass balancing cocks to 90% of maximum flow through coils that have three-way control valves.

10. Repeat the above procedures until all coils and units are operating ± 10% of design.

11. Measure and record final pump pressures, amperages, and voltages.

12. Measure and record all coil and unit pressure drops and entering and leaving water temperatures.

13. Measure and record data from all flow measuring devices.

1.7   Makeup Air System TAB Procedures

1. Follow items 1 through 16 listed in Section 1.5 where they apply to the installed makeup air system.

2. Confirm that all related system fans serving each area within the space being balanced are operating. If they are not, pressure differences and infiltration or exfiltration may adversely influence the balancing. Preliminary studies will have revealed whether or not the supply air quantity exceeds the exhaust air quantity from each area. Positive and negative pressure zones should be identified at the time.

3. In most building pressurization applications, space pressure may be a primary consideration. Pressure differentials may be as high as 0.25 inches water gauge (in.w.g.) (63 Pa) static pressure but normally are in the range of 0.05 to 0.10 in.w.g. (12.5 to 25 Pa). These differentials must be maintained during all airflow balancing. If the differential pressures were allowed to vary during the balancing or testing procedure, it would be difficult to repeat the test results, making the final TAB results unacceptable.

4. If the building is served by a primary system for airflow, air conditioning, and filtration, and a secondary system for makeup air and room pressurization, all systems should be in operation during all balancing work. If a system has a return air fan, the modulation of the dampers may adversely affect the TAB procedure.

5. Use an electronic manometer, inclined manometer, or differential pressure gauge to verify space pressure differentials with all HVAC and exhaust air systems operating and all doors and openings closed.

6. Report all measured pressure differentials to the nearest 0.01 in.w.g. (2.5 Pa).

7. Outdoor air intakes to makeup air units should be located as far as practical (on directionally different exposures whenever possible) but not less than 30 ft (9 m) from outlets of combustion equipment stacks, ventilation exhaust outlets, plumbing vent stacks, or from areas that may collect vehicular exhaust and other noxious fumes. The bottom of outdoor air intakes serving central systems should be located as high as practical but not less than 6 ft (2 m) above ground level, or if installed above the roof, 3 ft (1 m) above the roof level. Report any conditions that are questionable.

1.8   Final TAB Report

1. Prepare a final testing, adjusting and balancing report. All inlet and outlet device airflow quantities should be within ± 10% of design. Reasons for exceptions should be noted.

2. All equipment TAB report forms should be signed by the TAB technician taking the readings.

3. Review the contract documents and specifications to verify compliance.

4. Place "N/A" or a dash in all spaces on the report forms that do not apply to that project.

5. The final test report should also contain the name and signature of the TAB firm supervisor.

1.9   Sound (Optional)

1. Sound pressure level measurements and/or noise criterion (NC) levels or room criterion (RC) levels should be made in accordance with NEBB *Procedural Standards for the Measurement and Assessment of Sound and Vibration*.

2. Table 1.1 gives acceptable sound levels for various interior areas.

3. A type 1 sound level meter specified by American National Standards Institute (ANSI) Standard S1.4, *American National Standard Specifications for Sound Level Meters* with octave band filter sets and matching calibrator should be used for measurements.

4. NC levels and RC(N) levels must be obtained by plotting center frequency octave band sound pressure level readings on the appropriate graphs.

TABLE 1.1 Recommended NC and RC(N) Levels for Different Indoor Activity Areas

Type of area	NC or RC level	Approx. dB(A)	Type of area	NC or RC level	Approx. dB(A)
RESIDENCES			CHURCHES		
Private home (rural & suburban)	25–30	25–35	Sanctuaries	20–30	25–35
Private home (urban)	25–35	30–40	Libraries	30–40	35–45
Apartment house	30–40	35–45	Schools & classrooms	30–40	35–45
HOTEL			Laboratories	35–45	40–50
Individual rooms	30–40	35–45	Recreation halls	35–50	40–55
Ballroom, banquet room	30–40	35–45	Corridors & halls	35–50	40–55
Halls, corridors, lobbies	35–45	40–50	PUBLIC BUILDINGS		
Garages	40–50	45–55	Libraries, museums	30–40	35–45
Kitchens, laundries	40–50	45–55	Court rooms	30–40	35–45
HOSPITALS AND CLINICS			Post offices, lobbies	35–45	40–50
Private rooms	25–35	30–40	General banking areas	35–45	40–50
Operating rooms	30–40	35–45	Washrooms, toilets	40–50	45–55
Wards, corridors	30–40	35–45	RESTAURANTS, LOUNGES, AND CAFETERIAS		
Laboratories	30–40	35–45	Restaurants	35–45	40–50
Lobbies, waiting rooms	40–50	45–55	Cocktail lounges	35–45	40–50
Washrooms, toilets	40–50	45–55	Nightclubs	35–45	40–50
OFFICES			Cafeterias	40–50	45–55
Board rooms	20–30	25–35	RETAIL STORES		
Conference rooms	25–35	30–40	Clothing stores	35–45	40–50
Executive office	30–40	35–45	Department stores (upper floors)	35–45	40–50
General offices	30–45	35–50	Department stores (main floor)	40–50	45–55
Reception rooms	30–40	35–45	Small retail stores	40–50	45–55
General open offices	35–45	40–50	Supermarkets	40–50	45–55
Drafting rooms	35–45	40–50	INDOOR SPORTS ACTIVITIES		
Halls & corridors	40–55	45–60	Coliseums	30–40	35–45
Tabulation & computation areas	40–50	45–55	Bowling alleys	35–45	40–50
AUDITORIUMS AND MUSIC HALLS			Gymnasiums	35–45	40–50
Concert, opera halls	15–25	20–30	Swimming pools	40–55	45–60
Sound recording studios	15–25	20–30	TRANSPORTATION (RAIL, BUSES, AND PLANES)		
Legitimate theaters	25–35	30–40	Ticket sales offices	30–40	35–45
Multi-purpose halls	25–30	30–35	Lounges, waiting rooms	35–50	40–55
Movie theaters	30–35	35–40			
TV audience studios	30–35	35–40			
Amphitheaters	30–35	35–40			
Lecture halls	30–35	35–40			

5. The sound level readings used must be for HVAC noise only. Values should be corrected for ambient background noise.

6. Approximate room decibel-A scale [dB(A)] readings are not a substitute for NC levels or RC(N) levels.

7. Outdoor dB(A) readings at specified distances from equipment should be made in an acoustic free field.

Airflow Measurement Equations

2.1 Free Area and Duct Airflow

The basic airflow equation for any free area is found below. *Free area* is defined as the total minimum area of openings in an air outlet or air inlet device through which air can pass. Free area of return air grilles may be as low as 50% of the duct connection size. The free cross-sectional area of a duct normally is 100%. If other data are not available, it may be assumed that all similar return air grilles would have similar free areas when measured with the same instrument.

Equation 2.1 (U.S.) **Equation 2.1 (Metric)**

$$Q = AV$$ $$Q = 1000AV$$

Where: Q = airflow (cfm) Where: Q = airflow (L/s)
A = area (ft^2) A = area (m^2)
V = velocity (fpm) V = velocity (m/s)

The cross-sectional area of rectangular, round, and flat oval ducts may be calculated from equations found in Chapter 25.

Example 2.1 (U.S.) Find the velocity in a 24 × 12 in. duct handling 2000 cfm.

Solution

$$1 \text{ ft}^2 = 144 \text{ in.}^2; A = \frac{24 \times 12}{144} = 2 \text{ ft}^2$$

$$Q = AV \qquad \text{or} \qquad V = \frac{Q}{A}$$

$$V = \frac{2000 \text{ cfm}}{2 \text{ ft}^2} = 1000 \text{ fpm}$$

Example 2.1 (Metric) Find the velocity in a 600 mm × 300 mm duct handling 1500 L/s.

Solution 1 meter (m) = 1000 millimeters (mm);

$$\frac{600 \text{ mm}}{1000} = 0.6 \text{ m}; \ \frac{300 \text{ mm}}{1000} = 0.3 \text{ m}.$$

$$A = 0.6 \text{ m} \times 0.3 \text{ m} = 0.18 \text{ m}^2$$

$$Q = 1000 \ AV; \ V = \frac{Q}{1000 \ A}$$

$$V = \frac{1500 \text{ L/s}}{1000 \times 0.18 \text{ m}^2} = 8.33 \text{ m/s}$$

Example 2.2 (U.S.) A 48 × 36 in. return air grille has a measured average velocity of 370 fpm. A Pitot tube traverse of the connecting duct indicates an airflow of 2975 cfm. Find the return grille free area (percentage).

Solution

$$\frac{48 \times 36}{144} = 12 \text{ ft}^2; \text{ free area } A = Q/V = 2975/370 = 8.04 \text{ ft}^2$$

$$\% = \frac{8.04 \text{ ft}^2 \times 100}{12 \text{ ft}^2} = 67\%$$

Example 2.2 (Metric) A 1200 mm × 900 mm return air grille has a measured average velocity of 1.85 m/s. A Pitot tube traverse of the connecting duct indicates an airflow of 1340 L/s. Find the return grille free area (percentage).

Solution

$$1.2 \times 0.9 = 1.08 \text{ m}^2; A = Q/1000 \ V = 1340/1000 \times 1.85 = 0.724 \text{ m}^2$$

$$\% = \frac{0.742 \text{ m}^2 \times 100}{1.08 \text{ m}^2} = 67\%$$

Example 2.3 (U.S.) Find the nearest standard size round duct to handle 4600 cfm at a velocity of 1070 fpm.

Solution

$$Q = AV; A = \frac{Q}{V} = \frac{4600 \text{ cfm}}{1070 \text{ fpm}} = 4.3 \text{ ft}^2; 4.3 \times 144 = 619 \text{ in.}^2$$

Using Equation 25.2 from Chapter 25:

$$A = \pi R^2, \text{ where } R \text{ is radius.}$$

$$R = \sqrt{\frac{A}{\pi}} = \sqrt{\frac{619}{\pi}} = 14.04 \text{ in.}$$

$$D = 2R \text{ (where } D \text{ is diameter)}$$

$$D = 2 \times 14.04 = 28.08 \text{ in.; use standard size} = 28 \text{ in. diameter duct.}$$

Example 2.3 (Metric) Find the nearest standard size round duct to handle 2300 L/s at a velocity of 5.3 m/s.

Solution

$$Q = 1000 \, AV; A = \frac{Q}{1000 \, V} = \frac{2300 \text{ L/s}}{1000 \times 5.3 \text{ m/s}} = 0.434 \text{ m}^2 = 434{,}000 \text{ mm}^2$$

Using Equation 25.2 from Chapter 25:

$$A = \pi^2; R = \sqrt{\frac{A}{\pi}} = \sqrt{\frac{434{,}000}{\pi}} = 371.7 \text{ mm}$$

$D = 2R = 2 \times 371.7 = 743.4$ mm; use standard size = 750 mm diameter duct.

2.2   Air Changes

To find the amount of infiltration or ventilation air needed for room or space air changes per hour, the following equation may be used:

<table>
<tr><td align="center">**Equation 2.2 (U.S.)**</td><td align="center">**Equation 2.2 (Metric)**</td></tr>
<tr><td align="center">$$Q_o = \frac{Vol \times N}{60}$$</td><td align="center">$$Q_o = \frac{Vol \times N}{3.6}$$</td></tr>
</table>

Where: Q_o = outside air (cfm)   Where: Q_o = outside air (L/s)
      Vol = volume of space (ft^3) Vol = volume of space (m^3)
      N = number of air N = number of air
           changes/h   changes/h

Example 2.4 (U.S.) A 20 × 20 × 10 ft room requires one air change per hour. Calculate the amount of outside air one air change requires.

Solution

$$Q_o = \frac{Vol \times N}{60} = \frac{20 \times 20 \times 10 \times 1}{60} = 66.7 \text{ cfm}$$

Example 2.4 (Metric) A 6 × 6 × 3 m room requires one air change per hour. Calculate the amount of outside air one air change requires.

Solution

$$Q_o = \frac{Vol \times N}{3.6} = \frac{6 \times 6 \times 3 \times 1}{3.6} = 30 \text{ L/s}$$

Example 2.5 (U.S.) During the summer, a 100 × 60 × 10 ft space requires an air change every 3 min. Calculate the amount of outside air one air change requires.

Solution

$$N = \frac{60 \text{ min/h}}{3 \text{ min/air change}} = 20 \text{ air changes per hour}$$

$$Q_o = \frac{Vol \times N}{60} = \frac{100 \times 60 \times 10 \times 20}{60} = 20,000 \text{ cfm}$$

Example 2.5 (Metric) During the summer, a $30 \times 18 \times 3$ meter space requires an air change every 3 minutes. Calculate the amount of outside air one air change requires.

Solution

$$N = \frac{60 \text{ min/h}}{3 \text{ min/air change}} = 20 \text{ air changes per hour}$$

$$Q_o = \frac{Vol \times N}{3.6} = \frac{30 \times 18 \times 3 \times 20}{3.6} = 9000 \text{ L/s}$$

2.3   Percentage of Outdoor Air

To set outside air dampers to maintain a required percentage of outside air, the following equations may be used:

Equation 2.3 (U.S.)

$$T_m = \frac{X_o T_o + X_r T_r}{100}$$

Where: T_m = temperature of mixed air (°F)
X_o = % of outside air
T_o = temperature of outside air (°F)
X_r = % of return air
T_r = temperature of return air (°F)

Equation 2.3 (Metric)

$$T_m = \frac{X_o T_o + X_r T_r}{100}$$

Where: T_m = temperature of mixed air (°C)
X_o = % of outside air
T_o = temperature of outside air (°C)
X_r = % of return air
T_r = temperature of return air (°C)

Equation 2.4 (U.S.)

$$X_o = 100 \frac{(T_r - T_m)}{(T_r - T_o)}$$

Equation 2.4 (Metric)

$$X_o = \frac{(T_r - T_m)}{(T_r - T_o)}$$

Equation 2.5 (U.S.)

$$X_r = \frac{(T_m - T_o)}{(T_r - T_o)}$$

Equation 2.5 (Metric)

$$X_r = \frac{(T_m - T_o)}{(T_r - T_o)}$$

Example 2.6 (U.S.) Twenty percent of outside air is required. Find the dry bulb temperature of the mixed air when the outside air is at 95 °F and the return air is at 76 °F.

Solution

$$T_m = \frac{S_o T_o + X_r T_r}{100} = \frac{20 \times 95° + 80 \times 76°}{100}$$

$$T_m = \frac{1900 + 6080}{100} = \frac{7980}{100} = 79.8 \text{ °F}$$

Example 2.6 (Metric) Twenty percent of outside air is required. Find the drybulb temperature of the mixed air when the outside air is at 35 °F and the return air is at 24 °C.

Solution

$$T_m = \frac{X_oT_o + X_rT_r}{100} = \frac{20 \times 35° + 80 \times 24°}{100}$$

$$T_m = \frac{700 + 1920}{100} = 26.2 \text{ °C}$$

2.4   Measuring Airflow Through Orifice Plates

2.4.1   Orifice sizes

The use of sharp-edged orifice plates to balance airflow to outlets or branches induces a high level of accuracy, but loses the flexibility inherent in dampers. Where the flow is determined in advance, the following procedure can be used to determine the airflow and the total pressure loss accurately.

The sharp-edged orifice has more resistance to flow but is easily constructed. It also can be made readily interchangeable for several orifice sizes. The orifice is mounted between two flanged sections sealed with rubber gaskets (Figure 2.1). Three orifice sizes, 1.40 in.

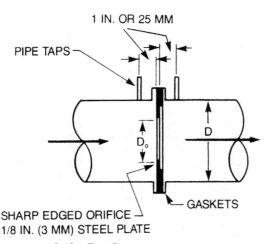

1 IN. OR 25 MM

PIPE TAPS

D_o

D

SHARP EDGED ORIFICE
1/8 IN. (3 MM) STEEL PLATE

GASKETS

Figure 2.1   Orifice Detail

(35.6 mm), 2.625 in. (66.7 mm), and 4.90 in. (124.5 mm) diameters, can be used to meter flow rates from 50 to 8000 fpm (0.25 to 40 m/s). If the orifice and pipe taps are made to exact dimensions, the calculated air volume will be within 1% of actual flow for standard air. Orifices for larger ducts can be sized using data found in Chapter 2 of the eighth edition of the *Fan Engineering Handbook* published by the Buffalo Forge Company, Buffalo, New York.

The orifice can be calibrated with a standard Pitot tube. A micromanometer is needed to read velocity pressures below 2000 fpm (10 m/s). At 2000 to 3000 fpm (10 to 15 m/s), with a 10:1 inclined manometer, an accuracy of ±0.3 to ±1.0% can be expected. At 3000 to 4000 fpm (15 to 20 m/s), an accuracy of ±0.25 to ±0.3% can be expected. If the orifice is made to the precise dimensions in Tables 2.1 and 2.2, no calibration is needed and the tabulated calculation can be used. Tables 2.1 and 2.2 give computer calculations for the three sizes of orifices listed above. The orifice sizes cover a range of measurable airflows from 18 to 1155 cfm (9 to 578 L/s).

TABLE 2.1   Orifice Flow Rate (Standard cfm) (U.S. Units)

	Orifice size				Orifice size		
ΔP (in.w.g.)	1.40 in.	2.625 in.	4.90 in.	ΔP (in.w.g.)	1.40 in.	2.625 in.	4.90 in.
0.10		29.3	121.5	2.60	41.7	147.8	596
0.20		41.3	169.6	2.70	42.5	150.6	607
0.30		50.5	206.5	2.80	43.3	153.3	618
0.40		58.3	237.5	2.90	44.0	156.0	629
0.50	18.5	65.1	264.9	3.00	44.8	158.7	639
0.60	20.2	71.3	289.6	3.20	46.2	163.8	660
0.70	21.8	76.9	312.3	3.40	47.6	168.8	680
0.80	23.3	82.2	333.5	3.60	49.0	173.7	700
0.90	24.7	87.2	353.3	3.80	50.3	178.4	718
1.00	26.0	91.9	372.1	4.00	51.6	183.0	737
1.10	27.3	96.3	390.0	4.20	52.9	187.5	755
1.20	28.5	100.6	407.0	4.40	54.1	191.9	772
1.30	29.6	104.7	423.4	4.60	55.3	196.2	789
1.40	30.7	108.6	439.1	4.80	56.5	200.4	806
1.50	31.8	112.4	454.3	5.00	57.6	204.4	822
1.60	32.8	116.1	468.9	5.50	60.4	214.3	862
1.70	33.8	119.6	483.1	6.00	63.1	223.8	899
1.80	34.8	123.1	496.9	6.50	65.6	232.8	935
1.90	35.7	126.4	510.4	7.00	68.1	241.4	970
2.00	36.6	129.7	523.4	7.50	70.4	249.9	1003
2.10	37.5	132.9	536.2	8.00	72.7	257.9	1036
2.20	38.4	136.0	548.6	8.50	74.9	265.8	1067
2.30	39.3	139.0	560.8	9.00	77.0	273.4	1097
2.40	40.1	142.0	572.6	9.50	79.1	280.8	1127
2.50	40.9	144.9	584.3	10.00	81.1	287.9	1155

TABLE 2.2   Orifice Flow Rate (Standard L/s) (Metric)

ΔP (Pa)	Orifice size 35.6 mm	66.7 mm	124.5 mm	ΔP (Pa)	Orifice size 35.6 mm	66.7 mm	124.5 mm
24.9		13.8	57.3	647.4	19.7	69.8	281.2
49.8		19.5	80.1	672.3	20.1	71.1	286.5
74.7		23.8	97.5	697.2	20.4	72.4	291.6
99.6		27.5	112.1	722.1	20.8	73.6	296.7
124.5	8.7	30.7	125.0	747.0	21.1	74.9	301.7
149.6	9.5	33.7	136.7	796.8	21.8	77.3	311.5
174.3	10.3	36.3	147.4	846.6	22.5	79.7	320.9
199.2	11.0	38.8	157.4	896.4	23.1	82.0	330.1
224.1	11.7	41.2	166.8	946.2	23.7	84.2	339.0
249.0	12.3	43.4	175.6	996.0	24.4	86.4	347.7
273.9	12.9	45.6	184.1	1046	25.0	88.5	356.4
298.8	13.5	47.5	192.1	1096	25.5	90.6	364.4
323.7	14.0	49.4	199.8	1145	26.1	92.6	372.4
348.6	14.5	51.3	207.3	1195	26.7	94.6	380.4
373.5	15.0	53.1	214.4	1245	27.2	96.5	388.0
398.4	15.5	54.8	221.3	1370	28.5	101.1	406.9
423.3	16.0	56.5	228.0	1494	29.8	105.6	424.3
448.2	16.4	58.1	234.5	1619	31.0	109.9	441.3
473.1	16.9	59.7	240.9	1743	32.1	113.9	457.8
498.0	17.3	61.2	247.0	1868	33.2	118.0	473.4
522.9	17.7	62.7	253.1	1992	34.3	121.7	489.0
547.8	18.1	64.2	258.9	2117	35.4	125.5	503.6
572.7	18.6	65.6	264.7	2241	36.3	129.1	517.8
597.6	18.9	67.0	270.3	2366	37.3	132.5	531.9
622.5	19.3	68.4	275.8	2490	38.3	135.9	545.2

2.4.2   Calculation of Flow

The airflow for a sharp-edged orifice with pipe taps located 1 in. (25 mm) on either side of the orifice (Figure 2.1) can be accurately computed from the following equations for standard air density of 0.075 lb/ft^3 (1.2041 kg/m^3) for ducts 2 to 14 in. (50 to 350 mm) in diameter.

Airflow measurements through orifice plates may be found in Equation 2.6 (for standard air):

Equation 2.6 (U.S.)

$$Q = 21.8 K D_0^2 \sqrt{h}$$

Where: Q = air volume (cfm)
  K = coefficient of airflow
  D_o = orifice diameter (in.)
  h = pressure drop across orifice (in.w.g.)

Equation 2.6 (Metric)

$$Q = 0.00101 K D_0^2 \sqrt{h}$$

Where: Q = air volume (L/s)
  K = coefficient of airflow
  D_o = orifice diameter (mm)
  h = pressure drop across orifice (Pa)

The constant K is affected by the Reynolds number, a dimensionless value expressing flow conditions in a duct. Equation 2.7 gives a sim-

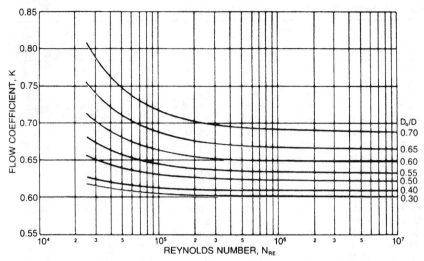

Figure 2.2   Flow coefficients K for square-edged orifice plates and flange taps in smooth pipe.

plified method of calculating the Reynolds number for standard air. The constant K can also be selected from Figure 2.2.

Equation 2.7 (U.S.)

$$N_{RE} = 8.56DV_o$$

Where: N_{RE} = Reynolds number
    D = duct diameter (inches)
    V_o = velocity of air through
        orifice (fpm)

Equation 2.7 (Metric)

$$N_{RE} = 66.4DV_o$$

Where: N_{RE} = Reynolds number
    D = duct diameter (mm)
    V_o = velocity of air through
        orifice (m/s)

3

Fan Equations

3.1 Fan Laws

The *fan laws* relate the performance variables for any similar series of fans. The variables are fan size, rotation speed (rpm), air density, volume flow rate, static pressure, power, and mechanical efficiency. The fan laws apply only to similar fans at the same point of rating on the performance curve. They can be used to predict the performance of any fan when test data are available for any fan of the same series.

Fan laws also may be used with a particular fan to determine the effect of speed change. However, caution should be exercised in these cases, since they apply only when all flow conditions are similar. Changing the speed of a given fan changes parameters that may invalidate the fan laws.

Equation 3.1 (U.S. & Metric)

$$\frac{Q_2}{Q_1} = \frac{\text{rpm}_2}{\text{rpm}_1}$$

Equation 3.2 (U.S. & Metric)

$$\frac{SP_2}{SP_1} = \left(\frac{\text{rpm}_2}{\text{rpm}_1}\right)^2 = \left(\frac{Q_2}{Q_1}\right)^2$$

Equation 3.3 (U.S. & Metric)

$$\frac{I_2}{I_1} = \frac{FP_2}{FP_1} = \left(\frac{\text{rpm}_2}{\text{rpm}_1}\right)^3 = \left(\frac{Q_2}{Q_1}\right)^3$$

Where: Q = airflow (cfm or L/s)
    rpm = fan revolutions per minute
    SP = static pressure (in.w.g. or Pa)
    FP = fan power [brake horsepower (BHP) or kW]
    I = current draw [Amps (A)]

Example 3.1 (U.S.) A fan is operating at 0.5 in.w.g. static pressure at a speed of 1000 rpm and moving 2000 cfm of air at 1.5 BHP. Using Equations 3.1, 3.2, and 3.3, find the new SP, rpm and FP if the design calls for 2200 cfm.

Solution

$$\text{rpm}_2 = \text{rpm}_1 \times \frac{Q_2}{Q_1} = 1000 \times \frac{2200}{2000} = 1100 \text{ rpm}$$

$$SP_2 = SP_1 \times \left(\frac{Q_2}{Q_1}\right)^2 = 0.5 \times \left(\frac{2200}{2000}\right)^2 = 0.61 \text{ in.w.g.}$$

$$FP_2 = FP_1 \times \left(\frac{Q_2}{Q_1}\right)^3 = 1.5 \times \left(\frac{2200}{2000}\right)^3 = 2.0 \text{ BHP}$$

Example 3.1 (Metric) A fan is operating at 125 Pa static pressure at a speed of 1000 rpm and moving 950 L/s of air at 1.1 kW fan power. Using Equations 3.1, 3.2, and 3.3, find the new SP, rpm, and FP if the design calls for 1050 L/s.

Solution

$$\text{rpm}_2 = \text{rpm}_1 \times \frac{Q_2}{Q_1} = 1000 \times \frac{1050}{950} = 1105 \text{ rpm}$$

$$SP_2 = SP_1 \times \left(\frac{Q_2}{Q_1}\right)^2 = 125 \times \left(\frac{1050}{950}\right)^2 = 153 \text{ Pa}$$

$$FP_2 = FP_1 \times \left(\frac{Q_2}{Q_1}\right)^3 = 1.1 \times \left(\frac{1050}{950}\right)^3 = 1.49 \text{ kW}$$

Rated fan capacities are based on air at sea level under standard conditions: 14.7 psi, 70 °F and a density of 0.075 lb/ft³ (101.3 kPa, 20 °C and a density of 1.2 kg/m³). In actual applications, the fan may be required to handle air or gas at some other density. The change in density may be caused by temperature, the composition of the gas, or altitude. As indicated by Equation 3.4, fan performance is affected by gas density. With constant size and speed, the horsepower and pressure vary in accordance with the ratio of gas density to the standard air density. **The airflow (cfm or L/s) remains constant because a centrifugal fan is a constant-volume machine.**

Density correction tables may be found in Tables 10.1 and 10.2 in Chapter 10 and in Appendix C, Table C-8 (U.S.) or Appendix D, Table D-6 (Metric). The correction factors in the tables substitute for d_2/d_1 in Equation 3.4.

Equation 3.4 (U.S. & Metric)

$$\frac{d_2}{d_1} = \frac{SP_2}{SP_1} = \frac{FP_2}{FP_1} \quad \text{when } Q_1 = Q_2 \text{ (cfm or L/s)}$$

Where: d = air density (lb/ft³ or kg/m³)
       SP = static pressure (in.w.g. or Pa)
      FP = fan power (BHP or kW)
(Fan size and fan speed (rpm) also are constant)

Example 3.2 (U.S.) A fan is rated to deliver 6000 cfm at 1.5 in.w.g. SP and it requires 3.1 BHP. Find the airflow, SP and BHP if the fan and system are moved to 7000 ft elevation and the air is at 40 °F.

Solution From Table 10.1 in Chapter 10 and Table C-8 in Appendix C, the correction factor (d_2/d_1) for 45 °F and 7000 ft. is 0.82. The cfm remains the same at 6000 cfm (constant volume). Using Equation 3.4:

$$SP_2 = SP_1 \times \frac{d_2}{d_1} = 1.5 \times 0.82 = 1.23 \text{ in.w.g.}$$

$$FP_2 = FP_1 \times \frac{d_2}{d_1} = 3.1 \times 0.82 = 2.54 \text{ BHP}$$

Example 3.2 (Metric) A fan is rated to deliver 3000 L/s at 375 Pa and it requires 2.3 kW. Find the airflow, SP, and kW if the fan and system is moved to a 2000 m elevation and the air is at 5 °C.

Solution From Table 10.2 in Chapter 10 and Table D-6 in the Appendix, the correction factor (d_2/d_1) for 5 °C and 2000 m is 0.84 (by interpolation). The airflow remains at 3000 L/s (constant volume).

$$SP_2 = SP_1 \times \frac{d_2}{d_1} = 375 \times 0.84 = 315 \text{ Pa}$$

$$FP_2 = FP_1 \times \frac{d_2}{d_1} = 2.3 \times 0.84 = 1.93 \text{ kW}$$

Additional equations and use of HVAC systems in other air density situations can be found in Chapter 10.

3.2 Fan Power

To find the theoretical fan power of a fan in brake horsepower [watts (W)], use Equation 3.5.

Equation 3.5 (U.S.)

$$BHP = \frac{Q \times SP}{6356 \times Eff}$$

Equation 3.5 (Metric)

$$W = \frac{Q \times SP}{1000 \times Eff}$$

Where: BHP = brake horsepower   Where: W = watts
$\quad$ Q = airflow (cfm) $\quad$ Q = airflow (L/s)
$\quad$ SP = static pressure (in.w.g.) $\quad$ SP = static pressure (Pa)
$\quad$ Eff = fan efficiency $\quad$ Eff = fan efficiency

Where the fan efficiency is not available, a good rule of thumb to use is a factor of 0.63. When the fan power is known, by rearranging Equation 3.5, the *fan static efficiency* may be calculated.

Equation 3.6 (U.S.)   **Equation 3.6 (Metric)**

$$Eff_{SP} = \frac{Q \times SP}{6356 \times BHP} \qquad Eff_{SP} = \frac{Q \times SP}{1000 \times W}$$

If total pressure (*TP*) is used in the above equations, then the *fan total efficiency* may be calculated.

Equation 3.7 (U.S.)   **Equation 3.7 (Metric)**

$$Eff_{TP} = \frac{Q \times TP}{6356 \times BHP} \qquad Eff_{TP} = \frac{Q \times TP}{1000 \times W}$$

Example 3.3 (U.S.)   A fan delivers 8000 cfm at 0.75 in.w.g. static pressure. If the fan efficiency is 65%, find the brake horsepower.

Solution

$$BHP = \frac{Q \times SP}{6356 \times Eff.} = \frac{8000 \times 0.75}{6356 \times 0.65} = 1.45 \; BHP$$

Example 3.3 (Metric)   A fan delivers 4000 L/s at 185 Pa static pressure. If the fan efficiency is 65%, find the fan power in watts and kilowatts.

Solution

$$W = \frac{Q \times SP}{1000 \times Eff} = \frac{4000 \times 185}{1000 \times 0.65} = 1138 \; W$$

$$kW = \frac{1138 \; W}{1000 \; W/kW} = 1.14 \; kW$$

3.3   Fan Tip Speed

The tip speed of a fan wheel may be found by using Equation 3.8.

Equation 3.8 (U.S.)   **Equation 3.8 (Metric)**

$$TS = \frac{\pi \times D \times rpm}{12} \qquad\qquad TS = \frac{\pi \times D \times r/s}{1000}$$

Where: TS = tip speed (fpm)   Where: TS = tip speed (m/s)
$\quad$ D = wheel diameter (in.) $\quad$ D = wheel diameter (mm)
$\quad$ rpm = revolutions per minute $\quad$ r/s = revolutions per second

Example 3.4 (U.S.) A 42-in. diameter fan is operating at 360 rpm. Determine the tip speed.

Solution

$$TS = \frac{\pi \times D \times \text{rpm}}{12} = \frac{\pi \times 42 \times 360}{12} = 3958 \text{ fpm}$$

Example 3.4 (Metric) A 1050 mm diameter fan is operating at 6 r/s. Determine the tip speed.

Solution

$$TS = \frac{\pi \times D \times \text{r/s}}{1000} = \frac{\pi \times 1050 \times 6}{1000} = 19.8 \text{ m/s}$$

3.4   Fan Speed/Coil Δt

Often there is a need to adjust the temperature drop (Δt) through a cooling coil or heating coil in HVAC units. An increase in fan speed (and airflow) will lower the temperature drop. The fan speed may be adjusted using the following equation:

Equation 3.9 (U.S. and Metric)

$$\frac{\text{rpm}_2}{\text{rpm}_1} = \frac{\Delta t_1}{\Delta t_2}$$

Where: rpm = fan revolutions per minute
Δt = coil temperature drop (°F or °C)

Example 3.5 (U.S.) The measured Δt for a coil is 24 °F that was specified to be 20 °F. The fan is running at 820 rpm. Find the required rpm.

Solution Using Equation 3.9:

$$\text{rpm}_2 = \text{rpm}_1 \times \frac{\Delta t_1}{\Delta t_2} = 820 \times \frac{24 \text{ °F}}{20 \text{ °F}} = 984 \text{ rpm}$$

The change in pulley size equations may be found in Chapter 5.

Example 3.5 (Metric) The measured Δt for a coil is 13.3 °C that was specified to be 11.1 °C. The fan is running at 820 rpm. Find the required rpm.

Solution Using Equation 3.9:

$$\text{rpm}_2 = \text{rpm}_1 \times \frac{\Delta t_1}{\Delta t_2} = 820 \times \frac{13.3 \text{ °C}}{11.1 \text{ °C}} = 983 \text{ rpm}$$

The change in pulley size equations may be found in Chapter 5.

4

Fan and System Curves

4.1 Centrifugal Fans and Curves

The first of the two basic types of fans normally encountered in HVAC systems is the *centrifugal fan*. Three variations of the centrifugal fan are commonly used: the forward curved fan, the backward inclined fan, and the airfoil fan.

4.1.1 Forward curved (FC) fans

The *forward curved* centrifugal fan travels at a relatively slow speed and generally is used for producing high volumes at low static pressures. The fan curve in Figure 4.1 shows that the FC fan may "surge" (see Section 4.1.5, Fan Surge, for an explanation of surge). The surge magnitude is less than for other types, because of the steep "hump" or peak of the fan curve.

The typical operating range of this type of fan is from 30 to 80% of wide open volume (see Figure 4.1). The horsepower curve has a continuous upward slope, so the FC fan is referred to as an *overloading type* fan.

4.1.2 Backward inclined (BI) fans

Backward inclined fans travel at about twice the speed of the forward curved fan. The normal selection range of the backward inclined fan is approximately 40 to 85% of wide open volume (Figure 4.2). Generally, the larger the fan, the more efficient for a given selection. The magnitude of surge for a BI fan is greater than for an FC fan, as shown by the gentle "hump" in the fan curve. The horsepower curve peaks

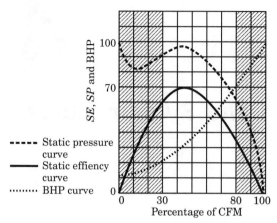

Figure 4.1 Characteristic curves for FC fans.

and then drops off with increased airflow, so the fan is a *nonoverloading* type.

4.1.3   Airfoil fans

The *airfoil fan* is a refinement of the flat-bladed backward inclined fan using airfoil-shaped blades. This improves the static efficiency, so that it is the highest of all centrifugal fans, and reduces the noise level slightly. The magnitude of surge also increases with the airfoil blades. Characteristic curves for airfoil fans (Figure 4.3) are similar to the BI fans shown in Figure 4.2. The horsepower curve peaks and then drops

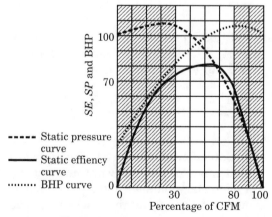

Figure 4.2 Characteristic curves for B1 fans.

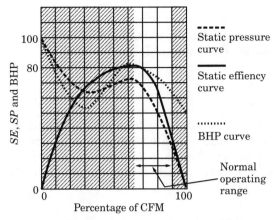

Figure 4.3 Characteristic curves for vaneaxial fans.

off with increased airflow, so the fan is a *nonoverloading* type. The airfoil fan operates at the highest speed of all centrifugal fans.

4.1.4 Tubular centrifugal fans

Tubular centrifugal fans generally consist of a single-width airfoil wheel arranged in a cylinder to discharge air radially against the inside of the cylinder. Air is then deflected parallel with the fan shaft to provide straight-through flow. Vanes are used to recover static pressure and to straighten air flow.

Characteristic curves are similar to the BI fan curves shown in Figure 4.2. The selection range is generally about the same as the scroll-type BI or airfoil bladed wheel—50 to 85% of wide open volume. However, because there is no housing of the turbulent airflow path through the fan, static efficiency is reduced and noise level is increased.

4.1.5 Fan surge

Fan surge may occur when a fan is operating at the peak or "hump" of the fan curve. Surging means that the airflow volume will rise and fall as the operating point moves back and forth across the hump of the fan curve to the same static pressure level on each side. For example, in Figure 4.1, the *SP* is at a 90% level at both 30% and 60% airflows. This back-and-forth rapid change of airflows causes the fan to become noisier and it may cause severe damage to the fan.

4.2   Axial Fans and Curves

The second basic type of HVAC fan is the *axial fan*. Axial fans are divided into three groups—propeller, tubeaxial, and vaneaxial.

4.2.1   Propeller fans

The *propeller fan* (Figure 4.4) is well suited for high volumes of air at little or no static pressure differential. With low efficiency, the fan is usually of inexpensive construction. The impeller is made of two or more blades of single-thickness metal attached to a relatively small hub.

4.2.2   Tubeaxial and vaneaxial fans

Tubeaxial and *vaneaxial* fans are simply propeller fans mounted in a cylinder with the vane-type straighteners. These vanes remove much of the swirl from the air and improve the efficiency. A vaneaxial fan is more efficient than a tubeaxial and can reach higher pressures. Note that with axial fans, the fan power is maximum at the blocktight static pressure. With centrifugal fans, the fan power is minimum at blocktight static pressure.

Tubeaxial and vaneaxial fans generally are used for handling large volumes of air at low static pressures. They have higher noise levels and lower efficiencies than centrifugal fans.

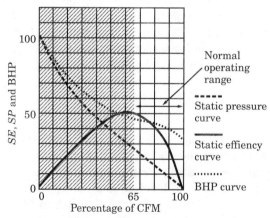

Figure 4.4 Characteristic curves for propeller and tubeaxial fans.

4.3   System Curves

4.3.1   Plotting a system curve

The *system resistance curve* for any HVAC system is represented by a single curve that is a plot of the static pressure required to move a specific amount of air through that system (Figure 4.5). For example, a system handling 1000 cfm (500 L/s) has a total measured resistance of 1 in.w.g. (250 Pa). If the airflow is doubled, the static pressure resistance will increase by that ratio squared to the 4 in.w.g. (1000 Pa) shown in Figure 4.5 (by using Equation 3.2 from Chapter 3).

4.3.2   System operating point

When a system curve is plotted on a fan curve of the fan being used (Figure 4.6) for the system, the *operating point* at which the fan and system will perform is determined by the intersection of the system curve and the fan performance curve. Every fan operates only along its performance curve for a given rpm. If the designed system resistance is not the same as the resistance in the installed system, the *operating point will move along the fan curve* and the static pressure and volume of air delivered will not be as calculated.

In Figure 4.6, the actual system pressure drop may be greater or less than that predicted by the system designer, so a new system curve indicates a lower or higher airflow volume. The point of operation would move along the fan curve to the new system curve. When a

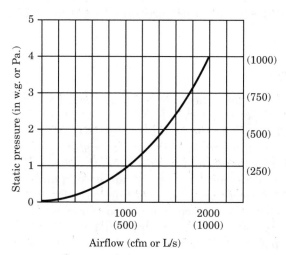

Figure 4.5   System resistance curve.

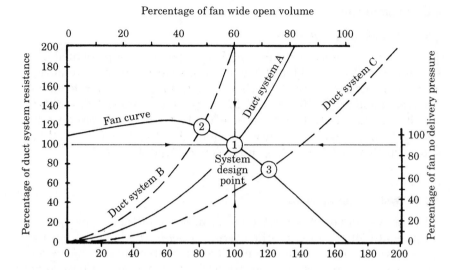

Figure 4.6 Interaction of system curves and fan curve. *(Reproduced with permission from AMCA)*

typical fan power curve is included, it would indicate a change in fan power (BHP or kW).

Duct system B has a greater static pressure resistance than designed; therefore, the volume of air is less (operating point 2). Duct system C has less static pressure resistance than designed, so it has more air being delivered at a lower static pressure as indicated by operating point 3.

4.4 Fan/System Curves

There is a different fan curve for every fan speed or rpm. Two fan curves for two different fan speeds are shown in Figure 4.7. Fan laws in Chapter 3 may be used to predict the results of changes. Increases or decreases in fan speed will alter the airflow rate through a system. Figure 4.7 illustrates the 10% increase in airflow when the speed of a fan is increased 10% to operating point 2. The 10% increase in airflow, however, creates a severe fan power penalty. According to the fan laws, the fan power increase is 33%. The static pressure also increased 21%.

On the other hand, a 10% reduction in fan speed has a somewhat similar decrease in fan power (25% decrease from the higher fan power).

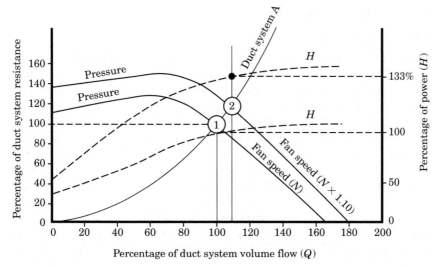

Figure 4.7 Effect of 10% increase in fan speed. *(Reproduced with permission from AMCA)*

Example 4.1 If the BHP of the fan (operating point 1) in Figure 4.7 is 10 HP (7.5 kW), find the new BHP and percentage increase if the fan speed is increased 10% (operating point 2).

Solution Using Equation 3.3 from Chapter 3:

$$\frac{FP_2}{FP_1} = \left(\frac{rpm_2}{rpm_1}\right)^3 ; \frac{rpm_2}{rpm_1} = 1.1 \text{ (a 10\% increase)}$$

U.S.: $FP_2 = 10(1.1)^3 = 13.31$ BHP

Metric: $FP_2 = 7.5(1.1)^3 = 9.98$ kW

Both are 33% increases in fan power.

Example 4.2 If the static pressure of the fan (operating point 1) in Figure 4.7 is 4 in.w.g. (1000 Pa), find the new SP and percentage increase if the fan speed is increased 10% (operating point 2).

Solution Using Equation 3.2 from Chapter 3:

$$\frac{SP_2}{SP_1} = \left(\frac{rpm_2}{rpm_1}\right)^2 ; \frac{rpm_2}{rpm_1} = 1.1 \text{ (a 10\% increase)}$$

U.S.: $SP_2 = 4(1.1)^2 = 4.84$ in.w.g.

Metric: $SP_2 = 1000(1.1)^2 = 1210$ Pa

Both are 21% increases in static pressure.

4.5 Multiple Fan Operation

4.5.1 Parallel fans

Many HVAC units contain multiple similar fans driven by a single motor. If the installation of the fans has unrestricted airflow, the total airflow is the total of all of the fans, but the static pressure does not change. Some fans have a "positive" slope in the pressure–volume curve to the left of the peak pressure point. If fans operating in parallel are selected in the region of this "positive" slope, unstable operation may result. Figure 4.8 shows the combined volume–pressure curve of two such fans in parallel.

The closed loop to the left of the peak pressure point is the result of plotting all the possible combinations of volume flow at each pressure. If the system curve intersects the combined volume–pressure curve in the area enclosed by the loop, more than one point of operation is possible. This may cause one of the fans to handle more of the air and could cause a motor overload if the fans are individually driven. This unbalanced flow condition tends to reverse readily, with the result that the fans will intermittently load and unload. This

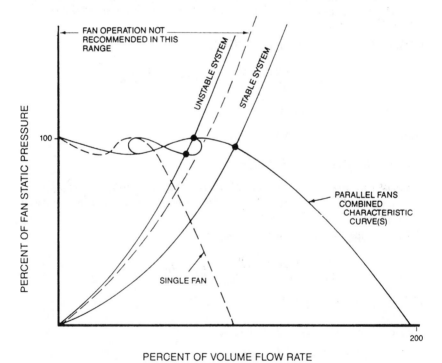

Figure 4.8 Parallel fan operation. *(Reproduced with permission from AMCA)*

"surging" often generates noise and vibration and may cause damage to the fans, ductwork, or driving motors (see Section 4.1.5).

4.5.2 Fans in series

Sometimes it is necessary to install two or more fans in series in the same system. When the fans make up a single assembly, the assembly is usually known as a *multi-stage fan*. This combination is seldom used in conventional HVAC systems.

In theory the combined volume–pressure curve of two fans operating in series is obtained by adding the fan pressures at the same airflow volumes (Figure 4.9). In practice, there is some reduction in airflow because of increased air density and a significant loss of performance caused by nonuniform flow into the inlet of the second-stage fan.

4.6 Fan Tables

Figure 4.10, copied from a fan catalog, shows both fan curves and a fan capacity table for a specific size and type of fan. Normally, fan curves are not shown, and sometimes not available if requested.

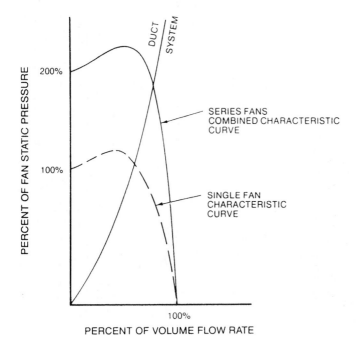

Figure 4.9 Fans operating in series. *(Reproduced with permission from AMCA)*

SWB-33

Performance Data

Wheel Diameter = 33 Inches
Outlet Area = 6.12 Square Feet
Maximum RPM = 1156
Tip Speed, FPM = 8.64 X RPM
Maximum BHP = $(RPM/452)^3$
Maximum Motor Frame = 286T

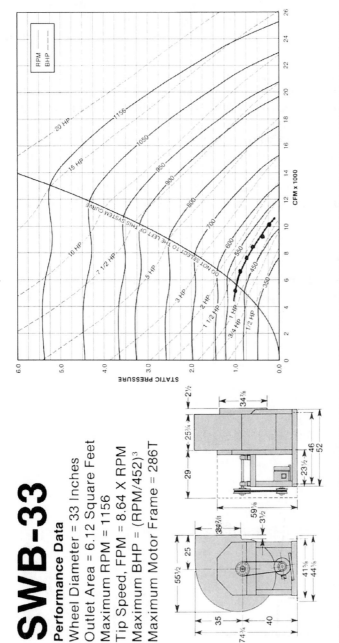

CFM	OV		¼"	⅜"	½"	⅝"	¾"	1"	1¼"	1½"	1¾"	2"	2½"	3"	3½"	4"
										STATIC PRESSURE						
5600	915	RPM	329	362	392	421	450	504								
		BHP	0.38	0.51	0.64	0.78	0.93	1.24								
6500	1062	RPM	361	391	419	445	471	521	567							
		BHP	0.50	0.65	0.79	0.95	1.11	1.44	1.79								

Figure 4.10 Fan curve and table. *(Courtesy of Greenheck Fan Corp.)*

CFM	Outlet Vel.															
7400	1209	RPM	395	422	448	472	496	541	585	626	667					
		BHP	0.64	0.80	0.97	1.14	1.31	1.67	2.05	2.45	2.87					
8300	1356	RPM	430	455	479	502	524	566	605	645	683	718				
		BHP	0.81	0.99	1.18	1.37	1.55	1.94	2.35	2.77	3.21	3.66				
9200	1503	RPM	466	489	511	533	553	593	630	666	701	737	801			
		BHP	1.01	1.21	1.41	1.62	1.83	2.25	2.68	3.13	3.59	4.08	5.07			
10100	1650	RPM	502	524	545	565	584	622	657	691	724	755	820	878		
		BHP	1.25	1.46	1.68	1.91	2.14	2.60	3.06	3.54	4.03	4.53	5.59	6.68		
11000	1797	RPM	539	559	579	598	617	652	685	717	749	779	838	897	951	
		BHP	1.53	1.76	1.99	2.23	2.48	2.99	3.48	3.99	4.51	5.04	6.14	7.30	8.49	
11900	1944	RPM	577	596	614	632	649	683	715	746	775	804	860	915	969	1019
		BHP	1.85	2.10	2.35	2.60	2.87	3.41	3.95	4.49	5.04	5.60	6.75	7.96	9.21	10.50
12800	2092	RPM	616	633	650	667	684	716	746	775	804	831	885	936	987	1037
		BHP	2.23	2.48	2.75	3.02	3.30	3.87	4.46	5.04	5.62	6.21	7.42	8.67	9.98	11.33
13700	2239	RPM	655	670	686	702	718	748	778	806	833	860	910	961	1008	1056
		BHP	2.66	2.92	3.21	3.49	3.79	4.39	5.01	5.64	6.26	6.88	8.15	9.46	10.80	12.21
14600	2386	RPM	695	707	723	738	753	782	811	838	864	889	939	986	1032	1076
		BHP	3.15	3.42	3.72	4.02	4.33	4.96	5.61	6.28	6.95	7.61	8.94	10.31	11.72	13.15
15500	2533	RPM	734	747	760	775	789	817	844	870	895	919	968	1012	1057	1101
		BHP	3.70	3.98	4.29	4.60	4.93	5.59	6.28	6.97	7.68	8.40	9.80	11.22	12.69	14.19
16400	2680	RPM	774	786	797	812	825	852	878	903	927	951	997	1041	1083	1126
		BHP	4.31	4.61	4.92	5.25	5.59	6.28	7.00	7.73	8.47	9.22	10.72	12.21	13.73	15.29
17300	2827	RPM	814	825	836	849	862	887	912	936	960	983	1027	1070	1111	1151
		BHP	4.99	5.31	5.62	5.96	6.32	7.04	7.79	8.55	9.32	10.11	11.71	13.26	14.85	16.46
18200	2974	RPM	854	865	875	886	899	923	947	970	993	1016	1058	1100	1140	
		BHP	5.74	6.07	6.40	6.75	7.12	7.87	8.65	9.44	10.25	11.06	12.73	14.39	16.04	
19100	3121	RPM	895	905	915	924	936	959	982	1005	1027	1048	1090	1130		
		BHP	6.57	6.91	7.26	7.61	7.99	8.78	9.58	10.40	11.24	12.09	13.82	15.59		
20000	3268	RPM	935	944	954	963	973	996	1018	1040	1061	1082	1123			
		BHP	7.47	7.83	8.20	8.56	8.94	9.76	10.59	11.44	12.31	13.19	14.99			

Performance shown is for Model SWB with outlet duct. BHP does not include drive losses.

Figure 4.10 (Continued)

Fan curves may be developed by plotting data from a fan capacity table, but often the exercise will result in inaccurate curves if sufficient plot points are not available. For example (using Figure 4.10), to plot the 500 rpm curve, the closest rpm values from the fan capacity table are 504, 521, 496, 502, 489 or 511, and 502 (5600 cfm to 10,100 cfm lines). Using the airflow and static pressure values as plot points, a rough 500 rpm fan curve could be drawn as noted by the points in Figure 4.10.

Fan Drives and Equations

5.1 Fan Drives—Pulleys

5.1.1 Pulley sizes

Since the motor pulley or sheave is usually the smallest of the two drive pulleys, it is more economical and common to change only this pulley if the desired results can be obtained. However, greater changes in fan speeds may require a change of both fan and motor pulleys. Hopefully, the motor and fan shaft center-to-center dimension can remain the same as the initial condition to prevent major revisions to the motor base and position. Drastic speed changes, requiring more powerful motors and larger electrical service, may complicate the physical modifications. Assuming that the fan pulley does not change in size, the following equations may be used to determine the new motor pulley pitch diameter. The pitch diameter (D) is smaller than the pulley outside diameter (D_o) as shown in Figure 5.1.

Equation 5.1 (U.S.)

$$\frac{d \text{ (motor)}}{D \text{ (fan)}} = \frac{\text{RPM (fan)}}{\text{rpm (motor)}}$$

Equation 5.2 (U.S.)

$$D \times \text{RPM} = d \times \text{rpm}$$

Where: d = motor pulley
        pitch diam. (in.)
    D = fan pulley pitch
        diam. (in.)
rpm = motor speed
RPM = fan speed

Equation 5.1 (Metric)

$$\frac{d \text{ (motor)}}{D \text{ (fan)}} = \frac{\text{R/S (fan)}}{\text{r/s (motor)}}$$

Equation 5.2 (Metric)

$$D \times \text{R/S} = d \times \text{r/s}$$

Where: d = motor pulley
        pitch diam. (mm)
    D = fan pulley pitch
        diam. (mm)
r/s = motor speed
R/S = fan speed

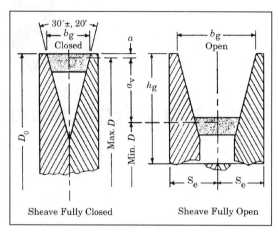

Figure 5.1 Adjustable sheave. D and d = pitch diameter, large and small sheave; D_o and d_o = outside diameter, large and small sheave.

Example 5.1 (U.S.) The fan motor speed is 1800 rpm, the fan pulley pitch diameter is 10.4 in. If the fan is to operate at 620 rpm, find the motor pulley pitch diameter.

Solution

$$d = \frac{D \times \text{RPM}}{\text{rpm}} = \frac{10.4 \times 620}{1800} = 3.58 \text{ in.}$$

Example 5.1 (Metric) The fan motor speed is 30 r/s, the fan pulley pitch diameter is 264 mm. If the fan is to operate at 10.3 r/s, find the motor pulley pitch diameter.

Solution

$$d = \frac{D \times \text{R/S}}{\text{r/s}} = \frac{264 \times 10.3}{30} = 90.6 \text{ mm}$$

5.1.2 Speed-O-Graphs

1. To use the Speed-O-Graphs in Figure 5.2, connect the known values of the two similar items forming a "box" and draw a diagonal line through that intersecting point parallel to the other diagonal lines.

2. Connect the other known item value to the diagonal line.

3. From that intersection, draw a line to obtain the missing value.

Example 5.2 (U.S.)
a. A fan drive has a motor running at 1800 rpm with a 4.5-in. pitch diameter pulley. The fan is to operate at 540 RPM. Determine the fan pulley pitch diameter size using the Speed-O-Graph in Figure 5.2.
b. Confirm the solution using the appropriate equation.

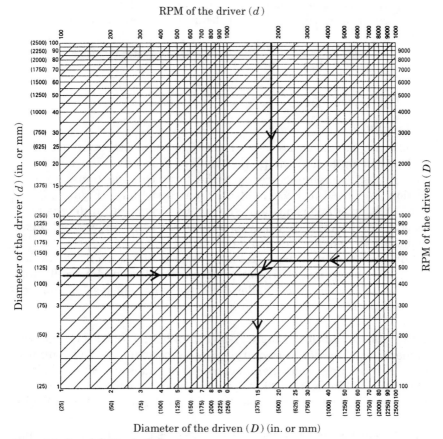

Figure 5.2   Speed-O-Graph: Pulley laws.

Solution

a. Intersect lines from 540 RPM and 1800 rpm (upper right-hand corner) and draw a diagonal line to intersect the line drawn from the 4.5 in. diameter *driver* (from the left side of the graph). Drawing a line downward from the intersection determines a *driven* pulley of 15 in. pitch diameter.

b.

$$D = \frac{d \times \text{rpm}}{\text{RPM}} = \frac{4.5 \times 1800}{540} = 15 \text{ in. pitch diameter}$$

Example 5.2 (Metric)

a A fan drive has a motor running at 1800 rpm with a 115 mm pitch diameter pulley. The fan is to operate at 540 RPM. Determine the fan pulley pitch diameter size using the Speed-O-Graph in Figure 5.2.

b. Confirm the solution using the appropriate equation.

Solution

a. Intersect lines from 540 RPM and 1800 rpm (upper right-hand corner) and draw a diagonal line to intersect the line drawn from the 112.5 mm diameter *driver* (from the left side of the graph). Drawing a line downward from the intersection determines a *driven* pulley of 375 mm pitch diameter.

b.

$$D = \frac{d \times r/s}{R/S} = \frac{112.5 \times 1800}{540} = 375 \text{ mm pitch diameter}$$

5.2 Fan Drives—Belts

5.2.1 V-Belts

V-Belts are classed by cross-section size, outside circumference, nominal pitch length, and the power they can transmit (Figures 5.7 and 5.8). All V-belt manufacturers use the same classifications. Nominal section sizes used in HVAC industry may be found in Figure 5.3 and Table 5.1, but dimensions and shapes may vary with different manufacturers. However, they may operate on the same pulleys, but mixed manufacturer belts never should be mixed on the same drive.

Equipment, using fractional horsepower motors such as in residential HVACs, uses special L sizes, which are lighter and more flexible. They will fit around the smaller diameter motor sheaves used in residential work.

5.2.1.1 Single V-belts. Single V-belt numbers consist of a letter–numeral combination designating cross-sectional size and length of belt. For example, one manufacturer designates an 8.5-in. belt as 2L085, a 52.0-in. belt as 3L520, and a 100.0-in. belt as 4L1000. The first digit indicates the cross-section size for light duty (L-type) belts.

5.2.1.2 Variable-speed V-belts. Variable-speed V-belts of the same manufacturer have a standard numbering system that indicates the nominal belt top width in sixteenths of an inch by the first two numbers. The third and fourth numbers indicate the angle of the groove in which the belt is designed to operate. The digits following the letter "V" indicate the pitch length to the nearest 1/10 in. A belt numbered 1530V450 is a V-belt of 15/16 in. nominal top width designed to op-

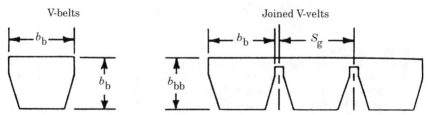

Figure 5.3 Standard V-belt sizes. See Table 5.1 for dimensions. (*ANSI / RMA IP-20-1888 Engineering Standard.*)

TABLE 5.1 V-Belt Nominal Dimensions (see Figure 5.3)

Cross section	b_b		h_b		h_{bb}		Std. S_g		Deep S_g	
	In.	mm	In.	mm	In.	mm	In.	mm	In.	mm
A, AX	13C, 13CX	0.50	12.7	0.31	7.9	0.41	10.4	0.625	15.9	—	—
B, BX	16C, 16CX	0.66	16.8	0.41	10.4	0.50	12.7	0.750	19.0	0.875	22.0
C, CX	22C, 22CX	0.88	22.4	0.53	13.5	0.66	16.8	1.000	25.4	1.250	30.0
D	32C	1.25	31.8	0.75	19.1	0.84	21.4	1.438	36.5	1.750	43.0

NOTE: Joined belts will not operate in deep groove sheaves.

b_b = belt width
h_b = belt height (single)
h_{bb} = belt height (joined)
S_g = sheave groove spacing

erate in a sheave of 30° (groove angle) and have a pitch length of 45.0 in. Other drive manufacturers use similar numbering systems.

For metric work, a soft conversion must be used, i.e., multiplying the data in inches by 25.4 millimeters per inch.

5.2.2 Changing drives

When it is necessary to change pulley sizes, the new pulley or sheave usually would be the same type. However, if a motor change is required, a new bore and keyway would be required, and the number of belts may increase to make the belt horsepower rating high enough, perhaps 150% of brake horsepower rating. Obviously, if the number of belts increases, the number of pulley grooves increases, and the motor *and* fan pulleys must be changed to suit.

To determine the required belt length, use Equation 5.3:

Equation 5.3 (U.S. & Metric)

$$L = 2C + 1.57\,(D + d) + \frac{(D - d)^2}{4C}$$

Where: L = Belt length (in. or mm)
$\quad\quad\;\; C$ = Shaft centerline to shaft center (in. or mm)
$\quad\quad\;\; D$ = Fan pulley pitch diam. (in. or mm)
$\quad\quad\;\; d$ = Motor pulley pitch diam. (in. or mm)

5.2.3 V-belt drive conditions

Regardless of whether drives consist of stock or special items, there are certain primary conditions to observe with respect to the installation of satisfactory drives. Those most commonly encountered are:

1. Drives should be installed with provisions for center distance adjustment. This is essential since all belts stretch.

2. Centers should not exceed 2½ to 3 times the sum of the pulley diameters nor be less than the diameter of the larger pulley.

3. Arc of contact on the smaller pulley should not be less than 120°.

4. Ratios should not exceed 8:1.

5. Never replace a belt with a smaller cross-section.

6. Replace multiple belts only with similar belts from the same manufacturer.

> **Example 5.3 (U.S.)** The center-to-center distance between shafts is 29 in. The fan pulley pitch diameter is 14.1 in. and the motor pulley has a pitch diameter of 4.3 in. Find the length of belt needed (to the nearest inch).

Solution

$$L = 2C + 1.57\,(D + d) + \frac{(D - d)^2}{4C}$$

$$L = (2 \times 29) + 1.57\,(14.1 + 4.3) + \frac{(14.1 - 4.3)^2}{4 \times 29}$$

$$L = 58 + 28.89 + \frac{96.04}{116}$$

$$L = 58 + 28.89 + 0.83$$

$$L = 87.72 \text{ in.}$$

Use an 88-in. belt or closest standard size.

Example 5.3 (Metric)   The center-to-center distance between shafts is 737 mm. The fan pulley pitch diameter is 358 mm, and the motor pulley has a pitch diameter of 109 mm. Find the length of belt needed.

Solution

$$L = 2C + 1.57\,(D + d) + \frac{(D - d)^2}{4C}$$

$$L = (2 \times 737) + 1.57\,(358 + 109) + \frac{(358 - 109)^2}{4 \times 737}$$

$$L = 1474 + 733.19 + \frac{62001}{2948}$$

$$L = 1474 + 733.2 + 21.0 = 2228.2 \text{ mm}$$

Use closest standard size to 2228 mm.

5.2.4 Multiple V-Belts

V-Belts are rated also by the power they transmit. To determine the number of V-Belts needed in a specific drive, use Equation 5.4.

Equation 5.4 (U.S. and Metric)

$$\text{No. of belts} = \frac{\text{motor power} \times \text{service factor}}{\text{belt power rating}}$$

Example 5.4 (U.S.)   A fan has a 30 HP motor. The V-Belt drive has a service factor of 1.5. Determine how many belts with a 16 HP rating are needed.

Solution

$$\text{No. of belts} = \frac{\text{motor power} \times \text{service factor}}{\text{belt power rating}}$$

$$\text{No. of belts} = \frac{30 \times 1.5}{16}$$

No. of belts = 2.8 (three belts are needed)

Example 5.4 (Metric)   A fan has a 22.5 kW motor. The V-Belt drive has a service factor of 1.5. Determine how many belts with a 12 kW rating are needed.

Solution

$$\text{No. of belts} = \frac{\text{motor power} \times \text{service factor}}{\text{belt power rating}}$$

$$\text{No. of belts} = \frac{22.4 \times 1.5}{12} = 2.8 \text{ belts (three belts are needed)}$$

5.2.5   V-belt tension—general rules

Tension of the belts on a V-belt drive is usually not critical. A few simple rules about tensioning will satisfy most requirements.

1. The best tension for a V-belt drive is the lowest tension at which the belts will not slip under the highest load condition, usually 1/64-in. (0.4-mm) deflection as shown in Figure 5.4.

2. Check the tension on a new drive frequently during the first day of operation.

3. Check alignment (Figure 5.5).

4. Check the drive tension periodically, thereafter.

5. Too much tension shortens belt and bearing life.

6. Keep belts and sheaves free from any foreign material that may cause slip.

7. If a V-belt slips, **tighten it**.

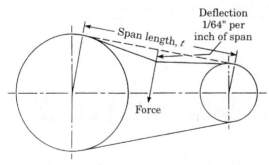

Figure 5.4   Belt tension. *(Gates Rubber Company.)*

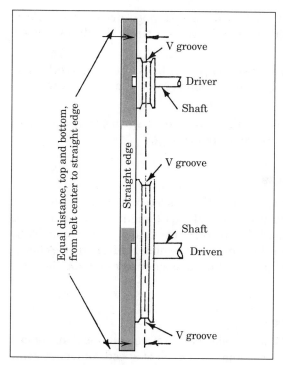

Figure 5.5 Drive Alignment.

5.2.6 V-belt tension—numerical method

Many users of V-belt drives rely on their experience and the above general rules for tensioning drives, but it has become common practice actually to measure the tension in a drive. Numerical methods for measuring tension have several advantages. For example, they prevent inexperienced personnel from drastically overtensioning or undertensioning a drive, thus preventing possible bearing or belt damage. Even with experienced personnel, it helps the individual get a feel for the tension needed in a particular drive. This is especially important with modern drives, where each V-belt is rated for higher horsepower than previous belts have been. If a belt is to carry more horsepower, it must be installed proportionally tighter. Experience with older drives may lead to undertensioning of modern drives unless tension is measured at least once to help get the feel for correct tension.

The procedure in numerically tensioning a drive is:

1. Check alignment (see Figure 5.5).

2. Determine the correct tension for the stopped drive, called *static tension,* so that the tension will be correct when the drive is operating.

3. Measure the static tension so that it can be set at the correct value with a tension tester (Figure 5.6) following the manufacturer's instructions.

4. Do not use this method if the drive uses a spring-loaded idler or other means of automatic drive tensioning.

5.3 Variable Speed Drives

The wide use of variable air volume (VAV) systems and variable-speed pumps has increased the need for efficient methods of controlling larger motor speeds. Small fan motors are easily and economically controlled, but using fan inlet vanes, adjustable blades, and discharge dampers to control large fan air volumes has not proved to be the most efficient method.

5.3.1 Variable sheave drives

Variable-pitch sheave drives are relatively simple and inexpensive and can provide a satisfactory, economical means of speed control in some applications. Speed ranges typically are limited to about 3 to 1, although ranges of up to 10 to 1 are possible. Belt wear is always a problem, the drive unit is relatively bulky, and they usually create more noise and vibration.

5.3.2 Multispeed motors

Multispeed induction motors can be used if universal speed control is not required, providing the advantages inherent in squirrel-cage motors. The number of speeds that can be provided is limited by the number of windings that can be placed on the stator or the manner in which poles can be reconnected external to the motor. It is seldom practical to provide for more than two speeds. Top speed also is limited to 3600-rpm synchronous speed.

Wound-rotor induction motors can be used to provide stepless speed control over a fairly broad range. Motor cost is higher than that of an induction motor, especially in smaller sizes. Overall system cost may be lower than that of static AC or DC drives. On the minus side are maintenance problems and problems of environmental compatibility— and the exclusions on use in hazardous (classified) locations that are inherent with any wound-rotor machine.

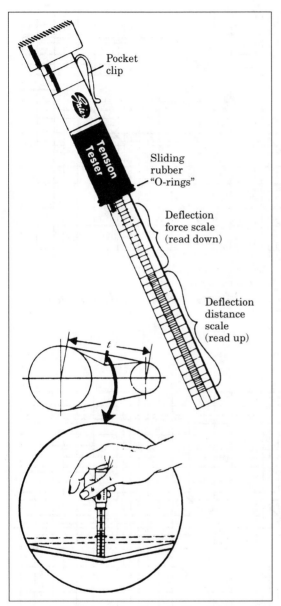

Pocket clip

Sliding rubber "O-rings"

Deflection force scale (read down)

Deflection distance scale (read up)

Tension Tester

t

Figure 5.6   Tension tester. *(Gates Rubber Company)*

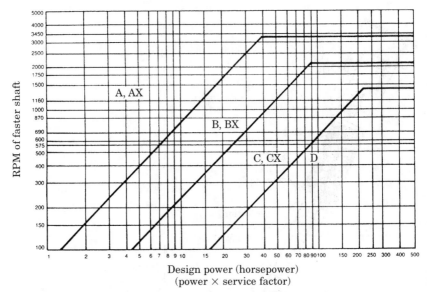

Figure 5.7 Selection of V-belt cross section (U.S.). *(Reprinted from ANSI / RMA Engineering Standard 1P-20-1988)*

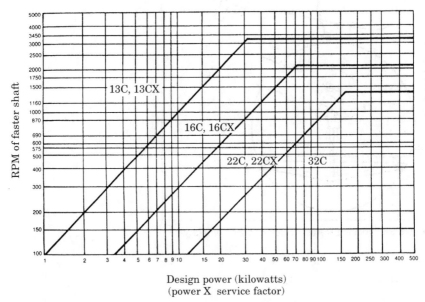

Figure 5.8 Selection of V-belt cross section (metric). *(Reprinted from ANSI / RMA Engineering Standard 1P-20-1988)*

5.3.3 Eddy-current clutches

Eddy-current clutches can provide economical stepless speed control over a broad speed range. Whereas fairly small in size, they must be interposed between the motor shaft and the load input shaft. First cost is fairly low, but efficiency is lower than for static AC and DC drives, especially at low speeds. Some units with large speed ranges require liquid cooling.

5.3.4 DC motors

In the past, the most common method of providing speed control with *DC motors* called for the motors to be served from a DC bus supplied by a motor generator set or from a static DC power supply. Few such systems are installed today. For most new installations, such fixed-bus DC systems have given way to packaged static DC drives dedicated to serving a single motor. For most types of variable-speed applications, when stepless control or very fine control is needed, the two most workable choices are packaged static DC drives and variable-frequency AC drives.

5.3.5 Variable-frequency drives (VFD)

If it were not for the fact that the *variable frequency drive* (VFD) can be used with a squirrel-cage inductor motor, the VFD would be hard pressed to compete with the static DC drive. In most applications, the VFD costs more initially and operates less efficiently. There are, however, many applications that can be served at lower first cost and higher efficiency by VFD.

Typical of applications that can be served at lower first cost with the VFD are those in which the speed of a number of motors must be synchronized. If used with synchronous-reluctance motors, a single VFD maintains all motors of the group in perfect synchronism. Control circuitry is simplified if speed is to be controlled automatically, because a single control feedback loop serves the entire operation.

A VFD decreases overall motor efficiency and increases motor heating. Fan cooled motors require a minimum speed to receive adequate cooling. Also, actual current flow to a VFD motor may be higher than at the constant-speed rating, but the voltage is lower so that the total power consumed by the motor is lower.

When checking motor performance, the current flow (amperage) to the drive must be checked at normal line voltage. Most VFDs have a manual override mode that allows the TAB technician to adjust the output speed. Section 19.7.3 also contains some practical information on amperage readings.

6

Duct System Pressure Losses

6.1 Duct Pressures

There are two sets of duct system pressures. One set is used to design the duct systems and to select the fans, and the other set is the actual system operating pressures.

6.1.1 Duct design pressures

The duct system designer calculates the static pressure losses of the straight sections of ductwork using engineering tables and charts. To these losses, the losses of all the duct fittings are calculated and added along with the pressure loss data of all manufactured items such as filters, coils, dampers, and diffusers or grills. The duct system fan(s) are selected from the total static pressure losses of the longest run(s) of the supply air and return air ducts connected to the fan(s).

6.1.2 Duct operating pressures

The three duct pressures that TAB technicians measure in the field are *total pressure (TP)*, *static pressure (SP)*, and *velocity pressure (V$_p$)*. The three pressures are related by the following equation:

Equation 6.1 (U.S. and Metric)

$TP = SP + V_p$

Where: TP = total pressure (in.w.g. or Pa)
$\quad\quad\ SP$ = static pressure (in.w.g. or Pa)
$\quad\quad\ V_p$ = velocity pressure (in.w.g. or Pa)

6.1.3 Total pressure

Total pressure (TP) of a duct is measured by the impact of the moving airstream on the end of a Pitot tube directly facing and perpendicular (Figure 6.1) to the airflow. Total pressure determines how much energy is in the airflow at any point in the system. Total pressures always decline in value from the fan to any terminal device (diffuser or grille).

6.1.4 Static pressure

Static pressure (SP) is exerted equally in all directions at any point or cross section of the duct (similar to the "blowup" or internal pressure of a balloon). It also is a measure of the potential energy to produce and maintain airflow against duct resistance. Static pressures may be positive or negative to the atmosphere.

6.1.5 Velocity pressure

Velocity pressure (V_p) is exerted only in the direction of airflow and is a measure of kinetic energy resulting from the airflow. Velocity pressure cannot be measured directly by a Pitot tube and pressure gauge or manometer, but it can be measured indirectly by subtracting the static pressure from the total pressure (see Figure 6.1) based on Equation 6.1, $V_p = TP - SP$. If the airflow velocity is known, the velocity pressure may be calculated using the following equations:

<table>
<tr><td align="center">Equation 6.2 (U.S.)</td><td align="center">Equation 6.2 (Metric)</td></tr>
<tr><td align="center">$V_p = (V/4005)^2$</td><td align="center">$V_p = 0.602V^2$</td></tr>
</table>

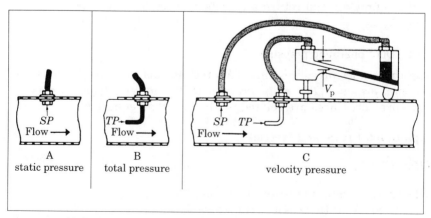

A
static pressure

B
total pressure

C
velocity pressure

Figure 6.1 Airflow pressure measurements.

Equation 6.3 (U.S.)

$$V = 4005 \sqrt{V_p}$$

Equation 6.3 (Metric)

$$V = \sqrt{1.66 V_p}$$

Where: V_p = velocity pressure (in.w.g. or Pa)
V = velocity (fpm or m/s)

(Equations 6.1 through 6.3 are for standard air at sea level.) Tables of velocities or velocity pressures in both U.S. and metric units may be found in Appendix A, Tables A-4 and A-5. Velocity pressures are always positive.

Example 6.1 (U.S.) A 48 × 16 in. duct handles 10,700 cfm at 2.3 in.w.g. static pressure. Find the total pressure of the duct.

Solution

Using Equation 2.1: $V = Q/A$ = 10,700/48 × 16/144 = 2006 fpm

Using Equation 6.2: $V_p = (V/4005)^2$ = 0.25 in.w.g.

Using Equation 6.1: $TP = SP + V_p$ = 2.3 + 0.25 = 2.55 in.w.g.

Example 6.1 (Metric) A 1200 × 400 mm duct handles 5030 L/s at 575 Pa static pressure. Find the total pressure of the duct.

Solution

Using Equation 2.1: $V = Q/1000A$ = 5030/1000 × 1.2 × 0.4 = 10.5 m/s

(1200 mm × 400 mm = 1.2 m × 0.4 m)

Using Equation 6.2: V_p = 0.602 (10.5)² = 66.4 Pa

Using Equation 6.1: $TP = SP + V_p$ = 575 + 66.4 = 641.3 Pa

6.2 Pressure Measurements

6.2.1 Supply ducts

A section of duct is capped at both ends and sealed tight (Figure 6.2). Pressurized at 0.06 psi (0.414 kPa), this pressure converts to 1.66 in.w.g. (414 Pa). Since the air is at rest, velocity equals 0, so V_p = 0. Both total pressure and static pressure would equal 1.66 in.w.g. (414 Pa).

If a fan is connected to one end of the duct and air flows through the duct at 4900 fpm (25 m/s), as seen in Figure 6.3, the airflow velocity would convert to 1.5 in.w.g. (375 Pa) velocity pressure using Equation 6.2 or the tables in Appendix A, Tables A-4 and A-5. Subtracting the velocity pressure from the total pressure:

$SP = 1.66TP - 1.5\ V_p$ = 0.16 in. SP (415 Pa − 375 Pa = 40 Pa SP)

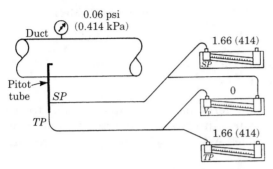

Figure 6.2 Pitot tube connections for supply air-stream.

The manometer does not sense the actual velocity pressure directly, but by using the Pitot tube hookup with the static opening connected to the low-pressure side of the gauge, and the total pressure opening connected to the high-pressure side of the gauge, the manometer will read the difference between the two, or the *velocity pressure*.

Velocity pressure and static pressure increase or decrease in the ductwork with every change in the duct configuration, but the total pressure remains in a constant decrease from the fan to the end of the system because of the fitting pressure losses and straight duct friction losses. Hence, at a point when the velocity pressure decreases, the static pressure increases, and vice versa, because the static pressure is always the difference between the total pressure and the velocity pressure. When the static pressure increases with a drop in velocity and velocity pressure, the change is called *static regain*.

6.2.2   Return or exhaust ducts

The static pressure in a return air or exhaust air system is always below atmospheric pressure, and it is customary among air ventilation

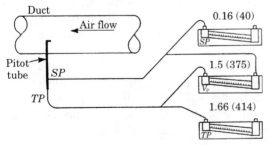

Figure 6.3 Pitot tube connections for supply air-stream.

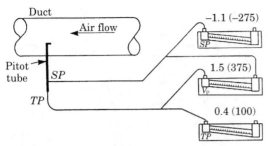

Figure 6.4 Pitot tube connections if airstream is exhausted from duct and *TP* is positive.

system designers to omit the minus sign affecting the static (gauge pressure). These system designers know that the total pressure is higher than the static pressure by the amount of the velocity pressure (Figure 6.6). But total pressure seldom is considered in the field of ventilation; most computation for design and testing being based on static pressure.

When the airstream is exhausted from the duct, the static pressure is negative and the hose connections to the manometer or gauge will depend on whether the velocity pressure is greater or smaller than the numerical value of the static pressure. If it is greater, the total pressure will be positive; if it is smaller, the total pressure will be negative.

In Figures 6.4 and 6.5. the ducts are connected to the inlet or suction side of the fan. With a negative static pressure of 1.1 in.w.g. (275 Pa) in Figure 6.4, the total pressure of 0.4 in.w.g. (100 Pa) will be positive $(TP = SP + V_p = -1.1 + 1.5 = 0.4$ in.w.g. or $TP = -275 + 375 = 100$ Pa).

In Figure 6.5, a negative static pressure of 2.0 in.w.g. (500 Pa) results in a negative total pressure of 0.5 in.w.g. (125 Pa); $(TP = SP + V_p = -2.0 + 1.5 = 0.5$ in.w.g. or $TP = -500 + 375 = -125$ Pa).

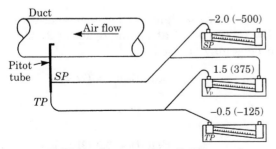

Figure 6.5 Pitot tube connections if airstream is exhausted from duct and *TP* is negative.

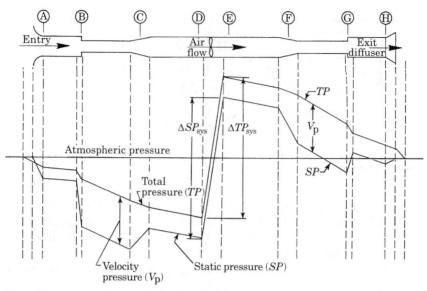

Figure 6.6 Pressure changes during flow in ducts.

6.2.3 Gauge connections

The various connections between the Pitot tube and the gauges or manometers frequently are made with rubber hose. Precaution must be taken so that all passages and connections are dry, clean, and free of leaks, sharp bends, and other obstructions. Pressure connection branches can be made by using T-fittings or using two-stem nipple adapters.

6.3 Air Density Changes

Chapter 10 discusses air density effects on HVAC systems in detail. The following equations, plus others, also may be found in Chapter 10.

When the airflow being tested is at a higher or lower temperature than used in the normal HVAC range, or at altitudes above 2000 ft (600 m), the following equations may be used:

Equation 6.4 (U.S.)

$$V_p = d \times \left(\frac{V}{1096}\right)^2$$

Equation 6.4 (Metric)

$$V_p = \frac{d}{2} \times V^2$$

Equation 6.5 (U.S.)

$$V = 1096 \sqrt{V_p/d}$$

Equation 6.5 (Metric)

$$V = 1.414 \sqrt{V_p/d}$$

Where: V_p = velocity pressure (in.w.g. or Pa)
        V = velocity (fpm or m/s)
        d = density (lb/ft^3 or kg/m^3)

The same velocity pressure (V_p) reading from a duct system airflow of a lesser density will result in a higher velocity.

Example 6.2 (U.S.)   The measured velocity pressure (V_p) for airflow at 250 °F and an altitude of 6000 feet is 0.32 in.w.g. Calculate the airflow velocity.

Solution   The air density correction factor from Table C-8 of the Appendix for 250 °F at 6000 feet is 0.60.

$$d = 0.60 \times 0.075 \text{ lb/ft}^3 = 0.045 \text{ lb/ft}^3$$

Using Equation 6.5:

$$V = 1096 \sqrt{0.32/0.045} = 2923 \text{ fpm}$$

Or, using Equation 6.3, corrected V_p = 0.32/0.6

$$V = 4005\sqrt{V_p} = 4005\sqrt{0.32/0.6} = 2925 \text{ fpm}$$

Example 6.2 (Metric)   The measured velocity pressure (V_p) for airflow at 125 °C and an altitude of 1750 m is 80 Pa. Calculate the airflow velocity.

Solution   The air density correction factor from Table D-6 of the Appendix for 125 °C and 1750 m is 0.60.

$$d = 0.60 \times 1.204 \text{ kg/m}^3 = 0.7224 \text{ kg/m}^3$$

Using Equation 6.5:

$$V = 1.414 \sqrt{V_p/d} = 1.414 \sqrt{80/0.7224} = 14.88 \text{ m/s}$$

Or, using Equation 6.3, corrected V_p = 80/0.6

$$V = \sqrt{1.66 \times 980/0.6} = 14.88 \text{ m/s}$$

6.4   Duct Air Leakage

The amount of air leakage from duct systems is not a mystery. The results of extensive leakage research by SMACNA and ASHRAE co-incides with similar research done in Europe. Tables 6.1 and 6.2 and Figure 6.7 are based on these results. Now TAB technicians can predict the amount of air leakage based on the duct plans and specifications. The calculated amount of air leakage should be added to the fan capacity unless so noted either in the specifications or on the drawings that estimated air leakage has been included. If air leakage has been added, the TAB technician can easily verify actual system conditions, and the specified fan drive sizes should be usable after completion of the TAB work.

Many fan drives, fan motors, and electrical services to the fans have had to be changed because duct system air leakage was ignored by

TABLE 6.1 Applicable Leakage Classes[a]

Duct class	½, 1, 2 in.w.g. (125, 250, 500 Pa)		3 in.w.g. (750 Pa)	4, 6, 10 in.w.g. (1000, 1500, 2500 Pa)
Seal class	None	C	B	A
Applicable sealing	N/A	Transverse joints only	Transverse joints and seams	All joints, seams, and wall penetrations
Leakage class (C_L) cfm/100 ft^2 (L/s per m^2) at 1 in.w.g. (250 Pa)				
Rectangular metal	48	24	12	6
Round and oval metal	30	12	6	3
Rectangular fibrous glass	N/A	6	N/A	N/A
Round fibrous glass	N/A	3	N/A	N/A

[a]The leakage classes listed in this table are averages based on tests conducted by ASHRAE. Leakage classes listed are not necessarily recommendations on allowable leakage. The designer should determine allowable leakage and specify acceptable duct leakage classifications.

the system designer. The *leakage classes* in Table 6.1 are based on seal classes (how the ductwork is sealed with mastic). The leakage class designation is based on the average amount of leakage (cfm or L/s) of 100 ft^2 (or m^2) of duct surface with an average internal static pressure of 1 inch water gauge (250 Pa).

For example, 5000 ft^2 of rectangular duct at 1 in.w.g. using seal class C (transverse joints only) would leak 1200 cfm (5000 × 24/100 = 1200 cfm).

Using seal class B, the same ductwork would leak 600 cfm (5000 × 12/100 = 600 cfm). Using seal class A, the same ductwork would leak 300 cfm (5000 × 6/100 = 300 cfm). The above ductwork also would be constructed in the related pressure classification gauges and reinforcement.

6.4.2 Leakage prediction

Table 6.2 allows a system designer to predict the percentage of duct air leakage by dividing the HVAC unit fan capacity by the total square feet (square meter) of duct surface, using the leakage class and then the average internal duct static pressure. For example, if the above 5000-ft^2 duct system had a 15,000-cfm fan, the cfm/ft^2 in the second column of Table 6.2 would be 3 (15,000/5000 = 3). A duct system with a leakage class of 6 (seal class A) would have 4.1% air leakage at 3 in.w.g. average or 3.1% leakage at 2 in.w.g. average. Table 6.2 does not include HVAC unit or terminal unit connection leakage. This leakage also must be added by the TAB technician.

TABLE 6.2 Leakage as a Percentage of System Airflow

| Leakage class | System airflow | | Average static pressure [in.w.g. (Pa)] | | | | | |
	cfm/ft^2	L/s per m^2	1/2 (125)	1 (250)	2 (500)	3 (750)	4 (1000)	6 (1500)
48	2	10	15	24	38			
	2.5	12.7	12	19	30			
	3	15	10	16	25			
	4	20	7.7	12	19			
	5	25	6.1	9.6	15			
24	2	10	7.7	12	19			
	2.5	12.7	6.1	9.6	15			
	3	15	5.1	8.0	13			
	4	20	3.8	6.0	9.4			
	5	25	3.1	4.8	7.5			
12	2	10	3.8	6	9.4	12		
	2.5	12.7	3.1	4.8	7.5	9.8		
	3	15	2.6	4.0	6.3	8.2		
	4	20	1.9	3.0	4.7	6.1		
	5	25	1.5	2.4	3.8	4.9		
6	2	10	1.9	3	4.7	6.1	7.4	9.6
	2.5	12.7	1.5	2.4	3.8	4.9	5.9	7.7
	3	15	1.3	2.0	3.1	4.1	4.9	6.4
	4	20	**1.0**	1.5	2.4	3.1	3.7	4.8
	5	25	**.8**	1.2	1.9	2.4	3.0	3.8
3	2	10	**1.0**	1.5	2.4	3.1	3.7	4.8
	2.5	12.7	**.8**	1.2	1.9	2.4	3.0	3.8
	3	15	**.6**	**1.0**	1.6	2.0	2.5	3.2
	4	20	**.5**	**.8**	1.3	1.6	2.0	2.6
	5	25	**.4**	**.6**	**.9**	1.2	1.5	1.9

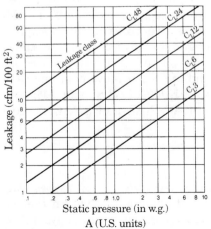

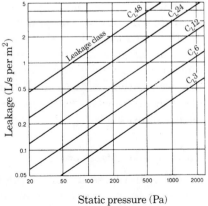

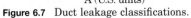

Figure 6.7 Duct leakage classifications.

6.4.3 "One percent leakage"

The often specified "maximum of 1% leakage" is almost impossible to attain under normal system design conditions (1% leakage or less are the bold numbers in Table 6.2). Two percent to 10% duct air leakage generally is found throughout the industry for average size HVAC systems. Note in Table 6.2 that the larger the system and/or the higher the average duct static pressure, the greater the duct leakage will be, even in the best, totally sealed duct systems.

Using the fan laws, a 3.1% air leakage will increase 10 fan brake horsepower to 10.96 BHP (using Equation 3.3 from Chapter 3):

$$\frac{FP_2}{FP_1} = \left(\frac{Q_2}{Q_1}\right)^3 ; \; FP_2 = 10 \left(\frac{1.031}{1.0}\right)^3 = 10.96 \text{ BHP}$$

So, 3.1% leakage causes almost a 10% increase in the fan power requirements. A more realistic 6% leakage would increase the 10 BHP requirements to 11.91 BHP, a 19% increase.

$$FP_2 = 10\left(\frac{1.06}{1.0}\right)^3 = 11.91$$

Unfortunately, many system designers do not include this extra airflow and energy requirements in their HVAC unit fan selection. However, TAB technicians should determine the amount of duct air leakage using field measurements.

6.5   System Effect

System effect is the derating or loss of capacity of a fan caused by poorly designed duct fittings at or close to the fan discharge and inlet. In addition to the generally unknown problem of system effect, **TAB technicians cannot measure system effect in the field.** Approximate fan capacity losses caused by system effect only can be calculated using dimensional measurements of the ductwork connections and data from tables and charts found in the Air Movement and Control Association (AMCA) Publication 201, *Fans and Systems*. This information also has been reprinted in ASHRAE and SMACNA duct design publications.

6.5.1   System effect curves

The diagonal lines in Figure 6.8 are called *system effect curves*. Any deviation from a piece of straight fan discharge duct within the "effective length" distance shown in Figure 6.9 may create a system ef-

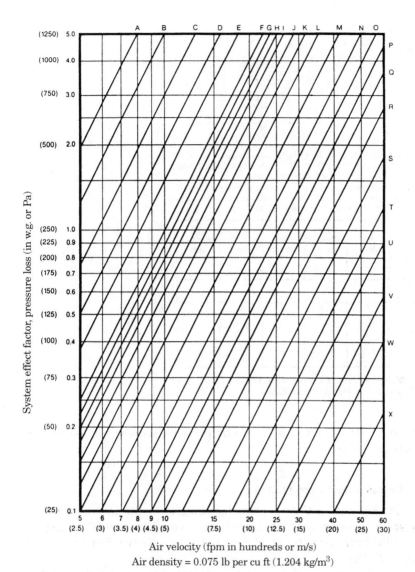

Air velocity (fpm in hundreds or m/s)
Air density = 0.075 lb per cu ft (1.204 kg/m³)

Figure 6.8 System effect curves. *(Reproduced with permission from AMCA)*

fect. The reason is that centrifugal fans are tested and rated by AMCA exactly as shown in Figure 6.9, i.e., no inlet duct and a straight discharge duct of at least 2½ duct diameters long.

For example, an exhaust fan on a roof without a discharge duct (assume a blast area ratio of 0.7 in Figure 6.9) would have an "S" system effect curve. Using Figure 6.8 [assume an outlet velocity of 2000 fpm (10 m/s)], a pressure loss of 0.2 in.w.g. (50 Pa) is obtained.

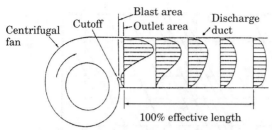

100% effective length

To calculate 100% effective duct length, assume a minimum of 2-1/2 duct diameters for 2500 fpm (12.5 m/s) or less. Add 1 duct diameter for each additional 1000 fpm (5 m/s).

Example: 5000 fpm (25 m/s) = 5 equivalent duct diameters. If the duct is rectangular with side dimensions a and b, the equivalent duct diameter is equal to $(4ab/\pi)^{0.5}$.

	No duct	12% effective duct	25% effective duct	50% effective duct	100% effective duct
Pressure recovery	0%	50%	80%	90%	100%
Blast area / Outlet area	System effect curve				
0.4	P	R-S	U	W	—
0.5	P	R-S	U	W	—
0.6	R-S	S-T	U-V	W-X	—
0.7	S	U	W-X	—	—
0.8	T-U	T-U	X	—	—
0.9	V-W	W-X	—	—	—
1.0	—	—	—	—	—

Figure 6.9   System effect curves for outlet ducts. *(Reproduced with permission from AMCA)*

That means that the rated fan static pressure capacity is *lowered* by 0.2 in.w.g. (50 Pa).

6.5.2   System effect from fittings

If an elbow (position C) was installed directly at the fan discharge, a "P" system effect curve would be obtained from Figure 6.10. Using the same 2000 fpm (10 m/s) in Figure 6.8 the system effect loss is 0.5 in.w.g. (125 Pa).

System effect losses also must be calculated for most fan inlet duct connections. Combined losses for both ends of the fan easily can exceed 1 in.w.g. (250 Pa) for each HVAC system fan.

6.5.3   Fan operation point

Figure 6.11 illustrates deficient fan or system performance caused by system effect. Curve A shows the HVAC system airflow capacity and pressure loss that has been calculated. The system designer selected the fan to operate at point 1 on system curve A. However, no allowance

Blast area Outlet area	Outlet elbow position	No outlet duct	12% effective duct	25% effective duct	50% effective duct	100% effective duct
0.4	A	N	O	P-Q	S	
	B	M-N	N	O-P	R-S	
	C	L-M	M	N	Q	
	D	L-M	M	N	Q	
0.5	A	O-P	P-Q	R	T	
	B	N-O	O-P	Q	S-T	
	C	M-N	N	O-P	R-S	
	D	M-N	N	O-P	R-S	
0.6	A	Q	Q-R	S	U	
	B	P	Q	R	T	
	C	N-O	O	Q	S	No system effect factor
	D	N-O	O	Q	S	
0.7	A	R-S	S	T	V	
	B	Q-R	R-S	S-T	U-V	
	C	P	Q	R-S	T	
	D	P	Q	R-S	T	
0.8	A	S	S-T	T-U	W	
	B	R-S	S	T	V	
	C	Q-R	R	S	U-V	
	D	Q-R	R	S	U-V	
0.9	A	T	T-U	U-V	W	
	B	S	S-T	T-U	W	
	C	R	S	S-T	V	
	D	R	S	S-T	V	
1.0	A	T	T-U	U-V	W	
	B	S-T	T	U	W	
	C	R-S	S	T	V	
	D	R-S	S	T	V	

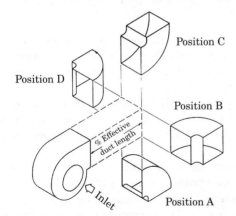

Position C

Position D

Position B

% Effective duct length

Position A

Inlet

For DWDI fans determine SEF using the curve for
SWSI fans. Then apply the appropriate multiplier from
the tabulation below

MULTIPLIERS FOR DWDI FANS

Elbow position	A	= $\Delta P \times 1.00$
Elbow position	B	= $\Delta P \times 1.25$
Elbow position	C	= $\Delta P \times 1.00$
Elbow position	D	= $\Delta P \times 0.85$

Figure 6.10 Outlet elbows on SWSI centrifugal fans.
(Reproduced with permission from AMCA)

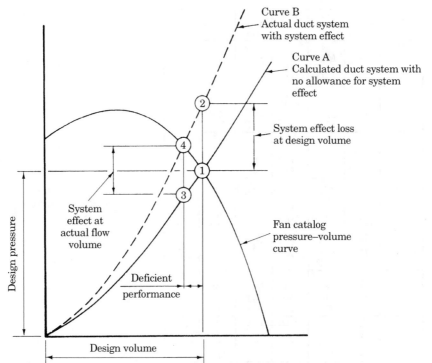

Figure 6.11 Effects of system effect. *(Reproduced with permission from AMCA)*

has been made for the effect of duct system connections on the fan performance. To compensate, a system effect factor must be added to the calculated system pressure losses to determine a new system curve that should be used to select the fan.

The point of intersection between the initially selected fan performance curve and this new "phantom" system curve B is point 4. Therefore, the actual system flow volume is deficient by the difference from point 1 to point 4.

To achieve the design airflow volume, a system effect factor equal to the pressure difference between point 1 and point 2 should be added to the calculated system pressure losses. The fan should be selected to operate at point 2, where a new, higher corrected rpm curve crosses system curve B. A higher fan brake horsepower also will occur.

When a TAB technician measures the actual HVAC system conditions with the corrected fan rpm, the airflow volume and static pressure will be established as point 1, because that is where the system actually is operating. The system is *not* operating on system curve B, which was used only to select the derated capacity fan rpm. System effect *cannot* be measured in the field, but can only calculated after a visual inspection is made to the fan and duct system connections.

Because system effect is velocity related, the difference between points 1 and 2 is greater than the difference between points 3 and 4. The system effect factor includes only the effect of the system configuration on the fan's performance. All duct fitting pressure losses were calculated as part of the HVAC system pressure losses and remain a part of system curve A.

6.6   Increased System Pressures

If inspections are not made during HVAC system installation, extra high-pressure drop fittings may be installed to avoid obstacles without the knowledge of either the design engineer or the TAB technician. Some sheetmetal contractors also may leave out every other turning vane in mitered 90° elbows.

Example 6.3   To avoid obstacles, eight 90° elbows are added to a primary air duct system designed with a 4.0 in.w.g. (1000 Pa) pressure loss, which includes six 90° elbows. Every other 4½ in., single-thickness turning vane is missing from all 14 of the installed elbows. The fitting loss coefficient with every other turning vane missing is 0.46. Find the new system pressure loss.

Solution   Use duct fitting pressure loss data from the design tables in Table A-9 of Appendix A, [assume a velocity of 2000 fpm (10 m/s)]. For fourteen 90° elbows, with every other vane missing:

Actual: $14 \times 0.25\ (V_p) \times 0.46\ (C) = 1.61$ in.w.g. (403 Pa)
Design: $6 \times 0.25\ (V_p) \times 0.23\ (C) = 0.80$ in.w.g. (200 Pa)

$$SP \text{ increase} \quad = 0.81 \text{ in.w.g. (203 Pa)}$$

Therefore the actual system loss is 4.81 in w.g. (1203 Pa), a system static pressure (*SP*) increase of 20%: 4.0 + 0.81 = 4.81 (1000 + 203 = 1203).

Use of TAB Instruments (Airflow)

7.1 TAB Instruments

Instruments for the measurement of airflow, water flow, rotation, temperature, humidity, and electric current are the "tools of the trade" of the testing, adjusting, and balancing (TAB) firm and its employees. Recommended uses and limitations for most of the TAB instruments may be found in Tables 7.1 to 7.4, which by experience, are reliable and accurate. Many of the newer electronic types also are accurate and reliable, and they may measure several different functions. Certified NEBB TAB firms are required to own, maintain, and periodically calibrate a required list of TAB instruments (see the NEBB *Procedural Standards for Testing Adjusting and Balancing of Environmental Systems*). Airflow TAB measuring instruments are discussed in this chapter and all other TAB instruments are discussed in Chapter 8.

7.2 Pitot Tube

As shown in Figure 7.1, a *Pitot tube* contains an impact tube for measuring total pressure, fastened concentrically inside a second tube of slightly larger diameter that senses static pressure through radially located sensing holes near the tip. The air space between the inner and outer tubes serves to transfer pressure from the sensing holes to these static pressure connection at the opposite end of the Pitot tube. When the static pressure outlet is connected to the low or negative pressure side of a manometer and the total pressure connection of the Pitot tube is connected to the high-pressure side of the manometer, velocity pressure is indicated directly on the instrument (see Figures 6.2 to 6.5 in Chapter 6). The Pitot tube is an important measuring

TABLE 7.1   Airflow-Measuring Instruments

Instrument	Recommended Uses	Limitations
U-Tube manometer	Measuring pressure of air and gas above 1.0 in.w.g. (250 Pa) Measuring low manifold gas pressures	Manometer should be clean and used with correct fluid. Should not be used for readings under 1 in.w.g. (250 Pa) of differential pressure.
Vertical inclined manometer	Measuring pressure of air and gas above 0.02 in.w.g. (5 Pa) Normally used with Pitot tube or static probe for determination of static, total, and velocity pressures in duct systems	Field calibration and leveling is required before each use. For extremely low pressures, a micromanometer or some other sensitive instrument should be used for maximum accuracy.
Micromanometer (electronic)	Measuring very low pressure or velocities Used for calibration of other instrumentation	Because some instruments utilize a time-weighted average for each reading, it is difficult to measure pressures with pulsations.
Pitot tube	Used with manometer for determination of total, static, and velocity pressures	Accuracy depends on uniformity of flow and completeness of duct traverse. Pitot tube and tubing must be dry, clean, and free of leaks and sharp bends or obstructions.
Pressure gauge (magnehelic)	Used with static probes for determination of static pressure or static pressure differential	Readings should be made in the midrange of the scale. Should be "zeroed" and held in same position. Should be checked against known pressure source with each use.
Anemometer rotating vane (mechanical and electronic)	Measurement of velocities at air terminals, air inlets, and filter or coil banks	Total inlet area of rotating vane must be in measured airflow. Correction factors may apply; refer to manufacturer data.
Anemometer deflecting vane	Measurement of velocities at air terminals and air inlets	Instruments should not be used in extreme temperature or contaminated conditions.
Anemometer thermal	Measurement of low velocities such as room air currents and airflow at hoods, troffers, and other low-velocity apparatus.	Care should be taken for proper use of instrument probe. Probes are subject to fouling by dust and corrosive air. Should not be used in flammable or explosive atmospheres. Temperature corrections may apply.

TABLE 7.1 *(Continued)*

Instrument	Recommended Uses	Limitations
Flow-measuring hood	Measurement of air distribution devices directly in CFM (L/s)	Flow-measuring hoods should not be used where the discharge velocities of the terminal devices are excessive. Flow-measuring hoods redirect the normal pattern of air diffusion, which creates a slight, artificially imposed pressure drop in the duct branch. Capture hood used should provide a uniform velocity profile at sensing grid or device.

TABLE 7.2 Hydronic Measuring Instruments

Instrument	Recommended Uses	Limitations
U-Tube manometer	Measuring pressure drops through heat exchange equipment, orifices, and venturi tubes	Manometer should be clean and used with correct fluid. Use collecting safety reservoirs on each side of a mercury manometer to prevent discharge of mercury into hydronic system, which can cause rapid deterioration of any copper it touches in the system. Should not be used for readings under one in.w.g. (250 Pa) of differential pressure.
Pressure gauge (calibrated)	Static pressure measurements of system equipment and/or piping	Pressure gauges should be selected so the pressure to be measured fall in the middle two thirds of the scale range. Gauge should not be exposed to pressures greater than or less than dial range. Pressures should be applied slowly to prevent severe strain and possible loss of accuracy of gauge.
Pressure gauge (differential)	Differential pressure measurements of system equipment and/or piping	Same as pressure gauge.
Flow-measuring devices	Used to obtain highly accurate measurement of volume flow rates in fluid systems	Must be used in accordance with recommendations of equipment manufacturer.

TABLE 7.3 Rotation Measuring Instruments

Instrument	Recommended Uses	Limitations
Revolution counter	Contact measurement of rotating equipment speed	Requires direct contact of rotating shaft. Must be used in conjunction with accurate timing device.
Chronometric tachometer	Contact measurement of rotating equipment speed	Requires direct contact of rotating shaft.
Contact tachometer	Contact measurement of rotating and linear speeds	Requires direct contact of rotating shaft or device to be measured.
Electronic tachometer (stroboscope)	Noncontact measurement of rotating equipment	Readings must be started at lower end of scale to avoid reading multiples (or harmonics) of the actual rpm.
Optical tachometer	Noncontact measurement of rotating equipment	Must be held close to object and at correct angle. Rotating device must use reflective markings.
Dual function tachometer	Contact or noncontact measurement of rotating equipment and linear speeds	Same as optical tachometer.

device and for accuracy, it must be carefully made. They are available in various lengths, normally from 8 to 60 in. (200 to 1500 mm).

Because the Pitot tube can be used to measure any one of three basic pressures (total pressure, static pressure, and velocity pressure), when used with the proper hose hookups between the Pitot tube and the manometer (as shown in Figures 6.2 to 6.5 in Chapter 6) it is accurate and reliable and is the preferred method of measuring air velocities and pressures in the field.

Smaller pocket-sized Pitot tubes can be used in ducts smaller than 8 in. (200 mm) in diameter. To ensure accurate sensing of total pressure, any size Pitot tube tip must be pointed directly into, or parallel with, the airstream. The Pitot tube tip is parallel with the static pressure outlet tube, and so the latter can be used as a pointer to align the tip properly.

The accuracy of the readings of the Pitot tube in a duct depend on the uniformity of the airflow in a cross section of the duct. Pitot tube traverses should be taken in a length of straight duct, preferably 6 to 10 duct diameters downstream from any elbows, branches, transitions, or other obstructions to uniform airflow and at least several duct diameters upstream.

Pitot tubes have smooth, well-rounded tips to minimize turbulence. They should be treated with care so that they do not become bent or

TABLE 7.4   Temperature Measuring Instruments

Instrument	Recommended Uses	Limitations
Glass tube thermometers	Measurement of temperatures of air and fluids	Ambient conditions may impact measurement of fluid temperature. Glass tube thermometers require immersion in fluid or adequate test wells. Some applications prohibit use of instruments containing mercury within the work area.
Dial thermometers	Measurement of temperatures of air and fluids	Ambient conditions may impact measurement of fluid temperature. Stem or bulb must be immersed a sufficient distance in fluid to record accurate measurement. Time lag of measurement is relatively long.
Thermocouple thermometers	Measurement of surface temperatures of pipes and ducts	Surface temperatures of piping and duct may not equal fluid temperature within because of thermal conductivity of material.
Electronic thermometers	Measurement of temperatures of air and fluids Measurement of surface temperatures of pipes and ducts	Use instrument within recommended range. Use thermal probes in accordance with recommendations of manufacturer.
Psychrometers	Measurement of dry and wet bulb air temperatures	Accurate wet bulb measurements require an air velocity between 1000 and 1500 fpm (5 to 7.5 m/s) across the wick, or a correction must be made. Dirty or dry wicks will result in significant error.
Electronic thermohygrometer	Measurement of dry and wet bulb air temperatures and direct reading of relative humidity	Accuracy of measurement above 90% relative humidity is decreased because of swelling of the sensing element.

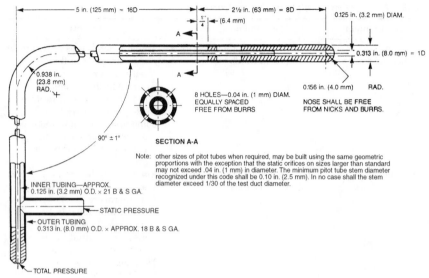

Figure 7.1 Pitot tube.

mashed; the noses should be protected so they do not become dented or otherwise roughened, and the small static pressure sensing holes should be kept open and clean.

7.2.1 Rectangular duct traverses

The primary use of the Pitot tube by the TAB technician will be measuring velocity pressures (velocities) in ducts to determine the duct airflow. Because the velocity pressures are not uniform across the duct, a series of readings must be made. The procedures for this are listed in the following sections.

7.2.1.1 Duct area. Measure the size of the duct. This means the free inside dimensions of the duct where the air is passing through. If the duct has a fibrous glass lining, the dimensions inside the insulation lining are what you use. From these dimensions, determine the cross-sectional area by multiplying the height in inches times the width in inches divided by 144. This will give you the duct cross-sectional area (A) in the required square feet. The duct cross-sectional area is in square meters in the metric system.

7.2.1.2 Method no. 1 (equal area). To perform a Pitot tube traverse of a duct, the readings must be taken in the duct at equal intervals (Figure 7.2). A rectangular duct, such as a 48 × 36 in. (1200 × 900 mm)

METHOD No. 1
(EQUAL AREA)

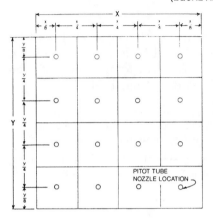

To determine the average air velocity in square or rectangular ducts, a Pitot tube traverse must be made to measure the velocities at the center points of equal areas over the cross section of the duct. The number of equal areas should not be less than 16 but need not be more than 64. The maximum distance between center points, for less than 64 readings, should not be more than 6 inches (150 mm). The readings closest to the ductwalls should be taken at one-half of this distance. For maximum accuracy, the velocity corresponding to each velocity pressure measured must be determined and then averaged.

METHOD No. 2
(LOG)

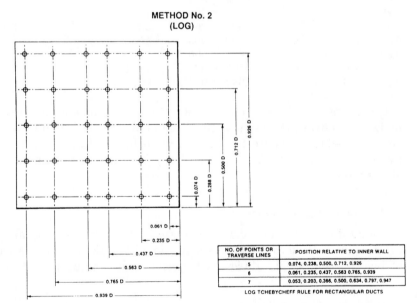

NO. OF POINTS OR TRAVERSE LINES	POSITION RELATIVE TO INNER WALL
5	0.074, 0.238, 0.500, 0.712, 0.926
6	0.061, 0.235, 0.437, 0.563 0.765, 0.939
7	0.053, 0.203, 0.366, 0.500, 0.634, 0.797, 0.947

LOG TCHEBYCHEFF RULE FOR RECTANGULAR DUCTS

Figure 7.2 Pitot tube traverse points (rectangular ducts).

duct, should be divided into equal areas and 48 readings taken. A minimum of 16 readings should be taken in small ducts and the readings should be no more than 6 in. (150 mm) apart for maximum accuracy. Rectangular duct traverse reports should be prepared in advance for each duct traverse to be made. Readings must be taken in the center of each area.

7.2.1.3   Method no. 2 (log).   To perform the log "Tchebycheff" Pitot tube traverse procedure, the traverse points shall be located at the intersection of the traverse lines (see Figure 7.2). The total number of readings would vary from 25 to 49 based on data from Table 7.5.

Example 7.1   A 36 × 24 in. (900 × 600 mm) duct would have seven traverse lines in the 36 in. (900 mm) dimension and five traverse lines in the 24-in. (600 mm) dimension. Determine the dimensions for the total of 35 readings.

Solution   Measured from the Pitot tube entry side, the dimensions for the readings [36 in. (900 mm)] depth would be (using the table in Figure 7.2):

U.S.	Inches	Metric	mm
0.053 × 36	= 1.91	0.053 × 900	= 47.7
0.203 × 36	= 7.31	0.203 × 900	= 182.7
0.366 × 36	= 13.18	0.366 × 900	= 329.4
0.500 × 36	= 18.0	0.500 × 900	= 450.0
0.634 × 36	= 22.82	0.634 × 900	= 570.6
0.797 × 36	= 28.69	0.797 × 900	= 717.3
0.947 × 36	= 34.09	0.947 × 900	= 852.3

From Example 7.1, it is easy to see that using metric dimensions makes the TAB work easier, as the decimal number in U.S. units must be converted to the nearest fraction of an inch. The dimensions for the five traverse lines [24 in. (600 mm)] depth would be calculated in a similar manner.

7.2.2   Round duct traverses

In round ducts, readings should be taken at the centers of equal concentric areas. Preferably 20 readings should be taken, 10 along each of two diameters. The divisions between each reading are to be of

TABLE 7.5   Duct Size and Traverse Lines

Duct side dimension	Number of traverse lines
<30 in. (750 mm)	5
30 to 36 in. (750 to 900 mm)	6
>36 in. (900 mm)	7

equal area. This means that the dimensions themselves will not be the same (Figure 7.3). Fewer readings may be taken with smaller ducts. A convenient chart (Table 1 in Figure 7.3) can be used for round ducts up to 36 in. (900 mm) in diameter, which gives the dimensions in inches from the pipe center for the test locations. For larger ducts, Table 2 in Figure 7.3 has the necessary constants, which are multiplied by the diameter of the duct being traversed. Once these dimensions have been determined, round duct traverse reports for each set of traverses being made can be readied.

Example 7.2 (U.S.) A 10-point traverse is made both horizontally and vertically across the duct. A 20 in. diameter duct requires 10 readings in each direction or 20 readings (see Table 2 in Figure 7.3). Calculate the dimensions for the traverses.

Solution Reading points can be calculated from Table 2 in Figure 7.3 as follows:

No. 1 = 20 × 0.1581 = 3.16 in. from the center
No. 2 = 20 × 0.2738 = 5.48 in. from the center
No. 3 = 20 × 0.3535 = 7.08 in. from the center
No. 4 = 20 × 0.4183 = 8.37 in. from the center
No. 5 = 20 × 0.4743 = 9.49 in. from the center

Example 7.2 (Metric) A 10-point traverse is made both horizontally and vertically across the duct. A 500 mm duct requires 20 readings (see Table 1 in Figure 7.3). Calculate the dimensions for the traverses.

Solution

No. 1 = 500 × 0.1581 = 79.1 mm from the center
No. 2 = 500 × 0.2738 = 136.9 mm from the center
No. 3 = 500 × 0.3535 = 176.8 mm from the center
No. 4 = 500 × 0.4183 = 209.1 mm from the center
No. 5 = 500 × 0.4743 = 237.2 mm from the center

Note that the dimensions in Example 7.2 check with those in Table 1 of Figure 7.3, and that the readings are numbered from the center of the duct. Reading point 1 is 3⅛ inches (79 mm) from the center of the duct. All of the number 5 readings are about ½ inch (13 mm) from the outside of the duct or 9½ inches (237 mm) from the center. These points will divide the round duct into equal areas and assure an accurate velocity pressure profile at five different points in each quadrant of the duct.

7.2.3 Using the Pitot tube

1. Once the Pitot tube traverse hole dimensions have been determined, the TAB technician can mark off the Pitot tube. Electrical

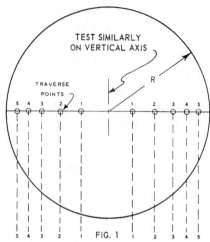

Fig. 1 shows the locations for Pitot tube tip making a 10-point traverse across one circular pipe diameter. In making two traverses across the pipe diameter, readings are taken at right angles to each other. The traverse points shown represent 5 annular zones of equal area.

TABLE 1

Duct diam.			Distances of Pitot Tube Tip From Pipe Center								
		Readings in one diam.	Point 1		Point 2		Point 3		Point 4		Point 5	
inches	(mm)		inches	(mm)	inches	(mm)	inches	(mm)	inches	(mm)	inches	(mm)
3	75	6	.612	15.3	1.061	26.5	1.369	34.2				
4	100	6	.812	20.4	1.414	35.4	1.826	45.6				
5	125	6	1.021	25.5	1.768	44.2	2.282	57.1				
6	150	6	1.225	30.6	2.121	53.0	2.738	68.5				
7	175	6	1.429	35.7	2.475	61.9	3.195	80.0				
8	200	6	1.633	40.8	2.828	70.7	3.651	91.3				
9	225	6	1.837	45.9	3.182	79.5	4.108	102.7				
10	250	8	1.768	44.2	3.062	76.6	3.950	98.8	4.677	116.9		
12	300	8	2.122	53.0	3.674	91.1	4.740	118.6	5.612	140.3		
14	350	10	2.214	55.3	3.834	95.8	4.950	123.7	5.857	146.4	6.641	166.0
16	400	10	2.530	63.2	4.382	109.5	5.657	141.4	6.693	167.3	7.589	189.7
18	450	10	2.846	71.1	4.929	123.2	6.364	159.1	7.530	188.2	8.538	213.4
20	500	10	3.162	79.1	5.477	136.9	7.077	176.8	8.367	209.1	9.487	237.2
22	550	10	3.479	87.0	6.025	150.6	7.778	194.4	9.203	230.1	10.435	260.9
24	600	10	3.795	94.9	6.573	164.3	8.485	212.1	10.040	251.0	11.384	284.6
26	650	10	4.111	102.8	7.120	178.0	9.192	229.8	10.877	271.9	12.222	308.3
28	700	10	4.427	110.7	7.668	191.7	9.900	247.5	11.713	292.8	13.282	332.0
30	750	10	4.743	118.6	8.216	205.4	10.607	265.1	12.550	313.7	14.230	355.7
32	800	10	5.060	126.5	8.764	219.0	11.314	282.8	13.387	334.6	15.179	379.4
34	850	10	5.376	134.4	9.311	232.7	12.021	300.5	14.233	355.6	16.128	403.2
36	900	10	5.692	142.3	9.859	246.4	12.728	318.2	15.060	376.5	17.176	426.9

For distances of traverse points from pipe center for pipe diameters other than those given in Table No. 1, use constants in Table No. 2.

TABLE 2

Readings in one Diameter	Constants To Be Multiplied By Pipe Diameter For Distances of Pitot Tube Tip From Pipe Center				
	P.1	P.2	P.3	P.4	P.5
6	.2041	.3535	.4564		
8	.1768	.3062	.3953	.4677	
10	.1581	.2738	.3535	.4183	.4743

Figure 7.3 Pitot tube traverse points (round ducts).

plastic tape will work well and can usually be reused several times.

2. Using the dimensions determined in Example 7.2, mark off the locations on the duct where you will need holes for the Pitot tube. The holes should be located in a section of straight duct from 6 to 10 duct diameters long. They should be nearer to the downstream end of a straight section of duct. Quite often, this much straight duct is not available. Therefore, you will have to use the best or longest straight section available, possibly with a sacrifice in accuracy.

3. Never traverse in an elbow, offset, or transition. Allow room to swing the Pitot tube while inserting and removing it from the duct. Many times you may need to use Pitot tubes of different lengths in the same hole because of a lack of room beside the duct to move the tube in and out. Do not locate holes in the bottom of ducts that carry moisture- or grease-laden air; instead, put these holes in the side of the duct where possible.

4. Drill holes in the duct slightly larger than the Pitot tube diameter, usually $\frac{3}{8}$ in. (9.5 mm) diameter. Where the duct is insulated make provision for neat repair of insulation.

5. The holes will have to be capped when the testing is complete. Snap-in plastic and rubber caps are available. Some specifications will call for metal test ports with a screw-on cap. These are available in various lengths for use with insulated ducts.

6. Set up the manometer in a convenient location. After connecting the tubing from the manometer to the Pitot tube, be sure to level and zero the manometer if required. Use a good grade of tubing that is thick enough that it will not kink, but thin enough to be flexible. Clear plastic or rubber tubing that has a $\frac{3}{16}$-in. (5 mm) inside diameter [usually $\frac{5}{16}$-in. (8 mm) outside diameter] works well.

7. Take velocity pressure readings at each location and record the readings on test report forms in the appropriate places. It is important that the Pitot tube be held straight and directly into the airstream. Notice that the static pressure port faces the same direction as the Pitot tube and can be used as a guide. Be sure to hold the Pitot tube at each position long enough to let the manometer fluid settle into position. If the fluid does not stay steady while the Pitot tube is at any one location, air turbulence may be indicated. Record the average reading midway between the points.

8. Turbulence should be noted on the test report form because turbulent readings will not be accurate. It is always advisable to take

static pressure readings at the Pitot tube duct opening. Some specifications require it, and it can be very useful for future adjustments or troubleshooting.

9. Once the readings are recorded they will have to be converted to duct velocities before they can be totaled and averaged. When adding the readings, the values must be velocities in feet per minute or in meters per second. **Velocity pressures cannot be totaled and averaged.** Unless every reading in the traverse has been identical (a very unlikely condition) adding and averaging the velocity pressures will result in erroneous results. Since many inclined manometers have scales that also read velocities in feet per minute or in meters per second, you can record these figures instead of the velocity pressures. These figures can be added and averaged; otherwise it will be necessary to convert the velocity pressures to velocities.

10. When measuring internally insulated or fiberglass ducts, it is wise to stop occasionally and blow out any glass fibers that collect in the Pitot tube to maintain as accurate a set of readings as possible. Passing the tube through the fiberglass does dislodge small amounts of fiberglass and will eventually clog the tube. Remove the plastic tube connections to the manometer and blow into the bottom of the inside tube to discharge any fiberglass fibers from the head end of the tube.

11. It is not unusual to get a negative pressure reading in ducts with considerable turbulence. The negative readings are added in at *zero* value but are counted in the number of readings to obtain the average velocity. Assume a duct with 16 positive pressure readings and four negative pressure readings. The 16 positive pressure readings would be added together and averaged by all 20 readings (16 positive plus 4 zero readings).

7.3   U-Tube Manometer

The *U-tube manometer* is a simple and useful means of measuring partial vacuums and pressures. It is so universally used that both the inch (millimeter) of water and the inch (millimeter) of mercury have become accepted units of pressure measurements. A manometer consists of a U-shaped glass or plastic tube partially filled with a liquid such as tinted water, oil, or mercury. The difference in height of the two fluid columns denotes the pressure differential. U-Tube manometers are made in different sizes and are recommended for measuring pressure drops above 1.0 in.w.g. (250 Pa or 25 mm w.g.) across filters, coils, fans, terminal devices, and sections of ductwork, but are not recommended for readings of less than 1.0 in.w.g. (250 Pa).

Mercury-filled manometers may not be used in many environments. Extreme care must be taken when they are used, following instructions for use of overpressure traps or three-valve cluster manifolds. Also see subsection 8.21 in Chapter 8.

7.4   Inclined and/or Vertical Manometer

The *inclined and/or vertical manometer* (Figure 7.4) for airflow pressure readings is usually constructed from a solid transparent block of plastic. It has an inclined scale that provides accurate air pressure readings below 1.0 in.w.g. (250 Pa) and a vertical scale for reading greater pressures. Note that all air pressures are given in *inches of water* (millimeters or pascals). For example, 3.0 in. of water (75 mm

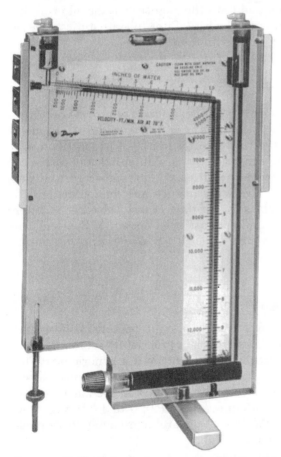

Figure 7.4 Inclined–vertical manometer. *(Courtesy of Dwyer Instruments Inc.)*

of water or 750 Pa) means that the air pressure on one end of a U-shaped tube is enough to force the water 3 in. (75 mm) higher in the other leg of the tube. Instead of water, this instrument uses colored oil, which is lighter than water. This means that although the scale reads in inches of water (millimeters), it is longer than a standard rule measurement. Whenever a manometer is used, the oil must be at normal room temperatures or the reading will not be correct. The manometer must be set level and mounted so it does not vibrate. Note the leveling screw and the magnetic clips.

Some manometers have two scales—one indicating some pressure in inches (millimeters) of water and the other indicating velocity in feet per minute (meters per second).

The manometer (or inclined draft gauge) is the standard in the industry. It can be read accurately down to approximately 0.03 in.w.g. (7 Pa) and contains no mechanical linkage. It is simple to adjust by setting the piston at the bottom until the meniscus of the oil is on the zero line. This instrument is used with a Pitot tube or static probe to determine pressures or air velocity in a duct.

7.5   Electronic Manometer

The *electronic manometer* (Figure 7.5) is designed to provide accurate readings at very low differential pressures. Some multimeters measure an extremely wide range of pressures from 0.001 to 60.00 in.w.g. (0.025 to 15,000 Pa). Airflow and velocity are automatically corrected for the density effect of barometric pressure and temperature. Readings can be stored and recalled with *average* and *total* functions. A specially designed grid enables the reading of face velocities at filter outlets, coil face velocities, and exhaust hood openings. Some multimeters provide additional functions such as temperature measurement.

7.6   Pressure Gauge (Magnehelic)

The *Magnehelic gauge* is a dry type diaphragm operated differential pressure gauge. Although *Magnehelic* is a proprietary name, its general use in the TAB industry has almost made it a generic one. The gauge is extremely sensitive and accurate, but it is resistant to shock and vibration. A zero calibration screw is located on the plastic cover.

Common ranges are 0 to 0.5 in.w.g. (125 Pa), 0 to 1.0 in.w.g. (250 Pa), and 0 to 5.0 in.w.g. (1250 Pa). There are approximately 30 pressure ranges of this instrument available.

When using the Magnehelic gauge:

Figure 7.5 Electronic manometer. *(Courtesy of Shortridge Instruments Inc.)*

- Readings should be made in midrange of the scale.
- The gauge should not be mounted on a vibrating surface.
- The gauge should be held in same position as *zeroed*, and in the vertical position.
- It should be checked against a known pressure source with each use.

7.7 Rotating Vane Anemometer

The basic propeller or *rotating vane anemometer* consists of a light-weight, wind-driven wheel connected through a gear train to a set of recording dials that read the linear feet of air passing through the wheel in a measured length of time. The instrument is made in various sizes, 3 in. (75 mm), 4 in. (100 mm), and 6 in. (150 mm) sizes being the most common.

At low velocities, the friction drag of the mechanism is considerable. To compensate for this, a gear train that overspeeds is commonly used. For this reason, the correction is often additive at the lower range and subtractive at the upper range, with the least correct in the middle of the 200 to 2000 fpm (1 to 10 m/s) range. Most older instruments are not sensitive enough for use below 200 fpm (1 m/s). Newer instruments can read velocities as low as 30 fpm (0.15 m/s).

The rotating vane anemometer reads in feet (meters), and so a timing instrument must be used to determine velocity. Readings are usually timed for 1 min, in which case the anemometer reading (when corrected according to a calibration curve) will give the result in feet per minute or meters per minute (divide by 60 for m/s). For moderate velocities, it may be satisfactory to use a ½ min timed interval, repeated as a check. A stop watch should be used to measure the timed interval, although a wristwatch with a sweep second hand may give satisfactory results for rough field checks.

In coils or filters, an uneven airflow is frequently found because of entrance or exit conditions. This variation is taken into account by moving the instrument in a fixed pattern to cover the entire amount of time over all parts of the area being measured so that the varying velocities can be averaged.

If the opening is covered with a grille, the instrument should touch the grille face but should not be pushed in between the bars. For a free opening without a grille, the anemometer should be held in the plane of the entrance edges of the opening. The anemometer always must be held so that the airflow through the instrument is in the same direction as was used for calibration (usually from the back toward the dial face). The manufacturer's recommendations must be followed very carefully when using this instrument.

Each reading must be corrected by a calibration chart, and the outlet correction factor (A_k or k factor) of a grille or register being measured must be used in calculating air quantities. Newer, electronic rotating vane anemometers (Figure 7.6) are battery operated, with direct digital or analog readouts with interchangeable remote rotating vane heads. The digital readout of the velocity is automatically averaged for a fixed time period depending on the measured velocity and the type of instrument. Analog instruments are direct readout with a choice of velocity scales.

7.8   Deflecting Vane Anemometer

The *deflecting vane anemometer* (Figure 7.7) operates by having a pressure exerted on a vane that causes a pointer to indicate the measured value. It is not dependent on air density because of the sensing of pressure differential to indicate velocities. The instrument is provided and always used with a dual hose connection between the meter and the probes, except as noted below.

A deflecting vane anemometer set traditionally has met the needs of TAB work, because most major air distribution device manufactur-

Figure 7.6 Electronic rotating vane anemometer. *(Courtesy of Alnor Instrument Company.)*

ers have published area (A_k) factors. However, many no longer furnish A_k factors. The set consists of the meter, measuring probes, range selectors, and connecting hoses. Meters are scaled through the following velocity ranges: 0 to 300 fpm (0 to 1.5 m/s); 0 to 1250 fpm (0 to 6.25 m/s); 0 to 2500 fpm (0 to 12.5 m/s); 0 to 5000 fpm (0 to 25.0 m/s); 0 to 10,000 fpm (0 to 50 m/s).

The deflecting vane anemometer may be used for the following:

Figure 7.7 Deflecting vane anemometer set. *(Courtesy of Alnor Instrument Co.)*

- Measurements of air velocity through both supply and return air terminals using the proper jet and the proper air terminal "k or A_k factors" (when available) for the airflow calculation

- Measuring some lower velocities where the instrument case is placed in the airstream

- Measuring static pressure

Equation 7.1

$$Q = V_c \times A_k$$

Where: Q = airflow (cfm or L/s)
V_c = velocity (or corrected velocity) (fpm or m/s)
A_k (or k) = outlet correction factor

Example 7.3 (U.S.) A manufacturer has published an A_k factor of 0.43 for the grille being measured. The corrected velocity reading for the anemometer is 987 fpm. Find the airflow.

Solution

$$Q = V_c \times A_k = 987 \times 0.43 = 424 \text{ cfm}$$

Example 7.3 (Metric) A manufacturer has published an A_k factor of 40.2 for the grille being measured. The corrected velocity reading for the anemometer is 4.96 m/s. Find the airflow.

Solution

$$Q = V_c \times A_k = 4.96 \times 40.2 = 199 \text{ L/s}$$

Example 7.4 (U.S.) A diffuser manufacturer states that four readings are to be taken equally spaced around the outer ring of the diffuser. The k factor for an 8-in. diameter diffuser with the cones up is 0.20. The readings are 920 fpm, 1160 fpm, 1200 fpm, and 1045 fpm. Find the diffuser airflow.

Solution Obtain the average velocity reading. Multiple the average velocity (fpm) times the k factor to obtain the cfm.

Reading	Velocity
1	920 fpm
2	1160 fpm
3	1200 fpm
4	1045 fpm
Total	4325 fpm/4 = 1081 fpm

$$Q = V_c \times A_k$$

$$Q = 1081 \text{ fpm} \times 0.20 = 216 \text{ cfm}$$

Example 7.4 (Metric) A diffuser manufacturer states that four readings are to be taken equally spaced around the outer ring of the diffuser. The k factor for a 200 mm diameter diffuser with the cones up is 18.8. The readings are 4.60 m/s, 5.78 m/s, 6.01 m/s, and 5.21 m/s. Find the diffuser airflow.

Solution Obtain the average velocity readings. Multiple the average velocity (m/s) times the k factor to obtain the L/s.

Reading	Velocity
1	4.60 m/s
2	5.78 m/s
3	6.01 m/s
4	5.21 m/s
Total	21.60 m/s/4 = 5.4 m/s

$$Q = V_c \times A_k = 5.4 \times 18.8 = 102 \text{ L/s}$$

7.9   Thermal Anemometer

The operation of a *thermal-type anemometer* depends on the fact that the resistance of a heated wire will change with its temperature. The probe of this instrument is provided with a special type of wire element that receives current from batteries contained in the instrument case. As air flows over the element in the probe, the temperature of the element is changed from that which exists in still air, and the resistance change is indicated as a velocity on the indicating scale of the instrument.

Most thermal anemometers can be used to measure very low airflow velocities and to make duct velocity traverses. The probe that is used with this instrument is quite directional and must be located at the proper point on the diffuser or grille as indicated by the manufacturer. They should not be used in flammable or explosive atmospheres. Corrections also must be made for the temperature of the measured air and the sensing tip must be kept clean to maintain accuracy.

7.10   Flow Measuring Hood

The *flow measuring hood* (Figure 7.8) rapidly is becoming the most popular instrument in the TAB industry for measuring the airflows of all types of registers, grilles and diffusers. They read directly in cfm (L/s), and they eliminate the need for calculations and for obtaining

Figure 7.8 Flow-measuring hood. *(Courtesy of Shortridge Instruments, Inc.)*

"k or A_k factors" and other exact data from the manufacturer. Because they are easy to use, many building owners and consulting engineers now have them and they are using them to verify the TAB contractor's readings.

The two main advantages of the flow measuring hood are speed of operation and accuracy. The hood is held up to the ceiling around the diffuser for several seconds, and then the airflow can be read and recorded directly in cfm with no further calculation. Many diffusers, particularly those with perforated faces, are very difficult to read accurately with any other instrument. Some diffuser manufacturers now have discontinued testing for and publishing "k or A_k factors." Instead they are recommending using a flow measuring hood.

The flow measuring hood consists of a measuring section that has an air-measuring device somewhat similar to those installed in ducts for monitoring cfm for control purposes. Inside is a series of Pitot tubes arranged in a pattern similar to a traverse. Pressures from these sensing elements are directed into a manometer or a deflecting vane anemometer that is calibrated directly in cfm (L/s). From the measuring section, a variety of sizes of collector skirts are available. A gasketed frame is provided on the larger end so all that has to be done is to hold it up tight against the surface around the diffuser and read the dial.

Most flow measuring hoods are available with from two to four ranges. A selector switch is provided to change ranges and switch from supply airflow to exhaust airflow. For very low airflows, some are equipped with a perforated plate or blank off panels that are inserted in the measuring section to reduce the area and increase the velocity of the air. There also are models that use a solid state digital electronic manometer that will read from 25 to 2500 cfm (12 to 1250 L/s) with no range switch required. The hood also will read supply or exhaust automatically. A minus sign appears on the display when a return or exhaust is being read. With the correct attachments, some hoods have manometers that can be removed and used with a Pitot tube. When reading higher airflows, the hood will create some static pressure in the system while being used. This will reduce the airflow coming from the outlet being tested. A curve is furnished so that a correction can be made if needed. Also, hoods should not be used where the discharge velocities from the diffusers are excessive.

Because of the weight and from having to hold it up tight, continuous use may cause fatigue. Inaccurate readings will result if the hood is not held up in a tight position. The hoods, even when packed for transport, are large and bulky. Electronic models will need to have their batteries charged or changed frequently.

7.11   Micromanometer (Hook Gauges)

Micromanometer (hook gauges) are manometers made to give very precise readings, accuracy within ±0.001 in. (0.025 mm) being possible. These are delicate instruments, in general being more adapted to precise industrial and laboratory testing rather than field testing. However, they have some application in commercial testing and balancing, where it may be necessary to measure very low pressures such as at exhaust hoods and air distributing ceilings. They are also useful for measuring air velocities below 600 fpm (3 m/s), where the corresponding velocity pressures are very low [approximately 0.025 in.w.g. (6 Pa) and less]. Using water as the liquid, these are primary measuring devices and can be used to calibrate other instruments.

Each of the two vials of this U-tube-type micromanometer contains a pointer (often called a *hook*) and a micrometer. To use the micromanometer, the vials are set at precisely the same level by using a gauge rod of precise length or using a longer gauge rod, the vials can be set at a precise distance, one above another, for a greater pressure range such as up to 4 in.w.g. (1000 Pa). The position of each hook is adjusted until it dimples the water surface, and its micrometer is then set to zero. When the pressure to be measured is imposed on one of the vials, the hooks are again adjusted to dimple the water surface. The pressure, as determined by the difference in height of the two water surfaces, is determined by reading the two micrometers and adding their readings together.

Although these are very precise instruments, there are disadvantages in using them. Both instruments must be very carefully leveled immediately before each reading, and they should be checked after being read to be sure they have not moved. Also, they are difficult to use if mounted on a surface that vibrates, or if there are pulsations in the pressure to be measured.

8

Use of TAB Instruments (Hydronic, Thermal, Rotational, Electrical)

8.1 Other TAB Instruments

Recommended uses and limitations of most of the normally used TAB instruments may be found in Chapter 7, Tables 7.1 to 7.4. Airflow measuring instruments are discussed in detail in Chapter 7. This chapter discusses the rest of the instruments generally used in HVAC TAB work.

As stated in Chapter 7, instruments for the measurement of airflow, water flow, rotation, temperature, humidity, and electric current are the *tools of the trade* of the testing, adjusting, and balancing (TAB) firm and its employees. Recommended uses and limitations for most of the TAB instruments may be found in Tables 7.1 to 7.4. Most have been proved by experience to be reliable and accurate. Many of the newer electronic types also are accurate and reliable, and they may measure several different functions. Certified NEBB TAB firms are required to own, maintain, and periodically calibrate a required list of TAB instruments (see the NEBB *Procedural Standards for Testing Adjusting and Balancing of Environmental Systems*).

8.2 Hydronic Measuring Instruments

8.2.1 U-tube manometer

Since the pressures to be measured in hydronic systems are usually considerably greater than those associated with airflow, *U-tube manometers* for hydronic use usually contain mercury rather than water or oil. A manometer of the type in Figure 8.1 is available in tube lengths up to 36 in. (900 mm). When filled with mercury, such a manometer can measure pressures up to 36 in.Hg, or 36×0.491 pounds per square inch = 17.7 psi, or 17.7 psi $\times 2.31 = 41$ ft.w.g. In the metric system, 36 in.Hg equals 914.4 mm Hg or 122.04 kPa. Manometers are therefore useful for measuring pressure drops through coils, chillers, condensers, and other heat exchangers, and also across orifices and venturis. They should not be used for measurements under 1 in. (25 mm) of water. One objection to the use of manometers is the possibility of excess pressure beyond the range or length of the manometer, which would blow the mercury out of the tube. A particularly annoying condition can easily occur when, even though the manometer selected may be of adequate range for the normal pressure to be measured, sudden surges that are caused by air in the system can momentarily exceed the range of the instrument and blow out the mercury.

Another case of blowout is the accidental disconnection of one of the two tubes or hoses that connect the manometer to the piping system. In such a case, the usual result is that static pressure in the pipe, being exerted against only one leg of the manometer, will be sufficient to blow out the mercury. Aside from the delay and expense of replacing the mercury, it is very objectionable for mercury to enter the water system because it can cause rapid deterioration of any copper (includ-

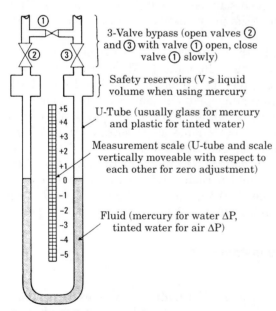

3-Valve bypass (open valves ② and ③ with valve ① open, close valve ① slowly)

Safety reservoirs (V ≥ liquid volume when using mercury

U-Tube (usually glass for mercury and plastic for tinted water)

Measurement scale (U-tube and scale vertically moveable with respect to each other for zero adjustment)

Fluid (mercury for water ΔP, tinted water for air ΔP)

Figure 8.1 U-Tube manometer.

ing copper alloys) with which it comes into contact in the system. Environmental laws also may restrict or prohibit the use of mercury in manometers.

8.2.2 Pressure gauge, calibrated

The *calibrated pressure gauge* (Figure 8.2) shall be of a minimum "grade A" quality; have a Bourdon tube assembly made of stainless steel, alloy steel, monel, or bronze; and have a nonreflecting white face with black letter graduations conforming to ANSI Specification U.S.A.S. B40-1. Test gauges are usually 3½ to 6 in. (90 mm to 150 mm) in diameter, with bottom or back connections. Many dials are available with pressure, vacuum, or compound ranges. Dial-type pressure gauges are used primarily for checking pump pressures; coil, chiller, and condenser pressure drops; and pressure drops across orifice plates, valves, and other flow calibrated devices.

1. Pressure ranges should be selected so the pressures to be measured fall in the middle two thirds of the scale range.

Figure 8.2 Calibrated Pressure Gauge. *(Courtesy of H. O. Trerice Co.)*

2. The gauge should not be exposed to pressures greater than the maximum dial reading. Similarly, a compound gauge should be used where exposed to vacuum.

3. Reduce or eliminate pressure pulsations by installing a needle valve between the gauge and the system equipment or piping. Under extreme pulsating conditions install a pulsation dampener or snubber (available from gauge manufacturers).

4. In using a gauge, apply pressure slowly by gradually opening the gauge cock or valve, to avoid severe strain and possible loss of accuracy that sudden opening of the gauge cock or valve can cause. Likewise, when removing pressure, slowly close the gauge cock or valve, to avoid a sudden release of pressure.

8.2.3 Pressure gauge, differential

In practically all cases of flow measurement, it will be necessary to measure a pressure differential, that is, a pressure drop across a piece of equipment, a balancing device, or a flow-measuring device. Normally, this requires two pressure measurements, one on the high-pressure side and one on the low-pressure side. The differential pressure, or pressure drop, is then the difference between the two pressure readings.

The *differential pressure gauge* is a dual-inlet, grade A dual–Bourdon tube pressure gauge with a single indicating pointer on the dial face which indicates the pressure differential existing between the two measured pressures. It can be calibrated in psi, in.w.g., ft.w.g., or in.Hg (Pa, kPa, mm w.g., or mm Hg). The differential pressure gauge will automatically read the difference between two pressures.

8.2.4 Differential pressure with calibrated gauge

When using a single calibrated pressure gauge, the gauge is alternately valved to the high-pressure side and the low-pressure side to determine the pressure differential. Such an arrangement eliminates any problem concerning gauge elevations and virtually eliminates errors resulting from gauge calibration.

Figure 8.3 illustrates the application of one type of gauge modification that uses a single standard gauge and eliminates the need for subtraction to determine differential. The gauge glass is calibrated to ft.w.g. (kPa) at its outer periphery. During operation, the gauge glass is left loose so it can be rotated. To measure a pressure differential, the high pressure is applied to the gauge by operating the valve to the high-pressure side, and the gauge glass is then rotated so that its *zero*

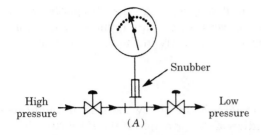

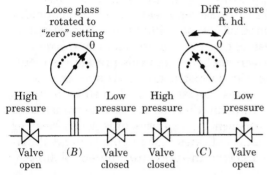

Figure 8.3 Single gauge for measuring differential pressures.

is even with the gauge pointer. Next, the high-pressure valve is closed and the valve to the low-pressure side is opened. The gauge pointer will now indicate a pressure that is directly equal to the pressure differential in ft. w.g. (kPa). If the gauge is of large diameter, such as 8-in. (200 mm) diameter, differential pressures can be read accurately to the order of 0.25 ft. w.g. (750 Pa).

8.3   Flow-Measuring Devices

8.3.1   Venturi tube and orifice plate

The *venturi tube* or *orifice plate* is a specific, fixed, area reduction in the path of fluid flow, installed to produce a flow restriction and a pressure drop. The pressure differential (the upstream pressure minus the downstream pressure) is related to the velocity of the fluid. The pressure differential also is equated to the flow in gallons per minute (gpm) (L/s), but the pressure drop is not equal to velocity pressure drop. By accurate measurement of the pressure drop with a manometer at flow rates from zero fluid velocity to a maximum fluid velocity established by a maximum practical pressure drop, a calibrated flow range may be established. The flow range may then be plotted on a graph that reads pressure drop versus flow rate (gpm or L/s) or the manometer scale may be graduated directly in the flow rate values.

8.3.2 Annular flow indicator

The *annular flow indicator* (Figure 8.4) is a flow sensing and indicating system that is an adaptation of the principle of the Pitot tube. The upstream sensing tube has a number of holes that face the flow and so are subjected to impact pressure (velocity pressure plus static pressure). The holes are spaced so as to be representative of equal annular areas of the pipe, in the manner of selecting Pitot tube traverse points. An equalizing tube arrangement within the upstream tube averages the pressure sensed at the various holes, and this pressure is transmitted to a pressure gauge. The downstream tube is similar to a reversed impact tube, and it senses a pressure equal to static pressure minus velocity pressure at this point; this pressure is also transmitted to a gauge. The difference between the two pressures, when referred to appropriate calibration data, indicates flow in gpm (L/s). A differential pressure gauge is used to read the pressure differential directly.

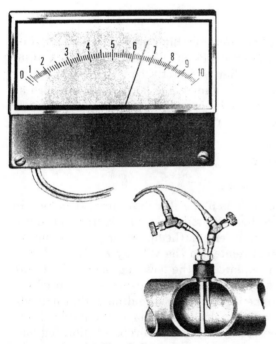

Figure 8.4 Annular Flow Indicator.

8.3.3. Calibrated balancing valves

Calibrated balancing valves (Figure 8.5) perform dual duty as flow measuring devices and as balancing valves, but the manufacturer has provided pressure taps into the inlet and outlet; and has calibrated the device by setting up known flow quantities while measuring the resistance that results from the different valve positions. These positions may be graduated on the valve body (as a dial) with a handle that has a pointer to indicate the reading. The manufacturer publishes a chart or graph that illustrates the percentage open to the valve (the dial settings), the pressure drop, and the resulting flow.

8.3.4 Location of flow devices

Flow measuring devices, including the orifice, venturi, and other types described above, give accurate and reliable readings only when fluid flow in the line is quite uniform and free of turbulence. Pipe fittings such as elbows, valves, etc., create turbulence and nonuniformity of flow. Therefore, an essential rule is that flow-measuring elements must be installed far enough away from elbows, valves, and other sources of flow disturbance to permit turbulence to subside and for flow to regain uniformity. This applies particularly to conditions upstream of the measuring element, and it also applies downstream, although to a lesser extent.

The manufacturers of flow measuring devices usually specify the lengths of straight pipe required upstream and downstream of the measuring element. Lengths are specified in numbers of pipe diameters, so that the actual required lengths will depend on the size of the pipe. Requirements will vary with the type of element and the types

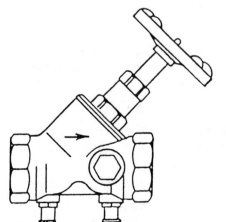

Figure 8.5 Calibrated balancing valve. *(Courtesy ITT Bell and Gossett.)*

of fittings at the ends of the straight pipe runs, ranging from about 5 to 25 pipe diameters downstream and 2 to 5 pipe diameters upstream from the fitting to the measuring element.

8.4   Temperature Measuring Instruments

8.4.1   Glass tube thermometers

Mercury-filled *glass thermometers* have a useful temperature range from minus 40 °F to over 220 °F (−40 °C to 105 °C). They are available in a variety of standard temperature ranges, scale graduations, and lengths.

8.4.1.1   Complete stem immersion.   The complete stem immersion calibrated thermometer, as the name implies, must be used with the stem completely immersed in the fluid in which the temperature is to be measured. If complete immersion of the thermometer stem is not possible or practical, then a correction must be made for the amount of emergent liquid column.

8.4.1.2   Partial stem immersion.   The thermometers calibrated for partial stem immersion are more commonly used. They are used in conjunction with thermometer test wells designed to receive them. No emergent stem correction is required for the partial stem immersion type.

8.4.1.3   Radiation effects.   When the temperatures of the surrounding surfaces are substantially different from the measured fluid, there is considerable radiation effect upon the thermometer reading, if left unshielded or otherwise unprotected from these radiation effects. Proper shielding or aspiration of the thermometer bulb and stem can minimize these radiation effects.

8.4.1.4   Test wells.   Thermometer test wells are used to house the test thermometer at the desired location and permit removal and insertion of a thermometer without requiring removal or loss of the fluid in the system.

8.4.1.5   Prohibitions.   Nuclear work and many cleanrooms prohibit the use of instruments containing mercury.

8.4.2   Dial thermometers

Dial thermometers are of two general types: the stem type and the flexible capillary type. They are constructed with various size dial heads, 1¾ to 5 in. (45 to 125 mm), with stainless steel encapsulated temperature sensing elements. Hermetically sealed, they are rust, dust, and leak proof and are actuated by sensitive bimetallic helix coils. Some can be field calibrated. Sensing elements range in length from 2½ to 24 in. (60 mm to 600 mm) and are available in many temperature ranges with and without thermometer wells.

8.4.2.1 Bimetallic type. The advantage of dial thermometers is that they are more rugged and more easily read than glass-stem thermometers, and they are fairly inexpensive. Small dial thermometers of this type usually use a bimetallic temperature sensing element in the stem. Temperature changes cause a change in the bend or twist of the element, and this movement is transmitted to the pointer by a mechanical linkage.

8.4.2.2 Capillary type. The flexible capillary type dial thermometer has a rather large temperature sensing bulb that is connected to the instrument with a capillary tube. The instrument contains a Bourdon tube, the same as in pressure gauges. The temperature sensing system, consisting of the bulb, capillary tube, and Bourdon tube, is charged with either liquid or a gas. Temperature changes at the bulb cause the contained liquid or gas to expand or contract, resulting in changes in the pressure exerted within the Bourdon tube. This causes the pointer to move over a graduated scale as in a pressure gauge, except that the thermometer dial is graduated in degrees. The advantage of this type thermometer is that it can be used to read the temperature in a remote location.

8.4.2.3 Use. In using a dial thermometer, the stem or bulb must be immersed a sufficient distance to allow this part of the thermometer to reach the temperature being measured. Dial thermometers have a relatively long time lag, so enough time must be allowed for the thermometer to reach temperature and the pointer to come to rest.

8.4.3 Thermocouple thermometers

Digital thermocouple thermometers (Figure 8.6) and analog or digital pyrometers normally used in measurements of surface temperatures in HVAC applications use a thermocouple as a sensing device and a millivoltmeter (or potentiometer) with a scale calibrated for reading temperatures directly.

Electronic-type thermometers have an instrument case containing items such as batteries, various switches, knobs to adjust variable resistances, and a sensitive meter. Thermocouple temperature sensing

Figure 8.6 Thermocouple thermometer. *(Courtesy of Cole-Parmer Instrument Company)*

elements are remote from the instrument case and connected to it by means of wire or cables. Electronic-type thermometers have advantages of remote-reading, good precision, and flexibility as to temperature range.

Additionally, some electronic-type thermometers have multiple connection points on the instrument case and a selector switch, enabling the use of a number of temperature sensors that can be placed in different locations and read one at a time by use of the selector switch.

In piping applications, it should be remembered that surface temperature of the conduit is not equal to the fluid temperature and that a relative comparison is more reliable than an absolute reliance on readings at a single circuit or terminal unit.

8.4.4 Electronic thermometers

There are many types of rugged, light weight, battery powered *digital electronic thermometers* that have precision accuracy with interchangeable probes and/or sensors. Types included are: resistance temperature detectors (RTD), thermistors, thermocouples, and diode sensors, with either liquid crystal or LED displays. Response time and ease of use will vary from model to model and type to type. Typical units are shown in Figures 8.7 and 8.8.

8.4.5 Psychrometers

The *sling psychrometer* can be used in determining the psychrometric properties of the conditioned spaces, return air, outdoor air, mixed air, and conditioned supply air. The readings taken from the sling psychrometer can be spotted on a standard psychrometric chart from which all other psychrometric properties of the measured air can be determined.

Figure 8.7 Resistance temperature detector. *(Courtesy of Cole-Parmer Instrument Company)*

Figure 8.8 Electronic thermometer. *(Courtesy of Cole-Parmer Instrument Company)*

The sling psychrometer (Figure 8.9) consists of two mercury filled thermometers, one of which has a wetted cloth wick or sock around its bulb. The two thermometers are mounted side by side on a frame fitted with a handle by which the device can be whirled with a steady motion through the surrounding air. The whirling motion is periodically stopped to take readings of the wet and dry bulb thermometers (in that order) until such time as consecutive readings become steady. Because of evaporation, the wet bulb thermometer will indicate a lower temperature than the dry bulb thermometer, and the difference is known as the *wet bulb depression.*

Accurate wet bulb readings require an air velocity of between 1000 and 1500 fpm (5 and 7.5 m/s) across the wick, or a correction must be made; therefore, an instrument with an 18 in. (450 mm) radius should be whirled at a rate of two revolutions per second. Significant errors will result if the wick becomes dirty or dry, so a constant supply of distilled water should be used.

Digital battery powered versions are available that blow the ambient air over the wetted wick. These instruments are accurate and they

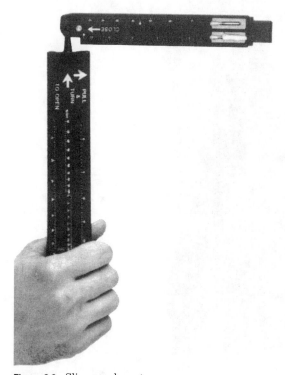

Figure 8.9 Sling psychometer.

can be placed into confined areas where there is insufficient room to whirl a sling psychrometer.

8.4.6 Electronic thermohygrometers

Unlike the psychrometer, the *thermohygrometer* (Figure 8.10) does not utilize the cooling effect of the wet bulb to determine the moisture content in the air. A thin film capacitance sensor is used as a sensing element in many instruments. As the moisture content and temperature change, the resistance in the sensor changes proportionally. Readout is normally in percent relative humidity. Because the instruments do not rely upon evaporation for measurement, the need for airflow across the wetted wick or sock is eliminated. The sensing element needs only to be held in the sampled air. Typical measuring ranges are 10 to 98% RH, 32 °F to 140 °F (0 °C to 60 °C).

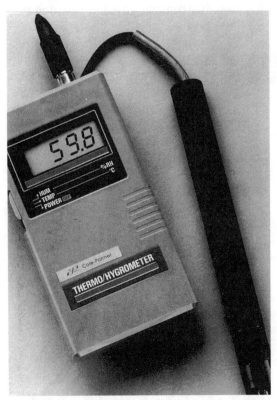

Figure 8.10 Thermohygrometer *(Courtesy of Cole-Parmer Instrument Company)*

The thermohygrometer can be used to determine the psychrometric properties of air in much the same way as the sling psychrometer. The reading can be spotted on a standard psychrometric chart from which all other psychrometric properties of the measured air can be determined.

It can be used for measuring and monitoring of areas sensitive to change in relative humidity such as cleanrooms, hospitals, museums, and paper storages. Continuous monitoring of conditions in areas sensitive to humidity is possible with a greater accuracy and ease of measurement. At relative humidities above 90%, the accuracy of the sensor is decreased because of swelling of the sensing element.

A multipurpose meter is shown in Figure 8.11 that measures air velocities and temperatures as well as humidities using various probes.

8.5   Rotation-Measuring Instruments

A *tachometer* is an instrument used to measure the speed at which a shaft or wheel is turning. The speed is usually determined in revolutions per minute (rpm), but some have many other ranges such as r/s, r/h, ft/s, in/s, cm/s, r/s, and r/m.

The several types of tachometers described below vary in cost, in dependability, and in accuracy of results obtainable. One basic differ-

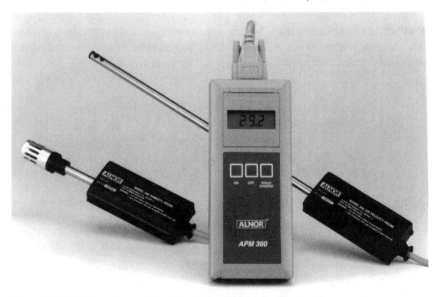

Figure 8.11   Multipurpose meter. *(Courtesy of Alnor Instrument Company)*

ence between the different types of tachometers is that many have digital readouts directly in revolutions per minute (rpm), whereas older types are primarily revolution counters that must be used with a timing device such as an accurate stop watch.

8.5.1 Tachometer, chronometric

The *chronometric tachometer* (Figure 8.12) combines a revolution counter and a stopwatch in one instrument. In using this type of tachometer, its tip is placed in contact with the rotating shaft. The tachometer spindle will then be turning with the shaft but the instrument will not be indicating. To take a reading, the push button is pressed and then quickly released. This sets the meter hand to zero, winds the stop watch movement, and then simultaneously starts both the revolution counter and the stopwatch. After a fixed time interval, usually 6 seconds, the counting mechanism is automatically uncoupled so that it no longer accumulates revolutions even though the instrument tip is still in contact with the rotating shaft. After the meter hands have stopped, the tachometer may be removed from the shaft and read. The meter face has two pointers and two dials, the smaller one indicating one graduation for each complete revolution of the larger pointer, and the reading will be directly in rpm (r/s). Some instrument spindles must be rotating in order to be reset without damage.

Since the timing is automatically synchronized with operation of the revolution counter, the human error that can occur when a revolution counter and separate stop watch are used is eliminated. In general, the chronometric tachometer is the preferred type of instrument when the shaft end is accessible and has a countersunk hole.

Figure 8.12 Chronometric tachometer.

There are new hand tachometers capable of producing instantaneous rpm measurement readings on a dial face (Eddy-current type) and solid-state instruments with a digital readout. Some have a "memory" button to recall the last reading as well as maximum and minimum readings.

8.5.2 Optical tachometer
(photo tachometer)

The *optical tachometer* or *photo tachometer* (Figure 8.13) uses a photocell that counts the pulses as the object rotates. Then, by use of a transistorized computer circuit, it produces a direct rpm (r/s) reading on the instrument dial that is either digital or analog. Several features make it adaptable for use in measuring fan speeds. It is completely portable and is equipped with long-life batteries for its light and power source. It has good accuracy and any error can be reduced by using more than one reflective marker on the rotating device. Its calibration can be checked continually on most jobs by directing its beam to a fluorescent light and comparing the indicated reading against 7200 on the rpm scale.

Figure 8.13 Digital optical tachometer.

The optical tachometer does not have to be in contact with the rotating device. It indicates instantaneous speeds, not average speed—whether constant or changing—thereby reading the speed as it is. It is easy to use: To read rpm, one need only place a contrasting mark on the rotating device by using chalk or reflective tape. It is a good instrument to use on inline fans and other such equipment where shaft ends are not accessible. It also has good application for use on equipment rotating at a high rate of speed.

8.5.3   Electronic tachometer (stroboscope)

The *Stroboscope* (Figure 8.14) is an electronic tachometer that uses an electrically flashing light. The frequency of the flashing light is electronically controlled and adjustable. When the frequency of the flashing light is adjusted to equal the frequency of the rotating machine, the machine will appear to stand still.

The Stroboscope (another proprietary name universally used in the TAB industry) does not need to make contact with the machine being checked but need only be pointed toward the machine so that a moving part is illuminated by the Stroboscope light and can be viewed by the operator. The light flashes are of extremely short duration, and their frequency is adjustable by turning a knob on the Stroboscope. When the frequency of the light flashes is exactly the same as the speed of the moving part being viewed, the part will be seen distinctly only once each cycle, and the moving part will appear to stand still. The

Figure 8.14   Stroboscope. *(Courtesy of Cole-Parmer Instrument Company)*

corresponding frequency, or rpm, can be read from an analog or digital scale on the instrument.

Care must be taken to avoid reading multiplies (or harmonics) of the actual rpm (r/s). Readings must be started at the lower end of the scale.

8.5.4 Dual function tachometer

The *dual function tachometer* (Figure 8.15) provides both optical and contact measurements of rotation and linear motions. Many allow a choice of 19 ranges depending on the application. A digital display always indicates the unit of measurement to identify the operating range. The "memory" button may be used to recall the last, maximum,

Figure 8.15 Dual-function tachometer.

minimum, and average readings. It measures rotation speeds by direct contact or by counting the speed of a reflective mark.

8.6 Electrical Measuring Instruments

8.6.1 Volt ammeter

The testing, adjusting, and balancing of mechanical systems requires the measurement of voltages and electrical currents as a routine matter. The clamp-on *volt ammeter* with digital readout (Figure 8.16) is one of the types used for taking field electrical measurements. The clamp-on volt ammeter shown has trigger operated, clamp-on transformer jaws that permit current readings without interrupting electrical service. Most meters have several scale ranges in amperes and volts. Two voltage test leads are furnished that may be quickly connected to the volt ammeter.

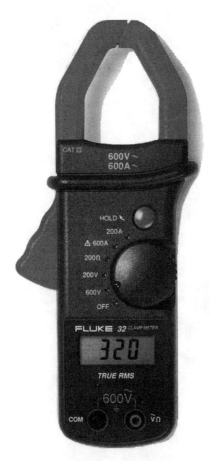

Figure   8.16 Clamp-on volt-ammeter. *(Courtesy of Fluke Corporation)*

Meters with "true RMS" (root-mean-square) on them can more accurately measure nonlinear loads such as personal computers, adjustable speed drives, and HID lighting. All of these draw current in short pulses, causing harmonics in the load current (see subsection 19.7.4 in Chapter 19).

8.6.2 Safety and use of volt ammeters

When using the volt ammeter, the proper range must be selected if the instrument does not do it automatically. When in doubt, begin with the highest range for both voltage and amperage scales and read the manufacturer's instructions.

Before using, be aware of the following safety precautions:

First. **Be careful not to make contact with an open electrical circuit. Hands should never be put into the electrical boxes. Do not attempt to pry wires over into position. Do not force the instrument jaws into position.** These precautions reduce the risk of causing a short circuit, which could injure both the equipment and personnel.

Second. **When taking amperage readings, do not attach the instrument and then start the motor. Position the instrument and read it after the motor is running at full speed.** The inrush current required to start a motor is from three to five times higher than the nameplate full-load current. Therefore, starting the motor with the instrument attached could damage the instrument.

Readings may be taken at the motor leads or from the load terminals of the starter. To determine the amperages of single-phase motors, place the clamp about one wire (often called a *leg*). When involved with three-phase current, take readings on each of the three wires and average the results. A somewhat higher amperage reading on one wire (leg) may cause the heater coils to trip.

To measure voltage with portable test instruments, set the meter to the most suitable range, connect the test lead probes firmly against the terminals or other surfaces of the line under test, and read the meter, making certain to read the correct scale if the meter has more than one scale. When reading single-phase voltage the leads should be applied to the two load terminals. The resulting single reading is the voltage of the current being drawn by the motor.

When reading three-phase voltage, it is necessary to apply the voltmeter terminals to pole 1 and pole 2; then to pole 2 and pole 3; and finally to pole 1 and pole 3. This will result in three readings, each of which will likely be a little different, but which should be close to each other.

If the average voltage delivered to the motor varies by more than a few volts from the nameplate rating of the motor, several things can occur. A rise in voltage may damage the motor and it will cause a drop in the amperage reading. A drop in the voltage will cause a rise in the amperage and can cause the overload protectors on the starter to "kick out." In either case, it is advisable to report high- or low-voltage situations promptly.

HVAC Unit Air Measurements

9.1 Introduction

In today's world, it almost is impossible to find either a factory-built or a field-built HVAC unit in which it is easy to make airflow or temperature measurements. The design of some rooftop HVAC units may not allow any accurate measurements to be made.

Determining the correct proportion of outside air to total supply air is basic to the proper balancing of any system. Because of close connections between the outside air duct and the equipment or the absence of an outside air duct, the Pitot tube traverse method of measuring outside air quantities is either extremely difficult or altogether impossible. Therefore, the most practical method of setting outside air proportional dampers is the *mixed air temperature* method. Air stratification within a rooftop HVAC unit may prevent the use of this method also.

9.2 Outside Air Measurements

9.2.1 Damper adjustments

Most systems are designed to operate with a minimum amount of outside air whenever the building is occupied. Everyone is more conscious of outside air quantities now that energy has become so expensive and indoor air quality has become a problem. The procedure for setting the outside air (OA) quantities will depend on the system and damper scheme.

Where a separate minimum and maximum OA damper are provided, start with the minimum OA dampers and the return air (RA) dampers open. The maximum OA dampers and exhaust air (EA) dampers should be closed. The airflow adjustment will be made by fan speed changes or minimum OA damper adjustment. Most systems use just one OA damper in conjunction with a minimum position controller, which will open the outside air and exhaust air dampers while closing the return air damper. Adjustment is made at the controller. Always ask the temperature control system technician to assist where possible.

9.2.2 Temperature method

The quantities of outside air should be tested by making a Pitot tube traverse of the OA duct where possible. Otherwise, calculate the amount of outside air by subtracting the actual return air flow volume from the actual supply air flow volume. If either actual flow volume test is not possible, the *temperature method*, using Equation 9.1, may be the only method left. Accurate temperature readings are essential to this method. (Equations 9.1 to 9.3 also may be found in Chapter 2 as Equations 2.3 to 2.5.)

Equation 9.1 (U.S.)

$$T_m = \frac{X_o T_o + X_r T_r}{100}$$

Where: T_m = temperature of mixed air (°F)
X_o = % of outside air
T_o = temperature of outside air (°F)
X_r = % of return air
T_r = temperature of return air (°F)

Equation 9.1 (Metric)

$$T_m = \frac{X_o T_o + X_r T_r}{100}$$

Where: T_m = temperature of mixed air (°C)
X_o = % of outside air
T_o = temperature of outside air (°C)
X_r = % of return air
T_r = temperature of return air (°C)

Equation 9.2 (U.S.)

$$X_o = 100 \frac{(T_r - T_m)}{(T_r - T_o)}$$

Equation 9.2 (Metric)

$$X_o = 100 \frac{(T_r - T_m)}{(T_r - T_o)}$$

Equation 9.3 (U.S.)

$$X_r = 100 \frac{(T_m - T_o)}{(T_r - T_o)}$$

Equation 9.3 (Metric)

$$X_r = 100 \frac{(T_m - T_o)}{(T_r - T_o)}$$

Example 9.1 (U.S.) The supply air fan of the HVAC system furnishes 12,500 cfm, and the specified outside air quantity is 2500 cfm. Measurements indicate OA = 92 °F and RA = 74 °F. Calculate the mixed air temperature that would allow the correct amount of outside air.

Solution

$$\% \text{ OA} = \frac{2500 \text{ cfm}}{12,500 \text{ cfm}} \times 100 = 20\%; \quad \% \text{ RA} = 100\% - 20\% = 80\%$$

Using Equation 9.1:

$$T_m = \frac{X_o T_o + X_r T_r}{100} = \frac{(20\% \times 92°) + (80\% \times 74°)}{100}$$

$$T_m = \frac{1840 + 5920}{100} = 77.6 \text{ °F}$$

Therefore, to obtain the correct amount of outside air (2500 cfm), the dampers will need to be adjusted to obtain a mixed air temperature of 77.6 °F.

Example 9.1 (Metric) The supply air fan of the HVAC system furnishes 5900 L/s, and the specified outside air quantity is 1180 L/s. Measurements indicate OA = 33 °C and RA = 23 °C. Calculate the mixed air temperature that would allow the correct amount of outside air.

Solution

$$\% \text{ OA} = \frac{1180 \text{ L/s}}{5900 \text{ L/s}} \times 100 = 20\%; \quad \% \text{ RA} = 100\% - 20\% = 80\%$$

Using Equation 9.1:

$$T_m = \frac{X_o T_o + X_r T_r}{100} = \frac{(20\% \times 33°) + (80\% \times 23°)}{100} = \frac{660 + 1840}{100}$$

$$T_m = 25.0 \text{ °C}$$

Therefore, to obtain the correct amount of outside air (1180 L/s), the dampers will need to be adjusted to obtain a mixed air temperature of 25 °C.

9.3   Temperature Measurements

Quite often, the TAB technician is going to find that the mixed-air temperature is very difficult to test accurately. **The duct configurations of many systems may create a considerable amount of airstream stratification**. Mixed-air temperatures will vary considerably, depending on where the readings are taken. In the case of air stratification, it will be necessary to take several temperature readings in the form of a careful thermometer traverse. Average the readings to obtain the correct mixed-air temperature. This can be a time-consuming process and a quick reading digital thermometer will speed up the process.

A helpful way to lay out the temperature traverse is to use the center of each filter section. Readings downstream of filters usually are acceptable. Mixed-air temperature readings should never be taken downstream of a coil (in use or not) or a fan, because the readings usually will be unreliable. Drastic air stratification also can cause other problems such as coil freezeup and bothersome freezestat tripping. If this is observed, it should be reported immediately so that the conditions may be corrected.

Wet bulb readings within HVAC units using wick soaked thermometers require velocities of 1000 to 1500 fpm (5 to 7.5 m/s). Small wet bulb temperature measurement errors of 3 °F (2 °C) may cause large errors in calculating total heat (enthalpy) performance of a cooling coil with the use of a psychrometric chart. (See Chapter 12 for use of psychrometric charts.)

Example 9.2 (U.S.) Outdoor air (OA) at 91 °F dry bulb (DB), 79 °F wet bulb (WB) is mixed with return air (RA) at 78 °F DB, 55% relative humidity (RH). Find the percentage of OA that will result in a 70 °F WB mixture.

Solution Plot the OA and RA on the psychrometric chart (Figure 9.1) and connect the slope. Now find 70 °F WB on the saturation curve and follow its line to the point where it intersects the OA/RA slope. From the point of intersection, drop down to the dry bulb scale and read 81.2 °F DB (see Figure 9.1).

Using Equation 9.2:

$$X_o = 100 \frac{(T_r - T_m)}{(T_r - T_o)} = 100 \frac{(78 - 81.2)}{(78 - 91)} = 100 \frac{(-3.2)}{(-13)} = 24.6\% \text{ OA}$$

Example 9.2 (Metric) Outdoor air (OA) at 33 °C DB, 26 °C WB is mixed with return air (RA) at 25.5 °C DB, 55% RH. Find the percentage of OA that will result in a 21 °C WB mixture.

Solution Plot the OA and RA on the psychrometric chart (Figure 9.2) and connect the slope. Now find 21 °C WB on the saturation curve and follow its line to the point where it intersects the OA/RA slope. From the point of intersection, drop down to the dry bulb scale and read 27.4 °C DB (see Figure 9.2).

Using Equation 9.2:

$$X_o = 100 \frac{(T_r - T_m)}{(T_r - T_o)} = 100 \frac{(25.5 - 27.4)}{(25.5 - 33)} = 100 \frac{(-1.9)}{(-7.5)} = 25.3\% \text{ OA}$$

Care also must be taken to avoid the radiation effect when taking dry bulb temperature readings, particularly on the heating cycle. Figure 9.3 illustrates the correct location for measuring the air temperature rise in a warm air furnace. The leaving bonnet temperature should be read as close to the heat exchanger as possible without being exposed to the radiation effect.

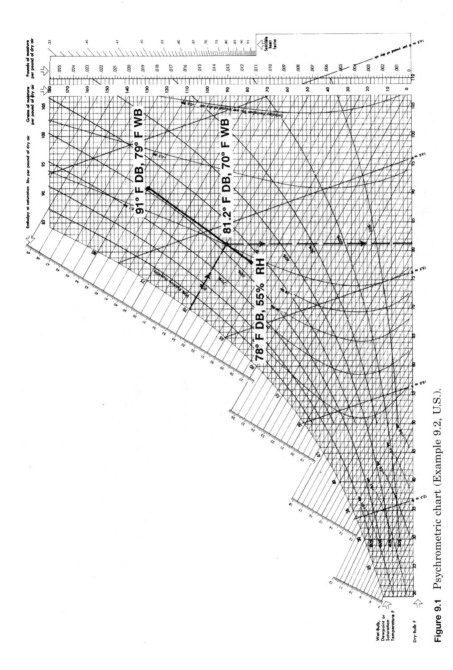

Figure 9.1 Psychrometric chart (Example 9.2, U.S.).

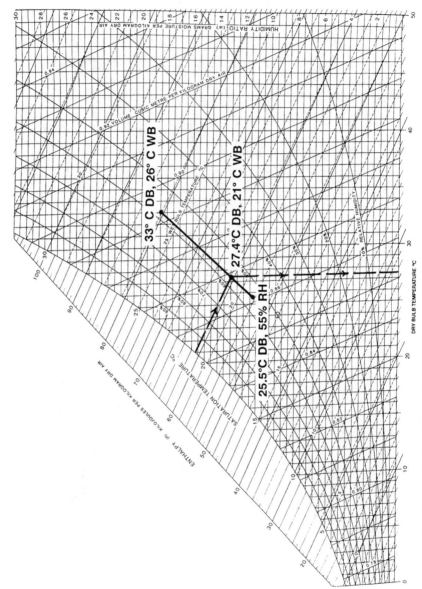

Figure 9.2 Psychrometric chart (Example 9.2, Metric).

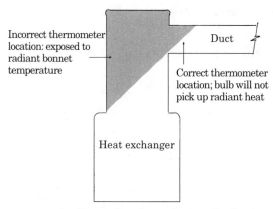

Incorrect thermometer
location: exposed to
radiant bonnet
temperature

Duct

Correct thermometer
location; bulb will not
pick up radiant heat

Heat exchanger

Figure 9.3 Avoiding radiant heat when reading leaving
air temperature.

The mixture conditions for any number of airstreams may be found
by using the following equation:

Equation 9.4 (U.S. and Metric)

$$T_m = \frac{X_oT_o + X_{r1}T_{r1} + X_{r2}T_{r2} + \cdots X_nT_n}{100}$$

Where:
The sum of $X_n = 100\%$.

Example 9.3 (U.S.) Conditions for a multizone unit are:

RA zone 1 = 1000 cfm at 78 °F   RA zone 3 = 2000 cfm at 77 °F
RA zone 2 = 1500 cfm at 76 °F   OA = 500 cfm at 91 °F

Find the temperature of the air mixture entering the coil.

Solution

1000 cfm + 1500 cfm + 2000 cfm + 500 cfm = 5000 cfm total airflow

1000/5000 = 20%   2000/5000 = 40%
1500/5000 = 30%   500/5000 = 10%

$$T_m = \frac{(20 \times 78°) + (30 \times 76°) + (40 \times 77°) + (10 \times 91°)}{100}$$

$$T_m = \frac{1560 + 2280 + 3080 + 910}{100} = 78.3 °F$$

Example 9.3 (Metric) Conditions for a multizone unit are:

RA zone 1 = 500 L/s at 16 °C   RA zone 3 = 1000 L/s at 25°C
RA zone 2 = 750 L/s at 24 °C   OA = 250 L/s at 33 °C

Find the temperature of the air mixture entering the coil.

Solution

$$500 + 750 + 1000 + 250 = 2500 \text{ L/s total air flow}$$

$$500/2500 = 20\% \qquad 1000/2500 = 40\%$$
$$750/2500 = 30\% \qquad 250/2500 = 10\%$$

$$T_m = \frac{(20 \times 26°) + (30 \times 24°) + (40 \times 25°) + (10 \times 33°)}{100}$$

$$T_m = \frac{520 + 720 + 1000 + 330}{100} = 25.7 \text{ °C}$$

10

Effects of Air Densities

10.1 Standard Conditions

Standard conditions for air in U.S. units are, at sea level, 70 °F, an air density of 0.075 lb/ft³ at a barometric pressure of 29.92 in.Hg, or 14.7 lb/in.². Standard conditions for air in metric units are at sea level, 20 °C, an air density of 1.2041 kg/m³ at a barometric pressure of 101.325 kPa or 760 mm Hg.

The sensible heat flow in HVAC work may be determined by the following equation:

Equation 10.1 (U.S.)   **Equation 10.1 (Metric)**

$$Q = 60 \times C_p \times d \times \text{cfm} \times \Delta t \qquad Q = C_p \times d \times \text{L/s} \times \Delta t$$

Where: Q = Heat flow (Btu/h or W)
  C_p = Specific heat [Btu/lb • °F or kJ/(kg • K)
  d = Density (lb/ft³ or kg/m³)
  Δt = Temperature differential (°F or °C)

The specific heat (C_p) for standard air is 0.24 Btu/lb • °F [1.005 kJ/ (kg • K)]. When the values for standard air specific heat and densities are substituted in Equation 10.1, the following equation is obtained:

Equation 10.2 (U.S.)   **Equation 10.2 (Metric)**

$$Q = 1.08 \times \text{cfm} \times \Delta t \qquad Q = 1.23 \times \text{L/s} \times \Delta t$$

This is the most used equation for air heat transfer.

10.2 Air Density Changes—Fans

If one multiplies "cfm × 60 × d" (L/s × d/1000) taken from Equation 10.1, an airflow rate in pounds per hour (kilograms per second) is obtained. If the air density is smaller, as it is when warmer or at a higher altitude, the mass or weight of air that the fan handles in the HVAC system is less. If the weight of the air is less, then the fan does not have to work as hard. This is reflected by the following equation (also found in Chapter 3) for fans:

Equation 10.3 (U.S. and Metric)

$$\frac{d_2}{d_1} = \frac{SP_2}{SP_1} = \frac{FP_2}{FP_1} \quad \text{(when } Q_1 = Q_2\text{)}$$

Where: Q = airflow (cfm or L/s)
  d = air density (lb/ft^3 or kg/m^3)
  SP = static pressure (in.w.g. or Pa)
  FP = fan power (BHP or kW)

The ratio d_2/d_1 is called the *density ratio* or *air density correction factors* (see Tables 10.1 and 10.2).

Example 10.1 (U.S.) A fan is rated to deliver 5000 cfm at 2.0 in.w.g. *SP* at 2.8 BHP. Find the airflow, *SP* and BHP at 5000 ft. elevation when handling 150 °F air.

Solution From Table 10.1 (or Table C-8 in Appendix C), the correction factor (d_2/d_1) for 5000 ft. and 150 °F is 0.72. The airflow remains constant at 5000 cfm. Using Equation 10.3:

$$SP_2 = SP_1 \times \frac{d_2}{d_1} = 2.0 \times 0.72 = 1.44 \text{ in.w.g.}$$

$$BHP_2 = BHP_1 \times \frac{d_2}{d_1} = 2.8 \times 0.72 = 2.02 \text{ BHP}$$

Example 10.1 (Metric) A fan is rated to deliver 2350 L/s at 500 pascals *SP* at 2.1 kW. Find the airflow, *SP*, and fan power at 1500 m elevation when handling 75 °C air.

Solution From Table 10.2 (or Table D-6 in Appendix D), the correction factor (d_2/d_1) for 1500 m and 75 °C is 0.71. The airflow remains constant at 2350 L/s. Using Equation 10.3:

$$SP_2 = SP_1 \times \frac{d_2}{d_1} = 500 \times 0.71 = 355 \text{ Pa}$$

$$FP_2 = FP_1 \times \frac{d_2}{d_1} = 2.1 \times 0.71 = 1.49 \text{ kW}$$

10.3   Air Density Changes—Systems

A common error in testing, adjusting, and balancing of HVAC systems is the neglect of correcting data for nonstandard air conditions. Pro-

TABLE 10.1 d_2/d_1, Air Density Correction Factors (U.S. Units)

Altitude (ft)	Sea level	1000	2000	3000	4000	5000	6000	7000	8000	9000	10,000
Barometer (in.Hg)	29.92	28.86	27.82	26.82	25.84	24.90	23.98	23.09	22.22	21.39	20.58
(in.w.g.)	407.5	392.8	378.6	365.0	351.7	338.9	326.4	314.3	302.1	291.1	280.1
Air temp. (°F)											
−40°	1.26	1.22	1.17	1.13	1.09	1.05	1.01	0.97	0.93	0.90	0.87
0°	1.15	1.11	1.07	1.03	0.99	0.95	0.91	0.89	0.85	0.82	0.79
40°	1.06	1.02	0.99	0.95	0.92	0.88	0.85	0.82	0.79	0.76	0.73
70°	1.00	0.96	0.93	0.89	0.86	0.83	0.80	0.77	0.74	0.71	0.69
100°	0.95	0.92	0.88	0.85	0.81	0.78	0.75	0.73	0.70	0.68	0.65
150°	0.87	0.84	0.81	0.78	0.75	0.72	0.69	0.67	0.65	0.62	0.60
200°	0.80	0.77	0.74	0.71	0.69	0.66	0.64	0.62	0.60	0.57	0.55
250°	0.75	0.72	0.70	0.67	0.64	0.62	0.60	0.58	0.56	0.58	0.51
300°	0.70	0.67	0.65	0.62	0.60	0.58	0.56	0.54	0.52	0.50	0.48
350°	0.65	0.62	0.60	0.58	0.56	0.54	0.52	0.51	0.49	0.47	0.45
400°	0.62	0.60	0.57	0.55	0.53	0.51	0.49	0.48	0.46	0.44	0.42
450°	0.58	0.56	0.54	0.52	0.50	0.48	0.46	0.45	0.43	0.42	0.40
500°	0.55	0.53	0.51	0.49	0.47	0.45	0.44	0.43	0.41	0.39	0.38
550°	0.53	0.51	0.49	0.47	0.45	0.44	0.42	0.41	0.39	0.38	0.36
600°	0.50	0.48	0.46	0.45	0.43	0.41	0.40	0.39	0.37	0.35	0.34
700°	0.46	0.44	0.43	0.41	0.39	0.38	0.37	0.35	0.34	0.33	0.32
800°	0.42	0.40	0.39	0.37	0.36	0.35	0.33	0.32	0.31	0.30	0.29
900°	0.39	0.37	0.36	0.35	0.33	0.32	0.31	0.30	0.29	0.28	0.27
1000°	0.36	0.35	0.33	0.32	0.31	0.30	0.29	0.28	0.27	0.26	0.25

Standard air density, sea level, 70 °F = 0.075 lb/ft³ at 29.92 in.Hg.

TABLE 10.2 d_2/d_1 **Air Density Correction Factors (Metric Units)**

Altitude (m)	Sea level	250	500	750	1000	1250	1500	1750	2000	2500	3000
Barometer (kPa)	101.3	98.3	96.3	93.2	90.2	88.2	85.1	83.1	80.0	76.0	71.9
Air temp. (°C) 0°	1.08	1.05	1.02	0.99	0.96	0.93	0.91	0.88	0.86	0.81	0.76
20°	1.00	0.97	0.95	0.92	0.89	0.87	0.84	0.82	0.79	0.75	0.71
50°	0.91	0.89	0.86	0.84	0.81	0.79	0.77	0.75	0.72	0.68	0.64
75°	0.85	0.82	0.80	0.78	0.75	0.73	0.71	0.69	0.67	0.63	0.60
100°	0.79	0.77	0.75	0.72	0.70	0.68	0.66	0.65	0.63	0.59	0.56
125°	0.74	0.72	0.70	0.68	0.66	0.64	0.62	0.60	0.59	0.55	0.52
150°	0.70	0.68	0.66	0.64	0.62	0.60	0.59	0.57	0.55	0.52	0.49
175°	0.66	0.64	0.62	0.62	0.59	0.57	0.55	0.54	0.52	0.49	0.46
200°	0.62	0.61	0.59	0.57	0.56	0.54	0.52	0.51	0.49	0.47	0.44
225°	0.59	0.58	0.56	0.54	0.53	0.51	0.50	0.48	0.47	0.44	0.42
250°	0.56	0.55	0.53	0.52	0.50	0.49	0.47	0.46	0.45	0.42	0.40
275°	0.54	0.52	0.51	0.49	0.48	0.47	0.45	0.44	0.43	0.40	0.38
300°	0.51	0.50	0.49	0.47	0.46	0.45	0.43	0.42	0.41	0.38	0.36
325°	0.49	0.48	0.47	0.45	0.44	0.43	0.41	0.40	0.39	0.37	0.35
350°	0.47	0.46	0.45	0.43	0.42	0.41	0.40	0.39	0.38	0.35	0.33
375°	0.46	0.44	0.43	0.42	0.41	0.39	0.38	0.37	0.36	0.34	0.32
400°	0.44	0.43	0.41	0.40	0.39	0.38	0.37	0.36	0.35	0.33	0.31
425°	0.42	0.41	0.40	0.39	0.38	0.37	0.35	0.34	0.33	0.32	0.30
450°	0.41	0.40	0.38	0.37	0.36	0.35	0.34	0.33	0.32	0.31	0.29
475°	0.39	0.38	0.37	0.36	0.35	0.34	0.33	0.32	0.31	0.29	0.28
500°	0.38	0.37	0.36	0.35	0.34	0.33	0.32	0.31	0.30	0.28	0.27
525°	0.37	0.36	0.35	0.34	0.33	0.32	0.31	0.30	0.29	0.27	0.26

Standard air density, sea level, 20 °C = 1.2041 kg/m³ at 101.325 kPa.

cess exhaust and drying systems, commercial kitchen exhaust, and warm air systems for thermal comfort demand exacting attention to problems of air density variation, as well as changes in elevation. At higher temperatures and elevations other than sea level, the velocity (fpm or m/s) and the volume of air (cfm or L/s) will increase as the reciprocal of the density ratio (d_2/d_1). Corrections normally are not made for elevations up to 2000 ft (600 m) and for the normal temperatures found in HVAC comfort systems.

The following equation may be used to find the actual dry air density.

Equation 10.4 (U.S.)   **Equation 10.4 (Metric)**

$$d = 1.325 \frac{P_b}{T} \qquad\qquad d = 3.48 \frac{P_b}{T}$$

Where: d = air density (lb/ft^3 or kg/m^3)
P_b = barometric pressure (in.Hg or kPa)
T = absolute temperature (°F + 460° or °C + 273°)

The *correction factor* (CF) used to correct HVAC airflow velocities and volumes at nonstandard air conditions may be calculated by Equation 10.5. It is the square root of the reciprocal of the d_2/d_1 correction factor.

Equation 10.5 (U.S.)   **Equation 10.5 (Metric)**

$$CF = \sqrt{\frac{0.075}{d}} \qquad\qquad CF = \sqrt{\frac{1.204}{d}}$$

Equation 10.6 (U.S.)   **Equation 10.6 (Metric)**

$$V = V_m \times CF \qquad\qquad V = V_m \times CF$$

or   or

Equation 10.7 (U.S.)   **Equation 10.7 (Metric)**

$$V = V_m \sqrt{\frac{0.075}{d}} \qquad\qquad V = V_m \sqrt{\frac{1.204}{d}}$$

Where: V = actual velocity (corrected)
(fpm or m/s)
V_m = measured velocity (fpm or m/s)
CF = correction factor
d = actual air density (lb/ft^3 or kg/m^3)

A quick, approximate correction for nonstandard air may be made by allowing a 2% increase in *velocity pressure* for each 10 °F above 70 °F (each 5 °C above 20 °C) and 4% for each 1000 ft (300 m) *altitude*

TABLE 10.3 Temperature Corrections (U.S.)

Temp. (°F)	d_2/d_1	CF	Temp. (°F)	d_2/d_1	CF
−40°	1.26	0.89	400°	0.62	1.27
0°	1.15	0.93	450°	0.58	1.31
40°	1.06	0.97	500°	0.55	1.35
70°	1.00	1.00	550°	0.53	1.39
100°	0.95	1.03	600°	0.50	1.41
150°	0.87	1.07	650°	0.48	1.44
200°	0.80	1.12	700°	0.46	1.47
250°	0.75	1.15	800°	0.42	1.54
300°	0.70	1.20	900°	0.39	1.60
350°	0.65	1.24	1000°	0.36	1.67

above sea level. When using Tables 10.3, 10.4 and 10.5 or Figures 10.1 and 10.2 for correcting for both temperature and altitude, the correction factors must be multiplied together.

To find the barometric pressure where the elevation is known, use the approximate correction of 0.1 in.Hg pressure reduction per 100 ft of elevation (1.13 kPa per 100 m).

Example 10.2 (U.S.)
a. Estimate the airflow velocity of a 0.20 in.w.g. velocity pressure (Vp) reading at 3000 ft. elevation and 70 °F.
b. Also calculate the answer using tables and equations.

Solution
a. 3000 ft. equals approximately 12% (3 × 4%)

$$0.20 \text{ in.w.g.} \times 1.12 = 0.224 \text{ in.w.g.}$$

$$V = 4005 \sqrt{V_p} = 4005 \sqrt{0.224} = 1896 \text{ fpm (Equation 6.3)}$$

b. $V = 4005 \sqrt{0.20} = 1791$ fpm

TABLE 10.4 Temperature Corrections (Metric)

Temp. (°C)	d_2/d_1	CF	Temp. (°C)	d_2/d_1	CF
0°	1.08	0.96	275°	0.54	1.36
20°	1.00	1.00	300°	0.51	1.40
50°	0.91	1.05	325°	0.49	1.43
75°	0.85	1.08	350°	0.47	1.45
100°	0.79	1.13	375°	0.46	1.47
125°	0.74	1.16	400°	0.44	1.51
150°	0.70	1.20	425°	0.42	1.54
175°	0.66	1.23	450°	0.41	1.56
200°	0.62	1.27	475°	0.39	1.60
225°	0.59	1.30	500°	0.38	1.62
250°	0.56	1.34	525°	0.37	1.64

TABLE 10.5   Altitude Corrections (U.S. and Metric)

Feet	Meters	d_2/d_1	CF
Sea level	Sea level	1.00	1.00
1,000	300	0.96	1.02
2,000	600	0.93	1.04
3,000	900	0.89	1.06
4,000	1200	0.86	1.08
5,000	1500	0.83	1.10
6,000	1800	0.80	1.12
7,000	2100	0.77	1.14
8,000	2400	0.74	1.16
9,000	2700	0.71	1.19
10,000	3000	0.69	1.20

From Table 10.5, CF for 3000 ft = 1.06

$V = V_\mathrm{m} \times \mathrm{CF} = 1791 \times 1.06 = 1898$ fpm

Example 10.2 (Metric)
a. Estimate the airflow velocity of a 50 Pa velocity pressure (Vp) reading at 900 m elevation and 20 °C.
b. Also calculate the answer using tables and equations.

Solution
a. 900 m equals approximately 12% (3 × 4%)

50 Pa × 1.12 = 56 Pa

$V = \sqrt{1.66\, V_\mathrm{p}} = \sqrt{1.66 \times 56} = 9.64$ m/s (Equation 6.3)

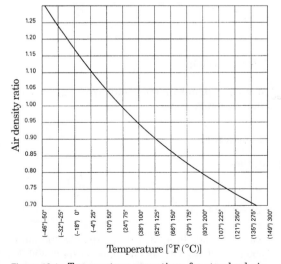

Figure 10.1   Temperature corrections for standard air.

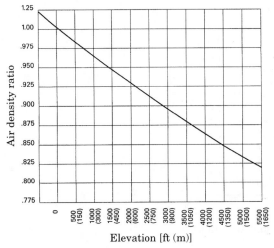

Elevation [ft (m)]

Figure 10.2 Altitude corrections for standard air.

b. $V = \sqrt{1.66 \times 50} = 9.11$ m/s

From Table 10.5, CF for 900 m = 1.06

$V = V_m \times CF = 9.11 \times 1.06 = 9.66$ m/s

Example 10.3 (U.S.) An exhaust air system at 4000 ft. with 250 °F air has an indicated velocity of 3000 fpm. Find the actual velocity.

Solution
From Table 10.3, CF = 1.15 (temperature)

From Table 10.5, CF = 1.08 (altitude)

Total CF = $1.15 \times 1.08 = 1.242$

$V = V_m \times CF = 3000$ fpm $\times 1.242 = 3726$ fpm

Example 10.3 (Metric) An exhaust air system at 1200 m with 125 °C air has an indicated velocity of 15 m/s. Find the actual velocity.

Solution
From Table 10.4, CF = 1.16 (temperature)

From Table 10.5, CF = 1.08 (altitude)

Total CF = $1.16 \times 1.08 = 1.253$

$V = V_m \times CF = 15$ m/s $\times 1.253 = 18.79$ m/s

Example 10.4 A Pitot tube traverse is made in a 36 × 24 in. (900 × 600 mm) duct located at an altitude of 5000 ft (1500 m) with an air temperature of 300 °F (150 °C). The Pitot tube traverse readings are found in Table 10.6 with the

TABLE 10.6 Pitot Tube Traverse Readings for
Example 10.4

Pitot tube traverse	Velocity pressure		Measured velocity	
	in.w.g.	Pa	fpm	m/s
1	0.056	13.9	946	4.80
2	0.068	16.9	1045	5.30
3	0.028	7.0	670	3.40
4	0.078	19.4	1119	5.68
5	0.081	20.2	1140	5.78
6	0.084	20.9	1161	5.89
7	0.086	21.4	1175	5.96
8	0.088	21.9	1188	6.03
9	0.088	21.9	1188	6.03
10	0.087	21.7	1181	5.99
11	0.083	20.7	1154	5.86
12	0.080	19.9	1133	5.75
13	0.072	17.9	1075	5.45
14	0.052	12.9	913	4.63
15	0.052	12.9	913	4.63
16	0.051	12.7	904	4.59

Grand total (velocities) = 16,905 fpm (85.77 m/s).
Average (velocities) = 1056 fpm (5.36 m/s).

average velocities calculated. NOTE: **Do not average velocity pressures.**
Find the actual airflow volumes and velocities (both U.S. and metric).

Solution (U.S.)

$$Q = AV = \frac{36 \times 24 \times 1056}{144} = 6336 \text{ cfm}$$

From Table 10.3, CF = 1.20 for 300 °F

From Table 10.5, CF = 1.10 for 5000 ft.

Total CF = 1.20 × 1.10 = 1.32

Actual velocity = 1056 × 1.32 = 1394 fpm

Actual airflow = 6336 × 1.32 = 8364 cfm

Solution (Metric)
$Q = 1000 \, AV = 1000 \times 0.9 \times 0.6 \times 5.36 = 2894 \text{ L/s}$

From Table 10.4, CF = 1.20 for 160 °C

From Table 10.5, CF = 1.10 for 1500 m

Total CF = 1.20 × 1.10 = 1.32

Actual velocity = 5.36 m/s × 1.32 = 7.08 m/s

Actual airflow = 2894 L/s × 1.32 = 3820 L/s

10.4   Air Density Changes

When the airflow being tested is at a higher or lower temperature than used in the normal HVAC range, or at altitudes above 2000 ft. (600 m), the following equations may be used:

Equation 10.8 (U.S.)   **Equation 10.8 (Metric)**

$$V_P = d \times \left(\frac{V}{1096} \right)^2 \qquad V_P = \frac{d}{2} \times V^2$$

Equation 10.9 (U.S.)   **Equation 10.9 (Metric)**

$$V = 1096 \sqrt{V_P/d} \qquad V = 1.414 \sqrt{V_P/d}$$

Where: V_P = velocity pressure (in.w.g. or Pa)
        V = velocity (fpm or m/s)
        d = density (lb/ft^3 or kg/m^3)

The same velocity pressure (V_P) reading from a duct system airflow of a lesser density will result in a higher velocity.

Example 10.5 (U.S.)   The measured velocity pressure (V_P) for airflow at 250 °F and an altitude of 6000 ft. is 0.32 in.w.g. Calculate the airflow velocity.

Solution   The air density correction factor from Table C-8 of Appendix C for 250 °F at 6000 ft is 0.60.

$$d = 0.60 \times 0.075 \text{ lb/ft}^3 = 0.045 \text{ lb/ft}^3$$

Using Equation 10.9:

$$V = 1096 \sqrt{0.32/0.045} = 2923 \text{ fpm, } or$$

Using Equation 6.3 from Chapter 6; corrected V_P = 0.32/0.6

$$V = 4005 \sqrt{V_P} = 4005 \sqrt{0.32/0.6} = 2925 \text{ fpm}$$

Example 10.5 (Metric)   The measured velocity pressure (V_P) for airflow at 125 °C and an altitude of 1750 m is 80 Pa. Calculate the airflow velocity.

Solution   The air density correction factor from Table D-6 of Appendix D for 125 °C and 1750 m is 0.60.

$$d = 0.60 \times 1.204 \text{ kg/m}^3 = 0.7224 \text{ kg/m}^3$$

Using Equation 6.5:

$$V = 1.414 \sqrt{V_P/d} = 1.414 \sqrt{80/0.7224} = 14.88 \text{ m/s}$$

11

Heat Transfer

11.1 Heat

11.1.1 Heat intensity

The intensity of heat of a substance is called *temperature* and is measured by a thermometer or other temperature indicating device. The Fahrenheit scale is used in the United States, whereas the Celsius (formerly called centigrade) scale is mostly used elsewhere (see Table 11.1).

The following equation can be used to convert temperature from the Celsius scale to the Fahrenheit scale:

Equation 11.1

$$°F = 1.8 \, °C + 32°$$

Equation 11.2 can be used to convert Fahrenheit scale temperatures to Celsius scale temperatures.

Equation 11.2

$$°C = \frac{(°F - 32°)}{1.8}$$

The temperature at which the continued removal of heat from a substance results in the substance having no molecular action is called *absolute zero*, which is $-460 \, °F$ on the Fahrenheit scale and $-273 \, °C$ on the Celsius scale. The thermodynamic absolute temperature (T) used in temperature–pressure calculations can be obtained in degrees Rankine by using Equation 11.3, and in degrees Kelvin by using Equation 11.4. The relationship between the temperature scales is shown in Figures 11.1 and 11.2. In metric equations where there is a temperature difference (Δt), degrees Kelvin (K) may be used instead of degrees Celsius (°C). Note that K does not use the degree symbol.

TABLE 11.1 Temperature Equivalents

Temperatures, 1° to 100°

°C	°F/°C	°F	°C	°F/°C	°F	°C	°F/°C	°F	°C	°F/°C	°F
−17.2	1	33.8	−3.3	26	78.8	10.6	51	123.8	24.4	76	168.8
−16.7	2	35.6	−2.8	27	80.6	11.1	52	125.6	25.0	77	170.6
−16.1	3	37.4	−2.2	28	82.4	11.7	53	127.4	25.6	78	172.4
−15.6	4	39.2	−1.7	29	84.2	12.2	54	129.2	26.1	79	174.2
−15.0	5	41.0	−1.1	30	86.0	12.8	55	131.0	26.7	80	176.0
−14.4	6	42.8	−0.6	31	87.8	13.3	56	132.8	27.2	81	177.8
−13.9	7	44.6	0.0	32	89.6	13.9	57	134.6	27.8	82	179.6
−13.3	8	46.4	0.6	33	91.4	14.4	58	136.4	28.3	83	181.4
−12.8	9	48.2	1.1	34	93.2	15.0	59	138.2	28.9	84	183.2
−12.2	10	50.0	1.7	35	95.0	15.6	60	140.0	29.4	85	185.0
−11.7	11	51.8	2.2	36	96.8	16.1	61	141.8	30.0	86	186.8
−11.1	12	53.6	2.8	37	98.6	16.7	62	143.6	30.6	87	188.6
−10.6	13	55.4	3.3	38	100.4	17.2	63	145.4	31.1	88	190.4
−10.0	14	57.2	3.9	39	102.2	17.8	64	147.2	31.7	89	192.2
−9.4	15	59.0	4.4	40	104.0	18.3	65	149.0	32.2	90	194.0
−8.9	16	60.8	5.0	41	105.8	18.9	66	150.8	32.8	91	195.8
−8.3	17	62.6	5.6	42	107.6	19.4	67	152.6	33.3	92	197.6
−7.8	18	64.4	6.1	43	109.4	20.0	68	154.4	33.9	93	199.4
−7.2	19	66.2	6.7	44	111.2	20.6	69	156.2	34.4	94	201.2
−6.7	20	68.0	7.2	45	113.0	21.1	70	158.0	35.0	95	203.0
−6.1	21	69.8	7.8	46	114.8	21.7	71	159.8	35.6	96	204.8
−5.6	22	71.6	8.3	47	116.6	22.2	72	161.6	36.1	97	206.6
−5.0	23	73.4	8.9	48	118.4	22.8	73	163.4	36.7	98	208.4
−4.4	24	75.2	9.4	49	120.2	23.3	74	165.2	37.2	99	210.2
−3.9	25	77.0	10.0	50	122.0	23.9	75	167.0	37.8	100	212.0

Temperatures, 100° to 1000°

°C	°F/°C	°F	°C	°F/°C	°F	°C	°F/°C	°F	°C	°F/°C	°F
38	**100**	212	160	**320**	608	288	**550**	1022	416	**780**	1436
43	**110**	230	166	**330**	626	293	**560**	1040	421	**790**	1454
49	**120**	248	171	**340**	644	299	**570**	1058	427	**800**	1472
54	**130**	266	177	**350**	662	304	**580**	1076	432	**810**	1490
60	**140**	284	182	**360**	680	310	**590**	1094	438	**820**	1508
66	**150**	302	188	**370**	698	316	**600**	1112	443	**830**	1526
71	**160**	320	193	**380**	716	321	**610**	1130	449	**840**	1544
77	**170**	338	199	**390**	734	327	**620**	1148	454	**850**	1562
82	**180**	356	204	**400**	752	332	**630**	1166	460	**860**	1580
88	**190**	374	210	**410**	770	338	**640**	1184	466	**870**	1598
93	**200**	392	216	**420**	788	343	**650**	1202	471	**880**	1616
99	**210**	410	221	**430**	806	349	**660**	1220	477	**890**	1634
100	**212**	414	227	**440**	824	354	**670**	1238	482	**900**	1652
104	**220**	428	232	**450**	842	360	**680**	1256	488	**910**	1670
110	**230**	446	238	**460**	860	366	**690**	1274	493	**920**	1688
116	**240**	464	243	**470**	878	371	**700**	1292	499	**930**	1706
121	**250**	482	249	**480**	896	377	**710**	1310	504	**940**	1724
127	**260**	500	254	**490**	914	382	**720**	1328	510	**950**	1742
132	**270**	518	260	**500**	932	388	**730**	1346	516	**960**	1760
138	**280**	536	266	**510**	950	393	**740**	1364	521	**970**	1778
143	**290**	554	271	**520**	968	399	**750**	1382	527	**980**	1796
149	**300**	572	277	**530**	986	404	**760**	1400	532	**990**	1814
154	**310**	590	282	**540**	1004	410	**770**	1418	538	**1000**	1832

NOTE: Read chart from bold column to left (°F to °C) or to right (°C to °F). Absolute temperature (K) = °C + 273.15° = °F + 459.67°/1.8. Temperature difference: K = °C = °F/1.8.

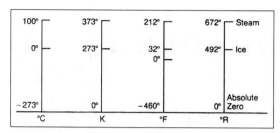

Figure 11.1 Relationship between temperature scales.

Equation 11.3 (U.S.)

°R = °F + 460

Where: °R = Absolute temperature (Rankine)
  °F = Fahrenheit temperature

Equation 11.4 (Metric)

K = °C + 273

Where: K = Absolute temperature (Kelvin)
  °C = Celsius temperature

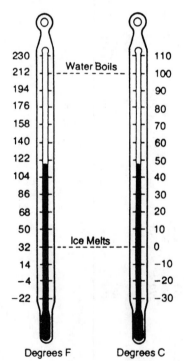

Figure 11.2 Comparison of Fahrenheit and Celsius thermometers.

11.1.2 Heat quantity

The quantity of heat is found by measuring the temperature and weight of a substance. In the United States the quantity or amount of heat in a substance is measured in *British thermal units* (Btu), which is defined as the amount of heat required to raise the temperature of one pound of water from 59 °F to 60 °F. In the metric system, the unit of measurement is called one calorie (also equal to 4.19 joules), which is the amount of heat required to raise the temperature of one gram of water from 4 °C to 5 °C. For electrical resistance heating, 1 W = 3.412 Btu/h, and in refrigeration work, 1 ton of cooling equals 12,000 Btu/h (3.517 kW).

The Btu (kilojoule) is seldom used without further definition either as flow rate with time or as a limited quantity of heat contained in some weight or volume of matter. By using the Btu per hour (watt), it is possible to determine the amount of heat flowing or heat transferred in a given amount of time. A thousand Btu per hour is abbreviated MBH.

11.1.3 Specific heat

As a thermodynamic process, heat transfer is a function of weight, specific heat, and temperature difference. It is expressed mathematically as: **Heat equals weight of the transfer medium times specific heat times temperature difference.**

Equation 11.5 (U.S. and Metric)

$$Q = M \times C_p \times \Delta t$$

Where: Q = heat (Btu or kJ)
M = weight (lb or kg)
C_p = specific heat [Btu/lb•°F [or kJ/(kg•K)]
Δt = temperature difference °F (°C or K)

Specific heat (C_p) is a reference index. It is used to measure the amount of heat necessary to raise the temperature of 1 lb (kg) of any substance 1 °F (°C or K). For the TAB technician, it is important to remember that the specific heat of standard air is 0.2388 Btu/lb•°F [1005 J/(kg•K)] and the specific heat of standard water is 1.0 Btu/lb•°F [4190 J/(kg•K)].

So for standard air, substituting in Equation 11.5:

$$(\text{U.S.}) \; Q = 60 \; \text{min/h} \times C_p \times d \times \text{cfm} \times \Delta t$$

$$Q = 60 \times 0.24 \times 0.075 \times \text{cfm} \times \Delta t$$

$$Q \; (\text{in Btu/h}) = 1.08 \times \text{cfm} \times \Delta t$$

$$(\text{Metric}) \; Q = C_p \times d \times \text{L/s} \times \Delta t = 1005 \times 1.204 \times \text{L/s} \times \Delta t$$

$$Q \; (\text{in watts}) = 1.23 \times \text{L/s} \times \Delta t$$

11.2 Heat Transfer

11.2.1 Methods of heat transfer

In HVAC systems, as in natural processes, heat is transferred by three means:

1. Radiation
2. Convection
3. Conduction

11.2.1.1 Radiation. *Radiation* is a form of energy transfer similar to that of light waves and radio waves, without heating the intervening space. Energy waves of the sun, for example, can be felt by a person until a heavy cloud layer passes in front of it. The change is felt immediately, but the air in the space in between was not heated directly by the sun's rays.

11.2.1.2 Convection. Most of the heat transfer in the HVAC industry is by *convection*. Convection is the transfer of heat by movement of a fluid such as air or water over a substance. The heat flow can be either to or from the substance or object.

When a fan is used to propel the air across a hot or cold surface, heat transfer generally increases with an increase in air velocity. *Forced convection* (air moved across the surface by a fan) is therefore a more efficient method of heat transfer and produces a greater volume of transferred heat. Major factors in the transfer of heat by convection are:

1. Temperature difference
2. Flow velocity
3. Type of fluid (or gas)
4. Conductivity of heat transfer material

5. Size and shape of the transfer surfaces

6. Condition of the transfer surfaces

11.2.1.3 Conduction. *Conduction* is the flow of heat through a substance or the flow of heat from one body to another when the bodies are in direct physical contact with one another. The heat of the handle of a poker placed in a fire is a good example.

11.2.1.4 Thermal conductivity (*k*). The ability of a substance to transfer heat by conduction is called *thermal conductivity* (*k*). Conductivity is defined as the amount of heat in Btu per hour flowing through 1 in. thickness of 1 ft^2 of a homogeneous material when the difference in temperature between the faces is 1 °F [W/(m•K)]. Therefore, materials having the lowest conductivity numerical values are the best insulators.

11.2.1.5 Thermal conductance (*C*). *Thermal conductance* (*C*) is a heat flow property of an object made of nonhomogeneous material, such as hollow clay tile or concrete blocks, wherein each succeeding inch of thickness is not identical with the preceding inch. Therefore, it is necessary to indicate the heat flow rate through the entire object. Conductance is defined as the heat flow rate in Btu per hour per 1 ft^2 of nonhomogeneous material of a certain specified thickness for a 1 °F difference in temperature between the two surfaces of the material. [W/(m^2•K)] Care should be taken not to confuse conductivity and conductance.

11.2.1.6 Thermal resistance (*R*). *Thermal resistance* (*R*) or resistivity is the reciprocal of the heat transmission coefficient (*U*). The overall resistance (*R$_t$*) is equal to the sum of the resistances and resistivities of the insulation and substances from which the wall, ceiling, floor, etc., is built. *R* values can be added together along with the *k* values and *C* values.

11.2.2 Coefficient of heat transfer (*U*)

The coefficient of heat transfer, *U*, can be obtained by taking the reciprocal of the resistance as shown in the following equation:

Equation 11.6

$$U = \frac{1}{R_t} = \frac{1}{R_1 + R_2 + R_3 \ldots + R_n}$$

Where: R_t = overall resistance total [ft²•°F/Btu/h or (m²•K)/W]
          U = coefficient of heat transfer [Btu/h/ft²•°F or W/(m²•K)]

$R_1 + R_2 ... R_n$ = individual resistances

Equation 11.7

$Q = A \times U \times \Delta t$

Where: Q = the rate of heat transfer (Btu/h or W)
       A = the area of a surface (ft² or m²)
       U = coefficient of heat transfer [Btu/h/ft²•°F or W/(m²•K)]
       Δt = °F (°C or K) temperature difference between the temperatures on each side of the surface

Example 11.1 (U.S.) An exposed wall of 800 ft³ in a building has a U factor or coefficient of heat transfer of 0.78. By insulating the wall, the U factor becomes 0.08. In a 10 hour period with 65 °F inside and 0 °F outside, find how many Btu would be saved.

Solution Using Equation 11.7:

$$Q = A \times U \times \Delta t$$

$$Q = 800 \times (0.78 - 0.08) \times (65° - 0°)$$

$$Q = 800 \times 0.7 \times 65°$$

$$Q = 36,400 \text{ Btu/h difference}$$

$$\text{Heat saved} = 10 \text{ hr} \times 36,400 \text{ Btu/h} = 364,000 \text{ Btu}$$

Example 11.1 (Metric) An exposed wall of 75 m² in a building has a U factor or coefficient of heat transfer of 6.01. By insulating the wall, the U factor becomes 0.62. In a 10 hour period with 18 °C inside and −18 °C outside, find how many kilojoules would be saved.

Solution Using Equation 11.7:

$$Q = A \times U \times \Delta t = 75 \times (6.01 - 0.62) \times [18 \text{ °C} - (-18 \text{ °C})]$$

$$Q = 75 \times 5.39 \times 36°C = 14553 \text{ W difference}$$

$$1 \text{ W} = 1 \text{ joule per second; } 1000 \text{ J/s} = 1 \text{ kJ/s;}$$

$$\text{Heat saved} = 10 \text{ h} \times 14.55 \text{ kJ/s} \times 3600 \text{ s/h}$$

$$\text{Heat saved} = 523,800 \text{ kJ}$$

Example 11.2 (U.S.) The R values for materials in a frame wall are 1.5, 3.2, 2.1, 9.0, and 0.8. Find the value of U.

Solution

$$R_t = 1.5 + 3.2 + 2.1 + 9.0 + 0.8 = 16.6$$

$$U = \frac{1}{R_t} = \frac{1}{16.6} = 0.06$$

Example 11.2 (Metric)   The R values for materials in a frame wall are 0.23, 0.48, 0.32, 1.35, and 0.12. Find the value of U.

Solution

$$R_t = 0.23 + 0.43 + 0.32 + 1.35 + 0.12 = 2.50$$

$$U = \frac{1}{R_t} = \frac{1}{2.5} = 0.4$$

11.2.3   Types of heat transfer

11.2.3.1   Sensible heat. *Sensible heat* is any heat transfer that causes a change in temperature that can be measured with a thermometer. Heating or cooling of air or water, measured with a thermometer, indicates an increase or decrease in sensible heat.

11.2.3.2   Latent heat. *Latent heat* is any heat transfer that causes a change of state from a solid to a liquid, a liquid to a gas, or vice versa. Evaporation of water is an example of a latent heat transfer. Latent heat transfer at terminal coils may be defined as any process that humidifies or dehumidifies the air. Both processes result in a change of actual moisture content in the air.

11.2.3.3   Total heat (enthalpy). *Total heat* is the sum of the sensible heat and latent heat in an exchange process. In many cases, the addition or subtraction of latent and sensible heat at terminal coils appears simultaneously. Total heat also is called *enthalpy*, both of which can be defined as the quantity of heat energy contained in that substance.

At any given time, a substance has only one value of enthalpy and a related specific temperature value on the thermometer. If the enthalpy is increased, the temperature increases. Conversely, if the temperature is decreased, the enthalpy decreases. The ability to increase or decrease enthalpy and temperature together is the basis for heat transfer in environmental systems and only *differences* in enthalpy and temperature are normally of importance.

11.3   Air Heat Flow Equations

11.3.1   Sensible heat

Sensible heat was defined as the heat associated with temperature differences as measured by a dry bulb thermometer. The sensible heat flow equations for air are:

Equation 11.8 (U.S)　　　**Equation 11.8 (Metric)**

$$Q_s = 1.08 \times cfm \times \Delta t \qquad Q_s = 1.23 \times L/s \times \Delta t$$

for standard air conditions

Where: Q_s = Sensible heat flow (Btu/h or watts)
　　　　cfm = Airflow (ft^3/min)
　　　　L/s = Airflow (liters per second)
　　　　Δt = Temperature difference (°F or °C or K)

Equation 11.9 (U.S.)　　　**Equation 11.9 (Metric)**

$$Q_s = 60 \times C_p \times d \times cfm \times \Delta t \qquad Q_s = C_p \times d \times L/s \times \Delta t$$

for non-standard air conditions

Where: C_p = specific heat [Btu/lb•°F or kJ/(kg•K)]
　　　　d = density lb/ft^3 or kg/m^3)

For standard air, C_p = 0.24 and d = 0.075 (C_p = 1.005 and d = 1.204).

11.3.2　Latent heat

Latent heat is the heat used to convert a liquid into a gas or vapor without a change in dry bulb temperature (such as water boiling at 212 °F or 100 °C) or the heat released when vapor condenses into a liquid, again without a change in dry bulb temperature.

Equation 11.10 (U.S.)　　　**Equation 11.10 (Metric)**

$$Q_L = 0.68 \times cfm \times \Delta W \text{ (gr.)} \qquad Q_L = 3.0 \times L/s \times \Delta W$$

or

$$Q_L = 4840 \times cfm \times \Delta W \text{ (lb.)}$$

Where: Q_L = latent heat flow (Btu/h or watts)
　　　　cfm = airflow (ft^3/min)
　　　　L/s = airflow (liters per second)
　　　　ΔW = humidity ratio (lb water/lb dry air or grains water/lb dry air in U.S. units, or (grams of water per kilogram of dry air in metric units)

The humidity ratio (ΔW) is obtained from a psychrometric chart (see Chapter 12).

11.3.3 Total heat (enthalpy)

Changes of enthalpy or the total heat content of air (obtained from a psychrometric chart) use the following equation:

Equation 11.11 (U.S.)   **Equation 11.11 (Metric)**

$$Q_T = 4.5 \times cfm \times \Delta h \qquad Q_T = 1.20 \times L/s \times \Delta h$$

Where: Q_T = total heat flow (Btu/h watts)
cfm = airflow (ft^3/min)
L/s = airflow (liters per second)
Δh = enthalpy difference (Btu/lb dry air or kJ/kg dry air)

A total of the heat flow answers obtained from Equations 11.9 and 11.10 should approximately equal the answer obtained when using Equation 11.11.

11.3.4 Hydronic heat flow equation

The heat flow equation used for water systems is:

Equation 11.12 (U.S.)

$$Q = 500 \times gpm \times \Delta t$$

Where: Q = heat flow (Btu/h)
gpm = gallons per minute
(water only)
Δt = temperature
difference (°F)

Equation 11.12 (Metric)

$$Q \text{ (in W)} = 4190 \times L/s \times \Delta t$$

or

$$Q \text{ (in kW)} = 4190 \times m^3/s \times \Delta t$$

Where: Q = heat flow (W or kW)
L/s = liters per second
(water only)
m^3/s = cubic meters per second
(water only)
Δt = temperature difference
(°C or K)

Example 11.3 (U.S.) Fifty gallons per minute of 200 °F water enters the coil of an HVAC unit and exits at 180 °F. If the unit fan handles 13,000 cfm of 70 °F entering air, find the leaving air temperature.

Solution Using Equation 11.12:

$$Q = 500 \times gpm \times \Delta t = 500 \times 50 \times (200° - 180°)$$

$$Q = 500,000 \text{ Btu/h}$$

Using Equation 11.8:

$$Q = 1.08 \times cfm \times \Delta t; \, \Delta t = Q/1.08 \times cfm$$

$$\Delta t = 500{,}000/1.08 \times 13{,}000 = 35.6 \text{ °F}$$

Leaving air temperature $= 70° + 35.6° = 105.6$ °F

Example 11.3 (Metric) A flow of 3.15 L/s of 93 °C water enters the coil of an HVAC unit and exits at 82 °C. If the unit fan handles 6100 L/s of 21 °C entering air, find the leaving air temperature.

Solution Using Equation 11.12:

$$Q = 4190 \times \text{L/s} \times \Delta t = 4190 \times 3.15 \times (93° - 82°)$$

$$Q = 145{,}184 \text{ W}$$

Using Equation 11.8:

$$Q = 1.23 \times \text{L/s} \times \Delta t; \Delta t = Q/1.23 \times \text{L/s}$$

$$\Delta t = 145{,}184/1.23 \times 6100 = 19.4 \text{ °C}$$

Leaving air temperature $= 21° + 19.4° = 40.4$ °C

Example 11.4 (U.S.) A condenser is operating at 100 gpm. Water enters at 60 °F and leaves at 70 °F. The condensing temperature is 80 °F and the condensing area is 300 ft². *Find:*

a. Btu/h capacity

b. Arithmetical mean temperature difference (AMTD)

c. Rate of heat transfer, Btu/h/ft²•°F

Solution

a. $Q = 500 \times \text{gpm} \times \Delta t = 500 \times 100 \ (70° - 60°) = 500{,}000$ Btu/h

b. 80 °F $-$ 60 °F $= 20$ °F water in

   80 °F $-$ 70 °F $= 10$ °F water out

   20 °F $+$ 10 °F $= 30$ °F

   Ave. $= 30$ °F/2 $= 15$ °F AMTD

c. Rate of heat transfer $= \dfrac{\text{Btu/h}}{\text{ft}^2 \times \Delta t} = \dfrac{500{,}000}{300 \times 15 \text{ AMTD}} = 111.1$ Btu/h/ft²/°F

Example 11.4 (Metric) A condenser is operating at 6.3 L/s. Water enters at 15.6 °C and leaves at 21.1 °C. The condensing temperature is 26.7 °C and the condensing area is 27.9 m². *Find:*

a. Watts capacity

b. Arithmetical mean temperature difference (AMTD)

c. Rate of heat transfer, W/(m²•°K)

Solution

a. $Q = 4190 \times L/s \times \Delta t = 4190 \times 6.3 \times (21.1° - 15.6°) = 145{,}184$ W

b. $26.7 °C - 15.6 °C = 11.1 °C$ water in

   $26.7 °C - 21.1 °C = 5.6 °C$ water out

   $11.1 °C + 5.6 °C = 16.7 °C$

   Ave. $= 16.7 °C/2 = 8.35 °C$ AMTD

c. Rate of heat transfer $= \dfrac{W}{m^2 \times \Delta t} = \dfrac{145{,}184}{27.9 \times 8.35° \text{ AMTD}} = 623.2$ W/(m²•K)

11.4   Log Mean Temperature Difference (LMTD)

Where two fluids are used in a heat transfer process the temperature difference at the end of the process will be less at the beginning. Thus a square foot of exchange surface at the end will do less work than an equal area at the beginning. The heat exchange will follow a logarithmic curve. The LMTD will give correct values for calculating heat transfer rates in double pipe energy recovery systems:

Equation 11.13 (U.S. and Metric)

$$\text{LMTD} = \frac{\Delta t_L - \Delta t_s}{\text{Log}_e (\Delta t_L/\Delta t_s)} = \frac{\Delta t_L - \Delta t_s}{2.3 \, \text{Log}_{10} (\Delta t_L/\Delta t_s)}$$

Where: Δt_L = the larger temperature difference (°F or °C)

   Δt_s = the smaller temperature difference (°F or °C)

   $\log_e$ = Napierian or natural logarithm

   Log_{10} = logarithm to base 10

Example 11.5 (U.S.)   Fluid enters a heat exchanger at 80 °F and leaves at 46 °F. Water enters in the opposite direction (counter flow) at 20 °F and leaves the other end of the exchanger at 40 °F. Figure 11.3 illustrates the log flow. *Find the logarithm mean temperature difference.*

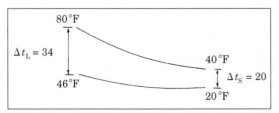

Figure 11.3 Mean temperature difference in heat exchanger.

Solution Using Equation 11.13:

$$\text{LMTD} = \frac{\Delta t_{\text{L}} - \Delta t_{\text{s}}}{\text{Log}_e (\Delta t_{\text{L}}/\Delta t_{\text{s}})} = \frac{(80° - 46°) - (40° - 20°)}{\text{Log}_e (34° - 20°)}$$

$$\text{LMTD} = \frac{14}{\log_e 1.7} = \frac{14}{0.53} = 26.4 \text{ °F}$$

Example 11.5 (Metric) The problem works the same as above except temperatures are in °C instead of °F.

11.5 Hot Water Coil Heat Transfer

Hot water coils in HVAC and fan coil units transfer most of the heat to the airstream even when the water flow rate is greatly reduced in low-temperature systems. Figure 11.4 shows the percentage of full rated heat flow into the airstream versus hydronic (water) flow capacity for 20 °F (11.1 °C) and 60 °F (33.3 °C) temperature drops.

For example in a typical 20 °F (11.1 °C) Δt heating coil, 50% of the rated water flow in the coil will still transfer about 90% of the heat to the airstream. To reduce heat output of the coil to 50% at a 20 °F (11.1 °C) Δt, water flow through the coil would need to be reduced to about 10%.

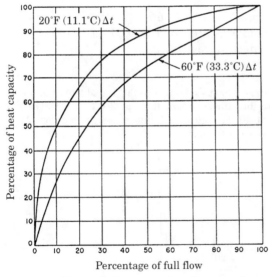

Figure 11.4 Heat emission versus flow, hot water heating coils.

11.6 Duct Heat Gain/Loss

Figures 11.5 and 11.6 contain duct heat transfer coefficients (U) for lined, insulated, and uninsulated rigid ducts and flexible ducts. Using these values and Equations 11.14, 11.15, and 11.16, duct heat transfer and entering or leaving duct air temperatures may be calculated.

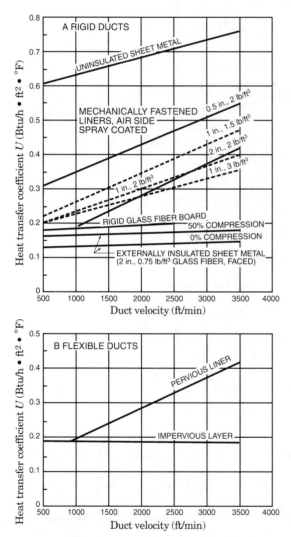

Figure 11.5 Duct heat transfer coefficients (U.S. units). *(Courtesy of ASHRAE.)*

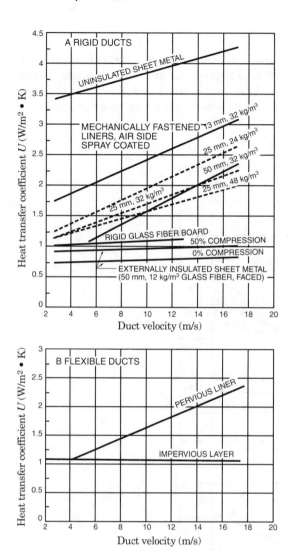

Figure 11.6 Duct heat transfer coefficients (metric units). *(Courtesy of ASHRAE.)*

Equation 11.14 (U.S.)

$$Q_1 = \frac{UPL}{12} \left[\left(\frac{t_e + t_1}{2} \right) - t_a \right]$$

Equation 11.15 (U.S.)

$$t_e = \frac{t_1(y + 1) - 2t_a}{(y - 1)}$$

Equation 11.14 (Metric)

$$Q_1 = \frac{UPL}{1000} \left[\left(\frac{t_e + t_1}{2} \right) - t_a \right]$$

Equation 11.15 (Metric)

$$t_e = \frac{t_1(y + 1) - 2t_a}{(y - 1)}$$

Equation 11.16 (U.S.)

$$t_1 = \frac{t_e(y - 1) + 2t_a}{(y + 1)}$$

Where: $y = 2.4AV\rho/UPL$ for
rectangular ducts
$y = 0.6DV\rho/UL$ for round
ducts
A = cross-sectional area of
duct (in.^{2})
V = average velocity (fpm)
D = diameter of duct (in.)
L = duct length (ft.)
Q_1 = heat loss or gain
through duct walls
Btu/h (negative for
heat gain)]
U = overall heat transfer
coefficient of duct wall
(Btu/h/ft^2•°F)
P = perimeter of bare or
insulated duct (in.)
ρ = density [lb/ft^3 (Std. =
0.075 lb/ft^3)]
t_e = temperature of air
entering duct (°F)
t_1 = temperature of air
leaving duct (°F)
t_a = temperature of air
surrounding duct (°F)

Equation 11.16 (Metric)

$$t_1 = \frac{t_e(y - 1) + 2t_a}{(y + 1)}$$

Where: $y = 2.01AV\rho/UPL$ for
rectangular ducts
$y = 0.5DV\rho/UL$ for round
ducts
A = cross-sectional area of
duct (mm^2)
V = average velocity (m/s)
D = diameter of duct (mm)
L = duct length (m)
Q_1 = heat loss/gain through
duct walls [W (negative
for heat gain)]
U = overall heat transfer
coefficient of duct wall
(W/m^2•K)
P = perimeter of bare or
insulated duct (mm)
ρ = density [kg/m^3 (Std. =
1.204 kg/m^3)]
t_e = temperature of air
entering duct (°C)
t_1 = temperature of air
leaving duct (°C)
t_a = temperature of air
surrounding duct (°C)

Example 11.6 (U.S.)   One hundred feet of 48 × 24 in. (inside dimensions) duct lined with 1 in., 1.5 lb/ft^3 duct liner containing an airflow of 16,000 cfm passes through a 32 °F attic. If 120 °F air is required at the end of the duct, find the entering air temperature and duct heat loss.

Solution   Using Equation 2.1 (Chapter 2):

$$V = \frac{Q}{A} = \frac{16,000}{4.0 \times 2.0} = 2000 \text{ fpm}$$

From Figure 11.5, $U = 0.35$; $P = 2(48 + 24) = 144$ in.

$$y = \frac{2.4\,AV\rho}{UPL} = \frac{2.4 \times 1152 \times 2000 \times 0.075}{0.35 \times 144 \times 100} = 82.29$$

Using Equation 11.15:

$$t_e = \frac{t_1(y + 1) - 2t_a}{(y - 1)} = \frac{120(82.29 + 1) - 2(32)}{(82.29 - 1)}$$

$$t_e = \frac{9994.8 - 64}{81.29} = 122.17 \; °F \text{ entering air}$$

Using Equation 11.14:

$$Q_1 = \frac{UPL}{12}\left[\left(\frac{t_e + t_1}{2}\right) - t_a\right]$$

$$Q_1 = \frac{0.35 \times 144 \times 100}{12}\left[\left(\frac{122.17 + 120}{2}\right) - 32\right]$$

$$Q_1 = 420\,(121.09 - 32) = 37{,}416 \text{ Btu/h heat loss}$$

Example 11.6 (Metric)   Thirty meters of 1200×600 mm (inside dimensions) duct lined with 25 mm, 24 kg/m^3 duct liner containing an airflow of 7500 L/s passes through a 0 °C attic. If 49 °C air is required at the end of the duct, find the entering air temperature and duct heat loss.

Solution   Using Equation 2.1 (Chapter 2):

$$V = \frac{Q}{1000A} = \frac{7500}{1000 \times 1.2 \times 0.6} = 10.42 \text{ m/s}$$

From Figure 11.6, $U = 2.0$; $P = 2\,(1200 + 600) = 3600$ mm

$$y = \frac{2.01\,AV\rho}{UPL} = \frac{2.01 \times 1200 \times 600 \times 10.42 \times 1.204}{2.0 \times 3600 \times 30} = 84.06$$

Using Equation 11.15:

$$t_e + \frac{t_1(y + 1) - 2t_a}{(y - 1)} = \frac{49(84.06 + 1) - 2(0)}{(84.06 - 1)}$$

$$t_e = \frac{4167.94 - 0}{83.06} = 50.18 \; °C \text{ entering air}$$

Using Equation 11.14:

$$Q_1 = \frac{UPL}{1000}\left[\left(\frac{t_e + t_1}{2}\right) - t_a\right]$$

$$Q_1 = \frac{2 \times 3600 \times 30}{1000}\left[\left(\frac{50.18 + 49}{2}\right) - 0\right]$$

$$Q_1 = 216\,(49.59 - 0) = 10{,}711 \text{ watts heat loss}$$

12

Psychrometrics

12.1 Moist Air

Psychrometrics is the study of the behavior of air–water mixtures. Moist air has five variable properties:

1. Dry bulb temperature
2. Wet bulb temperature
3. Dewpoint temperature
4. Relative humidity
5. Humidity ratio

If any two of the above properties are known and are plotted on a psychrometric chart, the other three can be determined (Figure 12.1).

12.1.1 Basic properties

12.1.1.1 Dry bulb temperatures. Dry bulb air temperatures are measured by an ordinary thermometer or temperature sensing device.

12.1.1.2 Wet bulb temperatures. Wet bulb temperatures are read from a thermometer that has a bulb covered by a wet wick. The wet bulb temperature is different from the dry bulb temperature, at the same ambient temperature conditions, because a cooling effect is produced by the evaporation of moisture from the wick, which reduces the temperature of the bulb and the temperature reading. The dryer the air, the faster the rate of evaporation and the lower the reading. Consequently, the difference in the dry and wet bulb temperature readings is a measure of the *dryness* of the air.

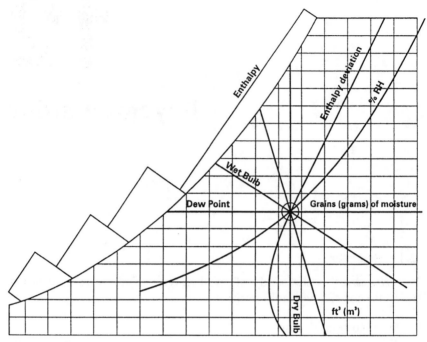

Figure 12.1 Properties of moist air.

12.1.1.3 Dewpoint. *Dewpoint temperature* is the temperature at which moisture leaves the air and condenses on objects, such as when dew forms on grass and other objects on a cool night. When the dewpoint, dry bulb, and wet bulb temperatures are the same, the air is saturated; it can hold no more moisture. When air is in a saturated condition, moisture entering the air displaces moisture within the air. The displaced moisture leaves the air in the form of fine droplets. This condition occurs in nature as fog. The curved top line on the left hand side of the psychrometric chart is a series of saturation points called the *saturation curve.*

12.1.1.4 Relative humidity. *Relative humidity* is a comparison of the amount of moisture within the air to the amount of moisture the same air (at the same dry bulb temperature) could hold if it were saturated. Relative humidity is given in percent. For example, if the relative humidity of the air is 50%, it contains one half the amount of moisture possible at the existing dry bulb temperature.

12.1.1.5 Humidity ratio. *Humidity ratio* describes the actual weight of water in a mixture of water vapor and air, expressed in grains of water per pound of dry air, or pounds of water per pound of dry air (1 lb =

7000 grains). In the metric system, grams of moisture per kilogram of dry air is used.

12.1.2 Other properties

Other properties and terms used in psychrometrics are listed below. Some have been discussed in Chapter 11.

12.1.3 Enthalpy

Enthalpy is a measure of the total heat energy of the air. Although there are some rather complex definitions of enthalpy, for practical purposes, enthalpy is the sum of the sensible heat and the latent heat content of the air. Changes in wet bulb temperature are indicative of changes in enthalpy.

In order to have specific values for practical use, enthalpy is stated as the amount by which the heat content is greater than the heat content that exists at some base point condition of the air. For general information, currently published data use a base point at which the enthalpy of dry air is taken to be zero at 0 °F (-17.8 °C), and the enthalpy of the water component is taken to be zero at 32 °F (0 °C).

12.1.2.2 Vapor pressure.
Vapor pressure is the pressure exerted by the water vapor contained in the air in inches (millimeters) of mercury.

12.1.2.3 Volume.
Volume is the cubic feet of the mixture per pound of dry air (m^3/kg of dry air).

12.1.2.4 Sensible heat factor.
Sensible heat factor is the ratio of sensible heat to total heat load.

12.1.2.5 Weight of dry air.
Pounds (kilograms) of dry air is the basis for calculations so that they would remain constant during all psychrometric processes.

12.1.2.6 Barometric pressures.
Atmospheric pressures differing from standard conditions of 29.92 in.Hg by 1 in. (760 mm Hg by 25 mm) of mercury or less may be assumed as standard in problems not requiring precise results, as in comfort HVAC work.

When dry bulb and dewpoint temperatures are known for air at nonstandard barometric pressures, values of percentage relative humidity and grains of moisture per cubic foot are correct as obtained from the standard chart. However, for any given dry bulb and wet bulb temperatures at nonstandard barometric pressures, all other properties of air must be corrected.

12.2 The Psychrometric Chart

12.2.1 What lines indicate

Sample psychrometric charts are shown in Figures 12.2 (U.S.) and 12.3 (metric). The direction to move on the chart from a reference point for a specific condition of air is shown in Figure 12.4 for each basic change to the condition of the air.

Psychrometric charts are graphic representations of psychrometric equations superimposed upon one another. A square grid is formed that is made up of air *dry bulb* (DB) *temperature lines* that run vertically with temperature values across the bottom, and the *moisture content of the air lines* that run horizontally with the values shown along the right side of the chart. *Wet bulb* (WB) *lines* run on a diagonal, upper left to lower right. The dry bulb lines and the moisture content lines are linear and each is the same distance from the adjacent line.

Wet bulb lines are not linear, but are straight lines unevenly spaced. There is less distance between the 25 °F and 30 °F (−4 °C and −1 °C) lines than there are between 80 °F and 85 °F (27 °C and 29 °C) lines, with one set being about three times as far apart as the other. Since wet bulb lines are normally coincident with enthalpy lines, then *enthalpy lines* are also not linear.

Curved *relative humidity* (RH) *lines* are superimposed over all of the other grids. *Air density lines*, although straight, are slanted differently.

Any point on a psychrometric chart indicates everything about the conditions of air at that point. If two different air measurements are noted on the chart, the *differences* in the conditions between those two points can be measured. For example, for a change in dry bulb temperature only, the two points would be on a horizontal line; for a change in moisture content of the air only, the two points would be on a vertical line.

A combination of the two changes in a particular ratio would result in a slanted line, which would then conform to some sensible heat factor line. Changes in enthalpy also can be determined between these two points by reading the difference between the values at the ends of the slanted wet bulb lines.

Notice that moisture content lines can be read on the right side in either pounds of moisture per pound of dry air or in grains of moisture per pound of dry air using a U.S. chart. Grams of moisture per kilogram of dry air are used on a metric chart.

12.2.2 Psychrometric chart processes

12.2.2.1 Sensible heating and cooling (U.S.).   When air at 70 °F DB and 20% RH is heated to 105 °F DB, this "heat-only" process establishes the conditions of the air at each point on the chart as shown in Figure

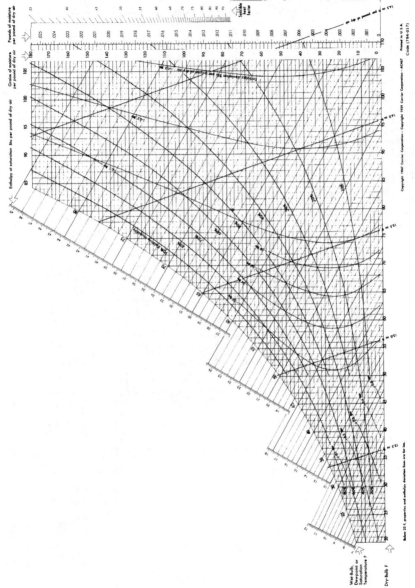

Figure 12.2 Psychrometric chart in U.S. units. *(Courtesy of Carrier Corporation.)*

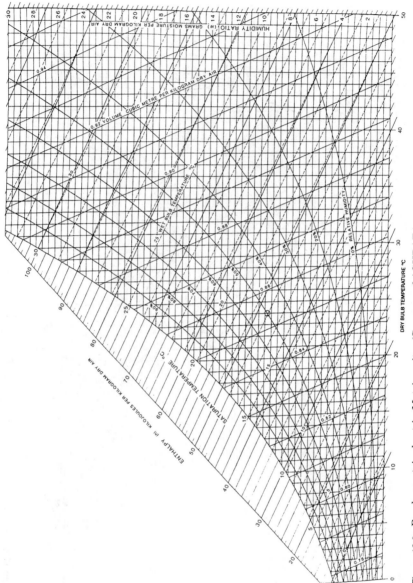

Figure 12.3 Psychrometric chart in Metric units. *(Courtesy of ASHRAE.)*

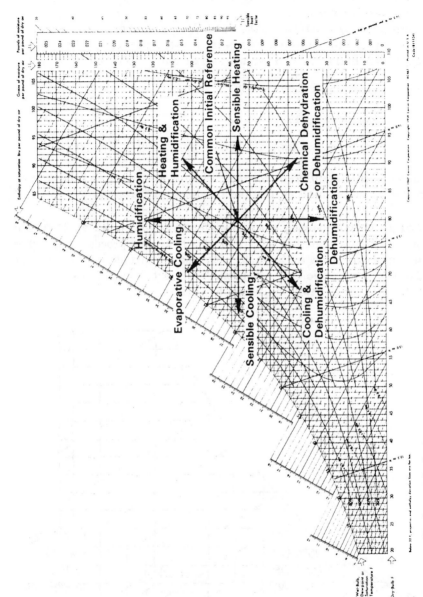

Figure 12.4 Psychrometric chart processes.

12.5. (Sensible cooling would reverse the process.) Using a U.S. unit psychrometric chart (as in Figure 12.2), all of the data for the two points, can be verified.

Normally $\Delta Q_T = \Delta Q_S + \Delta Q_L$; but $\Delta Q_L = 0$ as the moisture content of the air and the dewpoint temperature stay the same. Then $\Delta Q_T = \Delta Q_S$. Using Equation 11.8 (assume an airflow of 1000 cfm):

$$Q_T = 1.08 \times 1000 \times (105\ ^\circ F - 70\ ^\circ F) = 37{,}800\ \text{Btu/h}$$

Using Equation 11.11:

$$Q_T = 4.5 \times 1000 \times (29.0 - 20.3) = 39{,}150\ \text{Btu/h}$$

(Answers are 3% different because it is impossible to read the numerical values on a small psychrometric chart accurately).

12.2.2.2 Sensible heating and cooling (metric). When air at 21 °C DB and 40% RH is heated to 41 °C DB, this heat-only process establishes the conditions of the air at each point on the chart as shown in Figure 12.6. Using a metric unit psychrometric chart (as in Figure 12.3), all of the data for the two points can be verified.

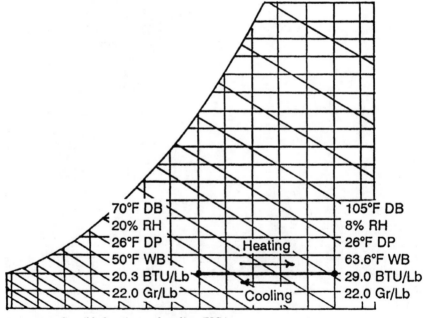

Figure 12.5 Sensible heating and cooling (U.S.).

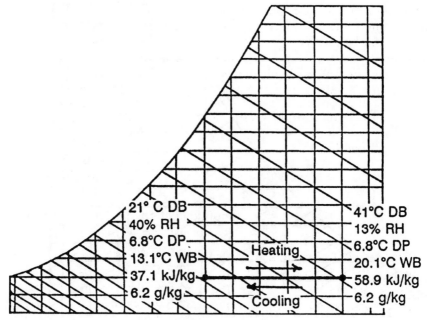

21° C DB
40% RH
6.8°C DP
13.1°C WB
37.1 kJ/kg
6.2 g/kg

Heating

Cooling

41°C DB
13% RH
6.8°C DP
20.1°C WB
58.9 kJ/kg
6.2 g/kg

Figure 12.6 Sensible heating and cooling (metric).

Normally $\Delta Q_T = \Delta Q_S + \Delta Q_L$; but $\Delta Q_L = 0$ as the moisture content of the air and the dewpoint temperature stay the same. Then $\Delta Q_T = \Delta Q_S$. Using Equation 11.8 (assume an airflow of 1000 L/s):

$$Q_S = 1.23 \times 1000 \times (41\ °C - 21\ °C) = 24{,}600\ W$$

Using Equation 11.11:

$$Q_T = 1.20 \times 1000 \times (58.9 - 37.1) = 26{,}160\ W$$

(Answers are 6% different; see explanation in Section 12.2.2.1).

12.2.2.3 Humidification. Air at 95 °F DB and 68.2 °F WB (41 °C DB and 20.1 °C WB) has moisture added by a humidifier until the final conditions are at 95 °F DB and 78 °F WB (41 °C DB and 28.2 °C WB). Using the proper psychrometric chart [Figure 12.2 (U.S.) or 12.3 (metric)], verify all of the data in Figures 12.7 (U.S.) and 12.8 (metric). Now $\Delta Q_S = 0$, so $\Delta Q_T = \Delta Q_L$. This may be verified by using Equations 11.10 and 11.11. (Note that the U.S. and metric conditions and temperatures used in this example are not conversions.)

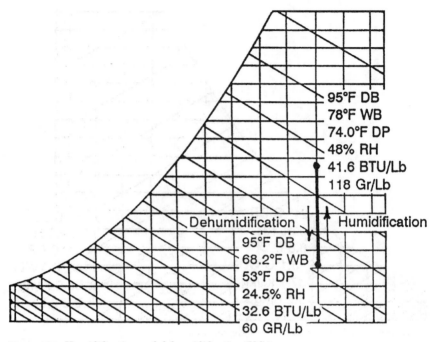

Figure 12.7 Humidification and dehumidification (U.S.)

12.2.2.4 Cooling and dehumidification. Figures 12.9 and 12.10 illustrate what happens when both sensible heat and latent heat changes are made. Equations 11.8, 11.10 and 11.11 may be used, because $\Delta Q_T = \Delta Q_S + \Delta Q_L$.

12.2.2.5 Chemical dehumidification. In most of the other processes, cooling accompanies dehumidification with the use of cooling coils. The chemical dehumidification "process line" on the chart is different in that dehumidification occurs, but the air gets warmer so that it is not a 90° vertical line. Many people have come in contact with this process without knowing it. When dryers are installed in refrigerant liquid lines, the refrigerant passes over a chemical, usually silica gel, to remove undesirable water vapor from the refrigerant. This moisture, which can be present in field-installed piping, generates enough heat during removal that a difference in the temperature between the dryer and the connecting piping can be noticed.

12.3 Using Psychrometric Charts

The charts in Figures 12.11 (U.S.) and 12.12 (metric) show two typical condition points, A and B. Once a point is located, everything can be

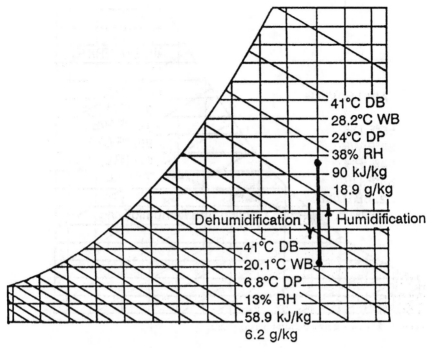

41°C DB
28.2°C WB
24°C DP
38% RH
90 kJ/kg
18.9 g/kg

Dehumidification   Humidification

41°C DB
20.1°C WB
6.8°C DP
13% RH
58.9 kJ/kg
6.2 g/kg

Figure 12.8 Humidification and dehumidification (metric).

determined about the conditions of the air at that point. The definitions of dry bulb lines, density lines, wet bulb lines, enthalpy, and relative humidity lines are given in Section 12.2.1.

If air is cooled without a change in moisture content, the relative humidity rises progressively until saturation is reached. This is the point, at which moisture begins to condense out of the air, called the *dewpoint* (DP). This also is the point where moisture condenses on the outside of a glass containing ice, because the air nearest the surface of the cold glass has become saturated and its dewpoint has been reached. Psychrometric charts in U.S. and Metric units are similar in layout and they work the same way. Only the values of the numbers and terms are different.

12.3.1 Heat flow example

12.3.1.1 U.S. units

a. If 4000 cfm is flowing through an HVAC duct system, the sensible heat change from point A to point B is 86,400 Btu/h (using Equation 11.8):

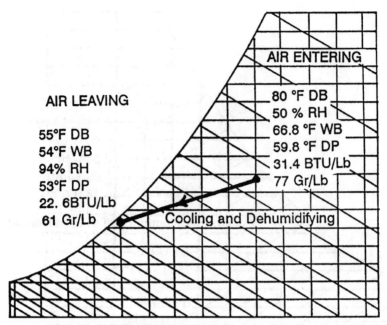

AIR ENTERING

AIR LEAVING

55°F DB
54°F WB
94% RH
53°F DP
22. 6BTU/Lb
61 Gr/Lb

80 °F DB
50 % RH
66.8 °F WB
59.8 °F DP
31.4 BTU/Lb
77 Gr/Lb

Cooling and Dehumidifying

Figure 12.9   Cooling and dehumidifying (U.S.).

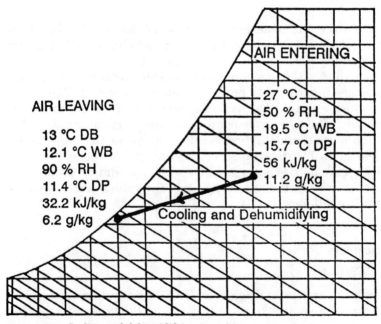

AIR ENTERING

AIR LEAVING

13 °C DB
12.1 °C WB
90 % RH
11.4 °C DP
32.2 kJ/kg
6.2 g/kg

27 °C
50 % RH
19.5 °C WB
15.7 °C DP
56 kJ/kg
11.2 g/kg

Cooling and Dehumidifying

Figure 12.10   Cooling and dehumidifying (metric).

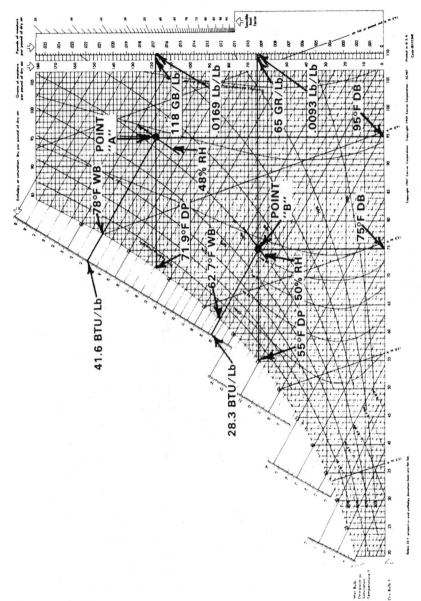

Figure 12.11 Psychrometric chart—Typical condition points (U.S.).

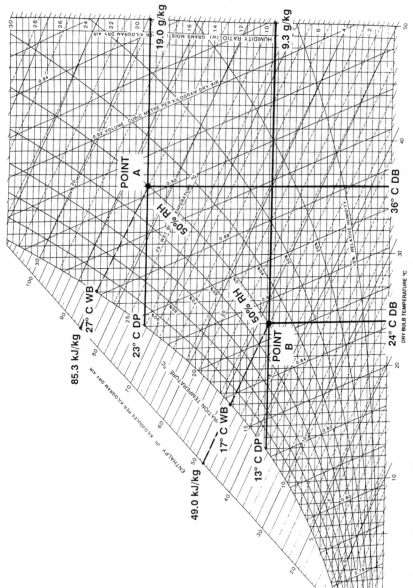

Figure 12.12 Psychrometric chart—Typical condition points (metric).

$$Q_S = 1.08 \times cfm \times \Delta t = 1.08 \times 4000 \times (95 \text{ °F} - 75 \text{ °F})$$

$$Q_S = 86,400 \text{ Btu/h}$$

b. The latent heat change is 147,136 Btu/h (using Equation 11.10):

$$Q_L = 4840 \times cfm \times \Delta W = 4840 \times 4000 \times (0.0169 - 0.0093)$$

$$Q_L = 147,136 \text{ Btu/h}$$

(Note that ΔW values are obtained from the psychrometric chart in Figure 12.11.)

c. The enthalpy or total heat change is 239,400 Btu/h (using Equation 11.11):

$$Q_T = 4.5 \times cfm \times \Delta h = 4.5 \times 4000 \times (41.6 - 28.3)$$

$$Q_T = 239,400 \text{ Btu/h}$$

(Note that Δh values are obtained from the psychrometric chart in Figure 12.11.)

d. To check the total heat change:

$$Q_T = Q_S + Q_L = 86,400 + 147,136$$

$$Q_T = 233,536 \text{ Btu/h}$$

The "check answer" is about 2½% lower because it is impossible to read the values closely on a small chart. However, all answers represent the accuracy normally found in TAB field work.

12.3.1.2 Metric units

a. If 2000 L/s is flowing through an HVAC duct system, the sensible heat change from point A to point B is 29,520 W (using Equation 11.8):

$$Q_S = 1.23 \times L/s \times \Delta t = 1.23 \times 2000 \times (36 \text{ °C} - 24 \text{ °C})$$

$$Q_S = 29,520 \text{ W (29.52 kW)}$$

b. The latent heat change is 58,200 W (using Equation 11.10):

$$Q_L = 3.0 \times L/s \times \Delta W = 3.0 \times 2000 \times (19.0 - 9.3)$$

$$Q_L = 58,200 \text{ W (58.2 kW)}$$

(Note that ΔW values are obtained from the psychrometric chart in Figure 12.12.)

c. The enthalpy or total heat change is 87,120 W (using Equation 11.11):

$$Q_T = 1.20 \times L/s \times \Delta h\ 0\ 1.20 \times 2000 \times (85.3 - 49.0)$$

$$Q_T = 87,120 \text{ W } (87.12 \text{ kW})$$

(Note that Δh values are obtained from the psychrometric chart in Figure 12.12.)

d. To check the total heat change:

$$Q_T = Q_S + Q_L = 29,520 + 58,200 = 87,720 \text{ W}$$

The "check answer" is 0.7% higher or very close. Because it is impossible to read the values closely on a small chart, all answers represent the accuracy normally found in TAB field work.

12.3.2   Mixing airstreams—dry bulb temperatures

When two airstreams are mixed (Figure 12.13) and are plotted in graph form on a psychrometric chart, the following steps should be used to avoid a common error.

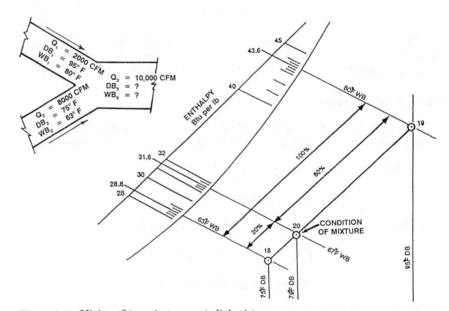

Figure 12.13 Mixing of two airstreams (adiabatic).

1. Assume that the airstreams are being mixed in a 50% ratio that causes the mixed point to be directly between points 18 and 19.

2. Then assume that a damper moves to a position to take in less hot outside air. This will cause the point to move away from 19 (because the distance from 19 to 20 gets longer when less air is taken in). By the time it reaches 18, 100% of the air of quality 18 will be used.

3. The mixed-air temperature will be closest to the air temperature of the largest airstream.

4. Only the dry bulb air temperature can be obtained accurately by using this method or by using Equation 9.1 (from Chapter 9):

$$T_m = \frac{X_o T_o + X_r T_r}{100}$$

12.3.3 Mixing airstreams—enthalpy

The wet bulb of the mixed airstream can be obtained by entering the values of the enthalpies of the two airstreams in Equation 12.1 and calculating the enthalpy of the mixed airstream. From this value, the mixed airstream wet bulb temperature can be obtained from the psychrometric chart.

Equation 12.1 (U.S. and Metric)

$$H_m = \frac{X_o H_o + X_r H_r}{100}$$

Where: H_m = mixed-air enthalpy (Btu/lb or kJ/kg)

X_o = percentage of outside air

H_o = outside air enthalpy (Btu/lb or kJ/kg)

X_r = percentage of return air

H_r = return air enthalpy (Btu/lb or kJ/kg)

Example 12.1 (U.S.) Calculate the dry bulb and wet bulb temperatures of the mixed airstream of Figure 12.13 using the equations in the text and by plotting in graph form.

Solution
a. Using Equation 9.1 (for dry bulb temperature):

$$T_m = \frac{X_o T_o + X_r T_r}{100}$$

$$T_m = \frac{20\% \times 95° + 80\% \times 75°}{100}$$

$$T_m = \frac{1900 + 6000}{100} = \frac{7900}{100} = 79 \text{ °F (DB)}$$

b. Using the enthalpy values from a psychrometric chart in Equation 12.1:

$$H_m = \frac{X_o H_o + X_r H_r}{100}$$

$$H_m = \frac{20\% \times 43.6 + 80\% \times 28.6}{100}$$

$$H_m = \frac{872 + 2288}{100} = 31.6 \text{ Btu/lb}$$

Using $H_m = 31.6$ Btu/lb and 79 °F DB, a wet bulb of approximately 67 °F may be determined using the chart. The solution also is shown in Figure 12.13 as plotted on a psychrometric chart.

12.3.4 Adiabatic saturation

Changes in airstream conditions caused by *adiabatic saturation* are shown in Figure 12.14 where the dry bulb can change drastically from 85 to 66 °F without a change in enthalpy. This change in airstream conditions occurs in cooling towers, when sprayed coils are used, and in evaporative cooling such as the fan–wet burlap combination or swamp cooler still found in Southwest desert areas. It does not occur when refrigerated cooling coils and dry-air evaporative coolers are used.

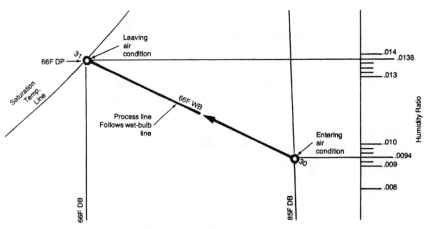

Figure 12.14 Example of adiabatic saturation.

12.3.5 Typical HVAC process example

Figure 12.15 shows a typical air conditioning application on a metric unit psychrometric chart. The path used to plot the cooling and dehumidification process does not matter. Consider what path air actually does take in cooling and dehumidification. As air leaves point C, it is sensibly cooled. This increases the relative humidity. As it approaches 70 to 90% relative humidity (depending on the conditions), the path begins to curve downward. Some dehumidification is now occurring. This happens even though the air is not at saturated temperatures, because the conditions in the air are not uniform. The air closest to the cold coils has moisture being condensed, whereas that further away has not reached condensing temperatures.

As air dry bulb temperatures become colder and colder and approach the "saturated line," more and more dehumidification takes place. The chart "air process line" is beginning to cross wet bulb lines. It can never cross them faster than dry bulb lines, but it is crossing them reasonably fast. The colder the air becomes, the closer it approaches the saturation line; however it never touches it. To touch it would mean that all the air across the coil has moisture condensing.

Notice how the "grams of moisture" on the chart drop rapidly. However, this process stops at the point where the air is discharged from the coil. That point has been labeled point A.

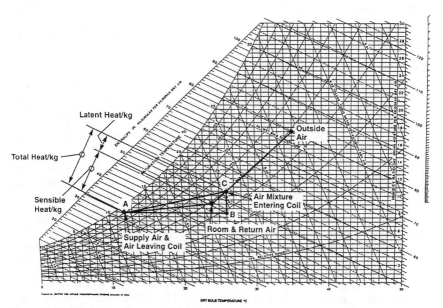

Figure 12.15 Typical cooling and dehumidifying process (metric).

The path that the process takes makes no difference to the total amount of heat that was transferred, only that it went from point C to point A. Point A is selected by the designer from experience as not being too cold for good air distribution. This temperature is approximately 12 °C when 24 °C is being maintained in the conditioned space. Notice that the line that leaves from point A and goes to that point labeled "room and return air" is not totally flat. This line is the line calculated by the heat loss and heat gain calculations, which have taken latent heat into account. The amount of latent heat from people, coffee pots, etc., keeps this line from being perfectly flat (or totally sensible heat).

Computer rooms have the flattest lines that may be found, whereas churches, auditoriums, and other places of assembly are frequently the highest in latent heat gain with the steepest lines. If point A of this room response line is carefully moved to the point at 26 °C and 50% relative humidity (being sure to keep the angle of the line the same), the sensible heat ratio can be read at the far right-hand scale. Notice that the line from the "room and return" point to the point labeled "outside air" is similar to the one in Figure 12.13. The room air is returned to the cooling coil mixed with the outside air in the proportions desired. It is emphasized that the processes of mixing and cooling of airstreams are most important. This typical example contains one set of each.

12.4 Air–Vapor Relationships

12.4.1 Atmospheric air

Dry air is an unequal mixture of gases consisting principally of nitrogen, oxygen, and small amounts of neon, helium, and argon. The percentage of each gas normally will be the same from sample to sample, although carbon dioxide, sulfur dioxide, and other pollutants may be present in varying quantities.

The air in our atmosphere, however, is not dry but contains small amounts of moisture in the form of water vapor. The amount of water vapor in atmospheric air normally represents under 1% of the weight of the moist air mixture. If the weight of air averages approximately 0.075 lb/ft^3, the moisture contained therein will weigh less than 0.00075 lb/ft^3. At first glance this would seem to be an insignificant amount to cause so much concern. Normally, atmospheric air contains only a portion of the water vapor it is able to absorb.

Air and water vapor behave as though the other were not present. Each acts as an independent gas and exerts the same pressure as if it were alone. The barometric pressure is the sum of the two pressures—the partial pressure of the air, plus the partial pressure of the water vapor.

The dry air component of the moist air exists as only a gas under all environmental conditions and cannot be liquefied by pressure alone; therefore, it acts as a perfect gas. However, water vapor does coexist with water as a liquid at all environmental temperatures, and it can be liquefied by pressure. Because it is not a perfect gas, the properties of water vapor are determined experimentally.

12.4.2 Condition changes

Although the following is not often used in HVAC calculations, a theoretically *perfect gas* is one in which the relationship of pressure, temperature, and volume may be defined and predicted by the following equation.

Equation 12.2

$$\frac{PV}{T} = R$$

Where: P = absolute pressure (psi + 14.7 or kPa + 101.3 or gauge pressure plus atmospheric pressure)
       V = volume of one pound (1 kg) in ft^3 (m^3)
       T = absolute temperature (460° + °F or 273° + °C)
       R = gas constant

An air and water vapor mixture behaves as a perfect gas provided that no condensation or evaporation takes place. As with the dual use of the symbol Q for airflow and heat flow, the symbol R is used to represent the gas constant when it is italic and also the absolute temperature in degrees Rankine when it is not italic.

The actual value of R, the gas constant, has little meaning in HVAC work except for the fact that it remains a constant for a perfect gas. For a given gas, Equation 12.3 is the most important equation to remember and use.

Equation 12.3

$$\frac{P_1 V_1}{T_1} = \frac{P_2 V_2}{T_2}$$

Where: P = absolute pressure (psia or kPa a)
       V = volume of air (ft^3 or m^3)
       T = absolute temperature (°R or K)

Example 12.2 (U.S.) A 100 ft^3 tank is indoors at 70 °F and at a pressure of 200 psi. If the pressure is released to a 50 ft^3 tank outside at 40 °F, calculate the new equalized gauge pressure found in both tanks (ignore piping volume).

Solution

$$\frac{P_1V_1}{T_1} = \frac{P_2V_2}{T_2}; P_2 = \frac{P_1V_1T_2}{T_1V_2}$$

$$P_2 = \frac{(200 + 14.7)(100)(460° + 40°)}{(460° + 70°)(50 + 100)} = \frac{214.7 \times 100 \times 500°}{530° \times 150}$$

$$P_2 = 135.0 \text{ psia} - 14.7 = 120.3 \text{ psi gauge pressure}$$

Example 12.2 (Metric) A 10 m³ tank is indoors at 20 °C and at a pressure of 1380 kPa. If the pressure is released to a 5 m³ tank outside at 4 °C, calculate the new equalized gauge pressure found in both tanks (ignore piping volume).

Solution

$$\frac{P_1V_1}{T_1} = \frac{P_2V_2}{T_2}; P_2 = \frac{P_1V_1T_2}{T_1V_2}$$

$$P_2 = \frac{(101.3 + 1380)(10)(273° + 20°)}{(273° + 4°)(10 + 5)} = \frac{1481.3 \times 10 \times 293°}{277° \times 15}$$

$$P_2 = 1044.6 \text{ kPa a} - 101.3 = 943.3 \text{ kPa gauge pressure}$$

13

Indoor Air Quality

13.1 Indoor Environment

Good indoor air quality (IAQ) is important for all types of buildings, from simple residences to complicated commercial and institutional buildings. The testing, adjusting, and balancing (TAB) technician plays an important part in maintaining quality air in the indoor environment. But many other persons with other skills may be involved.

The building indoor environment is a result of interactions between the building envelope, the building systems, the outdoor climate, and outdoor contaminant sources and potential interior contaminant sources from the building construction and furnishings, processes activities, and building occupants.

The five basic elements involved are:

1. Indoor or outdoor contaminant or pollution buildup

2. Poor HVAC system air distribution

3. Poor thermal and/or humidity comfort control

4. Building occupant perception and activities

5. Distribution of pollutants

13.1.1 Contaminant sources

Contaminants or pollutants may originate within the building or enter the building by infiltration or through the outdoor ventilation air intake. The TAB technician should recognize that these sources are numerous.

13.1.1.1 Outdoor air contaminants. A partial list of outdoor contaminants may include:

1. Pollen

2. Dust

3. Smog

4. Vehicle emissions

5. Odors from processes, dumps, or land fills

6. Re-entrained building exhaust air

13.1.1.2 HVAC systems. Contaminants from HVAC systems may include:

1. Dust or dirt from dirty filters or ductwork

2. Mold or microbiological growth from damp areas

3. Improper venting or combustion products

4. Induction of chemicals or odors into system

13.1.1.3 Building occupants and activities. A partial list of occupant developed pollutants include:

1. Smoking

2. CO_2 from respiration, body odors, perfumes

3. Cleaning materials and deodorizers

4. Dust, biocides, paint fumes

13.1.1.4 Building and furnishings. The IAQ problems of newly built buildings may decrease rapidly with proper system operations, but some problems can be ongoing:

1. Chemical outgassing from building materials, furnishings, and carpeting

2. Dust and mold

3. Emissions from trash or stored supplies

4. Housekeeping and maintenance activities

5. Unsanitary conditions and dry sewer traps

6. Food odors

7. Redecorating or repair activities

13.1.1.5 Source concentrations. Indoor air often contains a variety of contaminants at concentrations that are far below any standards or guidelines for occupational exposure. It is possible that the effects of these contaminants may be additive, or synergistic (interacting in a way that makes their combined effect stronger than their independent effects) under some circumstances. These factors often make it difficult to relate complaints of health effects to the concentration of a single specific pollutant.

13.1.2 HVAC systems

HVAC systems range in complexity from independently functioning units serving individual rooms to centrally controlled systems serving entire large buildings. Some buildings, such as warehouses and factories, use only natural ventilation or exhaust fans to remove odors and contaminants. In these buildings, indoor air quality problems may be associated with insufficient outdoor ventilation air, particularly during the heating season, when occupants tend to keep windows closed.

Public, institutional, and commercial buildings generally use mechanical ventilation systems to introduce outdoor air. Thermal comfort is maintained by mechanical equipment that distributes conditioned air, sometimes supplemented by piping systems that carry steam, hot water, or chilled water to perimeter areas of the building. Figure 13.1 shows the ASHRAE "thermal comfort zone" generally accepted for indoor environments.

13.1.2.1 Adequate IAQ. To provide adequate indoor air quality, properly functioning HVAC systems must:

1. Provide thermal comfort.

2. Distribute adequate ventilation air to all building occupants.

3. Exhaust or dilute contaminants to acceptable levels.

4. Control building pressurization.

13.1.2.2 Basic systems. Two of the most common HVAC designs used in modern public and commercial buildings are *constant volume* and *variable air volume* (VAV) systems. Constant volume systems provide a constant airflow and vary the air temperature to meet heating and cooling needs. The percentage of outside air may be held constant, but it is often controlled to vary with outside temperature, with a minimum setting that allows the system to meet ventilation guidelines.

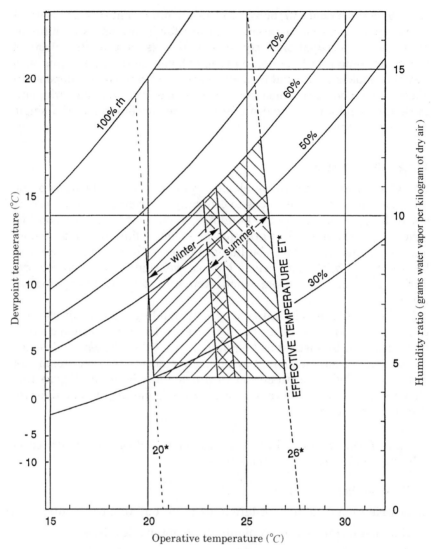

Figure 13.1 ASHRAE acceptable ranges for temperature (metric) and humidity for human occupancy. *(ASHRAE Standard 55.)*

Variable air volume (VAV) systems condition air to a constant temperature and vary the airflow to ensure thermal comfort. Most VAV systems do not allow control of the outside air quantity. Some more recent designs address this problem through the use of static pressure devices in the outside ventilation airstream.

13.1.2.3 Pressurization. Odors and contaminants may be controlled with the proper pressure relationships between rooms. This is accomplished by controlling the air quantities that are supplied to and removed form each room. If more air is supplied to a room than is exhausted, the excess air tends to leak out of the space and the room is said to be under *positive pressure*. If less air is supplied than is exhausted, adjacent room or outdoor air tends to leak into the space and the room is said to be under *negative pressure*. Control of pressure relationships is critically important in mixed-use buildings or buildings with special-use areas. Bathrooms, kitchens, and smoking lounges are examples of rooms that should be maintained under negative pressure.

13.1.3 Air movement

Air movement between zones and between the building's interior and exterior is intimately linked to the building structure and the functioning of the HVAC systems. Walls, ceilings, and floors divert or obstruct airflow, whereas openings provide pathways for air movement. Some openings are intentional (doors, windows, ducts); others are accidental (cracks, holes, utility chases). It is useful to think of the entire building, the rooms, and the connections such as chases, corridors, and stairways between them, as behaving like part of the duct system. Air must move from supply outlets through a room to return outlets, making the room channel the air. Any obstructions or openings in the room can affect the direction and amount of airflow.

13.1.3.1 Stack effect. Natural forces also exert an important influence. *Stack effect* is the pressure-driven flow produced by warm air rising and other factors, such as wind blowing across the top of a chimney. It draws outdoor air into openings at the lower levels of buildings and moves indoor air from lower to upper floors. Stack effect airflow can transport contaminants between floors by way of stairwells, elevator shafts, utility chases, or other openings. The stack effect is strengthened when indoor air is warmer than outdoor air.

13.1.3.2 Wind effects. *Wind effects* are transient, creating local areas of high pressure (on the windward side) and low pressure (on the leeward side) of buildings. Depending on the leakage openings in the building exterior, wind can affect the pressure relationships within and between rooms. **Both the stack effect and wind can overpower a building's mechanical system and disrupt air circulation and ventilation.**

13.1.3.3 Pressure differences. Indoor air contaminants are distributed within the buildings by pressure differences between rooms and between floors. Air moves from areas of higher pressure to areas of lower pressure through any available openings. Even if adjacent building zones are both under positive pressure relative to the outdoors, one of them is usually at a higher pressure than the other. Similarly, outdoor air contaminants are drawn into buildings by pressure differences between the outdoor air and the building interior. If the building or any zone within it is under negative pressure relative to the outdoors, any intentional or accidental opening can serve as an entry point. For example, a small crack or hole in a basement wall or floor can admit significant amounts of radon, if pressure differentials are large enough.

Although a building may be maintained under positive pressure, there is always some location, such as an outdoor air intake, that is under negative pressure relative to the outdoors. Entry of contaminants may be intermittent, occurring only when the wind blows from the direction of the pollutant source. If the pressure differential that brings the pollutant into the building is intermittent or acts upon different portion of the building, the IAQ problem may move from one location to another.

13.2 Occupant Response

The TAB technician must be aware that many other factors affect IAQ problems beside adequate ventilation and thermal comfort. Noise, excessive lighting, job stress, and personal problems are beyond the scope of TAB work but are some of the other factors that must be addressed in an IAQ audit.

13.2.1 Building occupants

Building occupants generally are those who spend a full workday in the facility, as opposed to *visitors,* who spend only minutes or a few hours in the facility.

Sometimes only one occupant is sensitive to a particular indoor air contaminant, whereas surrounding occupants have no ill effects. Symptoms that are limited to a single person also can occur when only one work station receives more of the pollutant. In other cases, complaints may be widespread throughout the building.

People often have different responses to the same pollutant. Further, different pollutants may cause similar physical reactions. Respiratory tract irritation can result from exposure to formaldehyde or other volatile organic compounds, dust, excessively dry air, or other influences.

Allergic reactions are caused by a wide array of materials. For the purposes of solving IAQ problems it is generally more useful to observe the pattern of symptom occurrence than to focus solely on the symptoms of one individual. However, it is worthwhile to collect symptom information in hopes that it will help solve the problem.

13.2.2 Acceptable IAQ

ASHRAE Standard 62-1989, *Ventilation for Acceptable Indoor Air Quality,* states that acceptable indoor air quality is "air in which there are no known contaminants at harmful concentrations as determined by cognizant authorities and with which a substantial majority (80 percent or more) of the people exposed do not express dissatisfaction."

13.2.3 Sick building syndrome

Sick building syndrome is a term that describes reactions that occupants may experience under certain circumstances. Although there is no universal agreement on exactly what "sick building syndrome" is, or even that such a condition exists, the following definition is used by ASHRAE. "Sick building syndrome exists when more than 20 percent of the occupants complain during a two week period of a set of symptoms, including headaches, fatigue, nausea, eye irritation, and throat irritation, that are alleviated by leaving the building and are not known to be caused by any specific contaminants." Sick building syndrome also sometimes is used to describe cases in which a significant number of building occupants experience acute health and comfort effects that are apparently linked to the time they spend in the building, but in which no specific illness can be identified. The complaints may be localized in a particular room or zone or may be widespread throughout the building. Analysis of air samples often fails to detect significant concentrations of any contaminants, so that the problem appears to be caused by the combined effects of many pollutants at low concentrations, with other environmental stresses, such as overcrowding and noise, as complicating factors.

13.2.4 Discomfort

Some complaints by building occupants are clearly related to discomfort rather than to health problems. The distinction is not always simple. One of the most common IAQ complaints is that "there's a funny smell in here." Odors often are associated with a perception of poor air quality, whether or not they cause symptoms. Environmental stresses such as over- or underheating, humidity extremes, drafts, lack of air circulation, noise, vibration, and overcrowding can produce symptoms that may be confused with the effects of poor air quality. For example, excessive heat can produce fatigue, stuffiness, and headache, whereas low temperatures can cause chills and flulike symptoms. Further, physical discomfort or psychosocial problems (such as job stress) can reduce tolerance for marginal air quality.

13.3   Solving IAQ Problems

Over recent years, over half of documented IAQ problems have been solved by increasing or redistributing outdoor ventilation air within the building. Generally, most IAQ problems are solved by:

1. Eliminating the contaminant source

2. Dilution with outdoor ventilation air

3. Filtration with high-efficiency particulate air (HEPA) filters and/ or activated carbon filters for odors

13.3.1   Source control

Some sources of pollution, such as smoking particulates, have been controlled by instituting "no smoking" areas or by banning smoking altogether. The need for such restrictions may be mitigated by improving the operation of the HVAC system, or by making modifications to the air distribution systems. Individual exhaust systems also can control smoke as well as fumes and gases from specific pieces of equipment such as copy machines. Sometimes new products that do not have offensive characteristics can be substituted for the offending products.

Radon gas concentrations can be controlled by sealing all cracks and openings in basement or lower level area floors, in walls, and wherever pipes, sumps, or other objects penetrate. Concrete block foundations must be thoroughly sealed. Vapor barriers can be used in new construction along with careful sealing procedures. Some crawl spaces and basements also may need the addition of outdoor air ventilation; and sumps, foundation drainage areas, and areas under floors may require exhaust systems.

HVAC ductwork and equipment may be a source of pollution. If air conditioning systems are not kept clean, they can become a source of biological contamination. Air conditioning units with poorly constructed drip pans often are sources of contamination, because stagnant water in the pans becomes an excellent breeding ground for bacteria. Ductwork, wet from condensation or moisture carryover, also may contribute to the pollution.

13.3.2 Dilution

Outdoor air ventilation is the main form of dilution control. ASHRAE standards have set the ventilation rates for various areas in different types of buildings. However, zoning of building ventilation creates problems unless multiple systems are used.

In homes or smaller buildings, insufficient outside air or inadequate combustion air may create a carbon monoxide problem from back drafts from fuel-burning furnaces and appliances. The best way to improve safety and air quality is to supply enough outside air so that air, high in oxygen, circulates throughout the building, and all gases or products of combustion are exhausted through chimneys or vents.

Many variable air volume (VAV) systems reduce airflow as occupancy increases during the heating season, because the heat generated by occupants reduces the thermal load. Reduced airflow is thus counter to the need for additional outdoor ventilation air under these conditions, and some type of compensation control is needed to increase the percentage of outdoor ventilation air.

13.3.3 Contaminant removal

Whereas dilution controls both particulates and gasses or vapors, energy for heating or cooling the outdoor ventilation air may be expensive. Air cleaning devices are highly effective in eliminating particles and require little additional energy (fan power) for HVAC equipment fan motors. A combination of dilution and removal usually is required for cost effective control of both gases and particles.

Air filters or air cleaning devices are a part of packaged HVAC units. However, built-up systems often require a thorough analysis of air filtration components and how the HVAC system is used. Because of energy conservation considerations, odor removal or control equipment may be used where excessive dilution of the air by ventilation would be too costly.

Atmospheric dust is a complex mixture of smokes, mists, fumes, dry granular particles, and fibers. When suspended in a gas, this mixture is generically called an *aerosol*. A sample of atmospheric dust gathered at any given point generally will contain materials common to the local environment, together with other components that originate at a distance but are transported by air currents or diffusion. These components vary with the geography of the locality, the season of the year, the direction and strength of the wind, and proximity of dust sources.

Different applications require different degrees of air cleaning effectiveness. Unfortunately, the smaller components of atmospheric dust are the worst offenders of smudging and discoloring building interiors. Electronic air cleaners or HEPA filters are required for small particle removal.

Cleaning efficiency is affected to some extent by the velocity of the airstream. The degree of air cleanliness required and aerosol concentration are both major factors influencing filter design and selection. Removal of particles becomes progressively more difficult and expensive as particle size decreases.

13.4   IAQ Audit

The investigation of a sick building usually is a team effort. In addition to the use of normal TAB instruments in establishing ventilation rates, temperature and humidity, and airflow distribution, the services of an industrial hygienist may be required to test for irritating or toxic gases and particulates.

13.4.1   Carbon dioxide levels

Carbon dioxide (CO_2) measuring instruments are accurate and easy to use. ASHRAE currently has set a recommended limit of 1000 parts per million (ppm) to satisfy comfort (odor) criteria. CO_2 is not a pollutant but an indicator of inadequate ventilating air. Outdoor air levels of CO_2 normally are near 350 ppm but may vary considerably. When CO_2 levels increase in a building, other contaminants also may increase.

13.4.2   Other contaminants

The following are some of the major contaminants that may require laboratory testing of samples of building air to determine if present in problem quantities.

13.4.2.1   Moisture. Moisture can be considered an IAQ contaminant in the sense that the presence of moisture promotes the growth of microorganisms such as mold and mildew. In addition to obvious reservoirs for microbiological growth such as stagnant water in drip pans, microorganism growth occurs where water vapor condenses onto surfaces. Problem sites include the relatively cool surfaces (and sometimes the interiors) of underinsulated exterior walls and areas around thermal breaks such as beams above windows, balconies. Carpeting is also a common site of biological growth.

13.4.2.2   Carbon monoxide. Carbon monoxide pollution occurs where combustion gases are not properly exhausted or are re-entrained into the building. Carbon monoxide should be measured if there are complaints of exhaust odors or if there is some other reason to suspect a problem with combustion gases.

13.4.2.3 Formaldehyde. Formaldehyde is a volatile organic compound (VOC) that is frequently emitted by new furnishings as well as by a wide variety of cleaning compounds, deodorants, adhesives, and other materials used in maintenance and housekeeping. IAQ complaints associated with formaldehyde or other VOCs are most commonly found under conditions in which there is new construction, new furnishings, or inadequate ventilation.

13.4.2.4 Tobacco smoke and other respirable particulates. Tobacco smoke, dust-producing processes, or inefficient or dirty filters can result in high levels of respirable particulates. The TAB technician should enquire about the smoking policy for the building and identify other internal sources of excessive dust. Examination of filters and dust accumulation in ductwork will reveal whether inefficient filtration seems to be a problem.

13.4.2.5 Nitrogen dioxide. Nitrogen dioxide, like carbon monoxide, is an air pollutant associated with combustion processes. It should be measured if outdoor sources of combustion gases are suspended (e.g., if the exhaust plume from a nearby power plant may be polluting air outside the building).

13.4.2.6 Ozone. Ozone problems may occur from outdoor ambient air or if there are unvented indoor sources such as some types of photocopiers or electrostatic precipitators. Ozone should be measured if someone has noticed its characteristic odor or if its presence is suspected because of identified equipment.

13.4.2.7 Volatile organic compounds (VOCs). Formaldehyde is the best known VOC causing indoor air quality problems, but many other related compounds are also emitted by building materials and furnishings. If specific sources are identified, test samples for VOCs such as styrene, paradichlorobenzene, or perchlorethylene may be required.

13.4.3 Audit procedures

IAQ audits generally are made at three levels depending on the complexity of the problems. Each is likely to improve the IAQ of the building, but a few IAQ problems have ended up without solution.

13.4.3.1 Level one—IAQ audit. A majority of problems are solved at this level by TAB technicians. First check to determine whether all HVAC systems are operating properly. After checking ventilation rates and system air distribution of the ventilation air (normal TAB work), make a checklist-type report of any system deficiencies or suspected sources of contaminants.

13.4.3.2   Level two—IAQ audit.   If the review of the HVAC systems and building indicates IAQ deficiencies present, possible pollution sources, lack of ventilation air, or other poor IAQ factors, the first step is to balance all HVAC systems for proper ventilation air distribution, after setting the required percentage of outdoor ventilation air.

Building pressurization levels should be checked and adjusted if possible. Suspected contaminant sources should be investigated and corrected or eliminated. Carbon dioxide (CO_2) measurements should be made to verify the ventilation air requirements.

13.4.3.3   Level three—IAQ audit.   Severe IAQ problems may require the help of an industrial hygienist to take and evaluate air samples, a consulting engineer to make changes to HVAC systems, relocation of employee work areas, and other more costly corrective actions beyond the scope of TAB work.

14

Energy Recovery

14.1 Recovery Fundamentals

14.1.1 Introduction

With the recent increase in outdoor air ventilation rates because of indoor air quality (IAQ) problems (see Chapter 13), the cost of energy to heat or cool this additional ventilation air will become more of a concern, particularly where there are extremes in climate. In Canada and parts of the northern United States, the winter use of heat recovery equipment in residences is growing rapidly.

The fundamentals of psychrometrics and heat transfer may be found in Chapter 11 and Chapter 12. The understanding of these fundamental concepts is necessary for calculating heat flow and equipment efficiencies of energy recovery devices.

14.1.2 Sensible heat devices

Energy recovery devices, which basically do not transfer moisture between the two airstreams, are called *sensible heat devices*. Most fixed-plate exchangers, heat pipe exchangers, run-around coil exchange systems, and some rotary wheel devices fall into this category. No latent heat (humidity) is exchanged between the airstreams except where the exhaust air is cooled below the exhaust air dewpoint in a rotary wheel drive. When this condensation occurs, some latent heat has been transferred.

14.1.3 Total heat devices

Total heat devices, such as some rotary wheel exchangers and twin tower enthalpy recovery loop systems, transfer both sensible and latent heat (called total heat or enthalpy) between the airstreams. Under typical summer conditions, a total heat device will transfer nearly three times as much energy as a sensible heat device. Under typical winter conditions, a total heat device recovers 25% more energy. Figure 14.1 is a graphic representation of total heat and sensible heat devices.

14.1.4 Effectiveness

ASHRAE Standard 84-91, *Method of Testing Air-to-Air Heat Exchangers*, defines *effectiveness* (*e*) as the ratio of actual transfer of energy and/or moisture divided by the maximum possible transfer of energy between airstreams.

The performance of air-to-air heat exchangers is usually expressed in terms of their effectiveness in transferring: (1) sensible heat (dry bulb temperature), (2) latent heat (humidity ratio), or (3) total heat (enthalpy). However, in the HVAC industry effectiveness (*e*) and efficiency generally are used interchangeably when evaluating the performance of air-to-air heat exchangers.

Equation 14.1

$$e = \frac{W_s(X_1 - X_2)}{W_{min}(X_1 - X_3)} = \frac{W_e(X_4 - X_3)}{W_{min}(X_1 - X_3)}$$

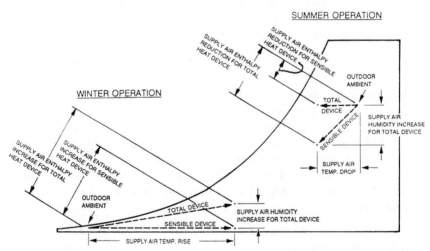

Figure 14.1 Comfort sensible heat versus total heat recovery devices *(Courtesy of ASHRAE.)*

Where (see Figure 14.2):

> e = sensible heat, latent heat, or total heat effectiveness
> X = drybulb temperature, humidity, ratio, or enthalpy at the locations indicated in Figure 14.2 (all differences are positive values)
> W_s = mass flow rate of supply (lb or kilograms of dry air per hour)
> W_e = mass flow rate of exhaust (lb or kilograms of dry air per hour)
> W_{min} = smaller of W_s and W_e

The leaving supply air condition is then:

Equation 14.2

$$X_2 = X_1 - e\frac{W_{min}}{W_s}(X_1 - X_3)$$

and the leaving exhaust air condition is:

Equation 14.3

$$X_4 = X_3 + e\frac{W_{min}}{W_e}(X_1 - X_3)$$

The above assumes that there is no heat transfer between the heat exchanger and its surroundings, and that there are no gains from cross leakage, fans, or frost control devices. This is generally true for larger commercial applications but not for heat recovery ventilators (HRVs). The equations used in Canada for HRVs determine energy recovery efficiency (i.e., the actual energy transfer efficiency) and the

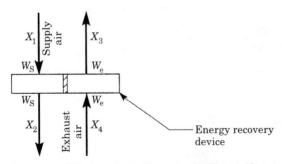

Figure 14.2 Effectiveness rating equation figure. *(Courtesy of SMACNA.)*

apparent sensible effectiveness (i.e., a measure of the temperature rise of the supply airstream, including that resulting from external gains).

A number of variables can affect heat exchanger effectiveness. These include whether the device is sensible or total, the humidity ratio of the warmer airstream, the heat transfer area, air velocities through the heat exchanger, the airflow arrangement, the supply and exhaust air mass flow rates, the supply to exhaust air mass flow ratio, and the method of frost control employed. Because of this, performance data must be established from each individual type of device. Manufacturers of each type may be contacted for detailed performance data.

The effectiveness of sensible heat recovery devices can be rated two ways:

1. The percentage ratio of the actual recovered portion of sensible heat to the total potentially transferable heat

2. The percentage ratio of the actual recovered heat to the maximum transferable sensible heat

The energy recovery device effectiveness numerical value, when computed with sensible heat transferred/total heat potential, will be smaller than when both values are sensible heat (sensible heat transferred/sensible heat potential).

Example 14.1 (U.S.)   Four thousand cfm of 40 °F outdoor (supply) air is being tempered to 56 °F by 3600 cfm of 80 °F exhaust air. Find the effectiveness of the heat recovery device and the leaving temperature of the exhaust air.

Solution   Using Equation 14.1 and Figure 14.2 (differences must be positive values):

$$e = \frac{W_s(X_1 - X_2)}{W_{min}(X_1 - X_3)}$$

$$e = \frac{4000 \text{ cfm} \times 0.075 \text{ lb/ft}^3 \times 60 \text{ min/h} \ (56° - 40°)}{3600 \text{ cfm} \times 0.075 \text{ lb/ft}^3 \times 60 \text{ min/h} \ (80° - 40°)}$$

$$e = \frac{18{,}000 \text{ lb/h} \times 16 \text{ °F}\Delta t}{16{,}200 \text{ lb/h} \times 40 \text{ °F}\Delta t} = \frac{288{,}000}{648{,}000}$$

$$e = 0.444 \text{ or } 44.4\%$$

Using Equation 14.3 and Figure 14.2:

$$X_4 = X_3 + e \frac{W_{min}}{W_e} (X_1 - X_3)$$

$$X_4 = 80 \text{ °F} - 0.444 \frac{3600}{3600}(80 \text{ °F} - 40 \text{ °F})$$

$$X_4 = 80 \text{ °F} - 17.8 \text{ °F} = 62.2 \text{ °F}$$

Example 14.1 (Metric)   Two thousand liters per second of 5 °C outdoor (supply) air is being tempered to 13 °C by 1800 L/s of 27 °C exhaust air. Find the effectiveness of the heat recovery device and the leaving temperature of the exhaust air. (1000 L/s = 1 m³/s)

Solution   Using Equation 14.1 and Figure 14.2 (differences must be positive values):

$$e = \frac{W_s(X_1 - X_2)}{W_{min}(X_1 - X_3)}$$

$$e = \frac{2.00 \text{ m}^3/\text{s} \times 1.204 \text{ kg/m}^3 \times 3600 \text{ s/h } (13° - 5°)}{1.80 \text{ m}^3/\text{s} \times 1.204 \text{ kg/m}^3 \times 3600 \text{ s/h } (27° - 5°)}$$

$$e = \frac{8669 \times 8 \text{ °C}\Delta t}{7802 \times 22 \text{ °C}\Delta t} = 0.404 \text{ or } 40.4\%$$

Using Equation 14.3 and Figure 14.2:

$$X_4 = X_3 + e\frac{W_{min}}{W_e}(X_1 - X_3) = 27 \text{ °C} - 0.404\frac{1800}{1800}(27° - 5°)$$

$$X_4 = 27 \text{ °C} - 8.9 \text{ °C} = 18.1 \text{ °C}$$

14.1.5   Use of effectiveness percentages

Care should be taken when working with effectiveness percentages. Mathematically, sensible heat effectiveness is not the same thing as total heat effectiveness. High sensible heat effectiveness for a device does not mean it has high total heat effectiveness and vice versa.

None of the energy recovery devices are justified on effectiveness alone. Rather, they are justified on their ability to recover usable energy and be adaptable to the many other conditions required by the system installation. Generally:

1. Sensible heat devices, including sensible heat wheels, usually are justified when most of the energy savings for an application are in the sensible heating of the supply airstream.

2. Total heat devices usually are justified when significant energy savings are obtained when it is possible to heat and humidify and/or cool and dehumidify supply airstreams.

14.2   Energy Recovery Equipment

14.2.1   Airflow configurations

Airflow arrangement refers to the relative flow directions of the supply and exhaust airstreams. Counterflow is the arrangement with the highest theoretical performance, but other arrangements are common

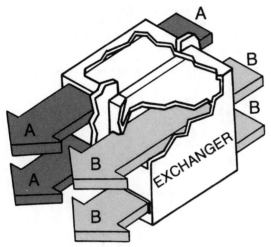

Figure 14.3 Parallel flow airstreams. *(Courtesy of SMACNA.)*

because of practical design or construction considerations. Figures 14.3, 14.4, and 14.5 illustrate common airflow arrangements.

14.2.1.1 Parallel flow   In parallel flow devices (Figure 14.3), both airstreams, upon entry, confront each other at maximum difference of temperature and humidity. Therefore, maximum heat transfer will take place at this point. In winter, the exhaust airstream progressively loses heat and the supply airstream absorbs it. The differences in tem-

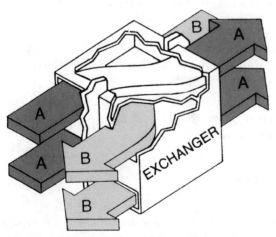

Figure 14.4 Counterflow airstreams. *(Courtesy of SMACNA.)*

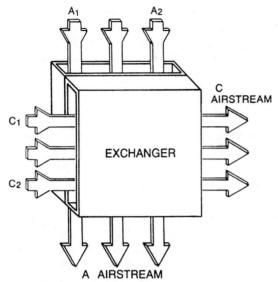

Figure 14.5 Crossflow airstreams. *(Courtesy of SMACNA.)*

perature and humidity will grow smaller, and in time, the two airstreams will become equal in temperature and humidity ratio. The maximum energy actually lost by the exhaust airstream and gained by the supply airstream will be only half of the potential transferable energy. The theoretical maximum effectiveness of the parallel flow device is 50%.

14.2.1.2 Counterflow In a counterflow pattern (Figure 14.4), the high–heat content moist air (hotter of two and having a higher humidity ratio) and the low–heat content air enter the device from opposite sides, with the exhaust air giving up some energy and the supply air gaining this energy. The exhaust airstream becomes colder and drier while the supply airstream becomes warmer and more humid in a total heat situation. Given opportunity and time, all potentially transferrable heat would be transferred. The theoretical maximum effectiveness of the counterflow device is 100%, although the practical effectiveness is much lower.

14.2.1.3 Crossflow In the crossflow configuration (Figure 14.5), the A_1 portion of airstream A will cross airflow C at its maximum value and will have excellent heat transfer. The same can be said about the C_1 portion of airstream C in relation to the airflow A. But airstream A_2 will just begin to transfer energy crossing airstream C after it is completely depleted and the same can be said about airstream C_2. Still, given opportunity and time, a 100% theoretical effectiveness is possible. Because of economic considerations, this maximum is never obtained and actual effectiveness falls short of that for counterflow devices.

14.2.2 Types of exchangers

Table 14.1 gives the essential data on HVAC system energy recovery equipment and systems. A brief description of each system follows.

14.2.2.1 Fixed-plate exchangers

Fixed surface plate exchangers have no moving parts. Alternate layers of plates, separated and sealed (referred to as the heat exchanger core), form the exhaust and supply airstream passages. Plate spacings range from 0.1 to 0.5 in. (2.5 to 13 mm), depending on the design and the application. Heat is transferred directly from the warm airstreams through the separating plates into the cool airstreams. Practical design and construction restrictions inevitably result in crossflow heat transfer, but additional effective heat transfer surface arranged properly into counterflow patterns can increase heat transfer effectiveness.

Normally, both latent heat of condensation (from moisture condensed as the temperature of the warm, exhaust airstream drops below its dewpoint) and sensible heat are conducted through the separating plates into the cool, supply airstream. Energy is transferred but moisture is not. A sensible effectiveness as high as 80% may be attained under ideal conditions.

14.2.2.2 Rotary wheel exchangers

Rotary wheel exchange devices (Figure 14.6) are porous discs, fabricated from some material having a high heat retention capacity, that rotate through two side-by-side ducts. The axis of the disc is located parallel to and on the partition between the two ducts. As the disc is slowly rotated, sensible heat (and often moisture containing latent heat) is transferred to the disc by the hot air from the first airstream and then from the disc to the second, cooler airstream. The pores of heat wheels carry a small amount of the exhaust stream into the supply side of the exchanger. Should this result in an undesirable contamination, a purge section may be added.

Rotary wheel exchangers are commercially available with a number of different types of discs. Some discs have a metal frame packed with a core of knitted stainless steel or aluminum wire mesh, not unlike common kitchen "steel wool." Other wheels are made of corrugated metal assembled to form many parallel passages (laminar wheels);

TABLE 14.1 Comparison of Energy Recovery Systems

	Fixed plate	Rotary wheel	Heat pipe	Coil loop	Multiple towers
Airflow arrangements	Counter flow Cross flow Parallel flow	Counter flow Parallel flow	Counter flow Parallel flow	Counter flow Parallel flow	Independent
Equipment size range (airflow)	50 cfm and up (25 L/s and up)	50–70,000 cfm (25–35,000 L/s)	100 cfm and up (50 L/s and up)	100 cfm and up (50 L/s and up)	Up to 100,000 cfm (up to 50,000 L/s)
Type of heat transfer (typical effectiveness)	Sensible (50–80%)	Sensible (50–80%) Latent (45–55%)	Sensible (45–65%)	Sensible (55–85%)	Sensible (40–60%) Latent (45–55%)
Face velocity [fpm (m/s)]	100–1000 (0.5–5)	500–1000 (2.5–5)	400–800 (2.0–4.0)	300–600 (1.5–3.0)	300–450 (1.5–2.3)
Pressure drop range [in.w.g. (Pa)]	0.02–1.8 (5–450)	0.4–0.7 (100–175)	0.4–2.0 (100–500)	0.4–2.0 (100–500)	0.7–1.2 (175–200)
Temperature range		−70 ° to 1500 °F (−60 ° to 800 °C)	−40 ° to 66 °F (−40 ° to 35 °C)		−40 °F to 115 °F (−40 °C to 46 °C)
Unique advantages	No moving parts No cross leakage All sizes Most materials Low pressure drop High effectiveness Easily cleanable	Latent transfer Compact large sizes Low pressure drop High effectiveness	No moving parts No cross leakage Most sizes Fan location not critical Allowable pressure differential to 60 in.w.g. (15 kPa)	Exhaust airstreams can be separated from supply air	Latent transfer from remote airstreams. Multiple units in a single system. Efficient microbiological cleaning of both supply and exhaust airstreams.
Limitations	Latent available only in special units	Cold climates may increase service Cross air contamination possible	Effectiveness limited by pressure drop and cost Few suppliers	Effectiveness may be limited by pressure drop Few suppliers	Few suppliers
Cross contamination	0–5%	1–10%	0%	0%	0.025%

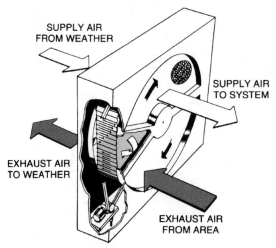

Figure 14.6 Rotary wheel exchanger. *(Courtesy of SMACNA.)*

ceramic laminar wheels for higher temperatures; and metal or composition wheels coated, impregnated, or treated with a hygroscopic material so that latent heat may be transferred.

An effectiveness of 80% may be obtained for sensible heat wheels and as high as 55% for latent heat wheels.

14.2.2.3 Heat pipe exchangers The *heat pipe exchanger* (Figure 14.7), although appearing as one large coil, is face divided to provide separate exhaust air and supply air sides connected by sealed tubes containing an appropriate heat transfer fluid. The effectiveness of this fluid depends on the anticipated temperatures involved, because evaporation of the fluid takes place on the hot side and condensation takes place on the cold side.

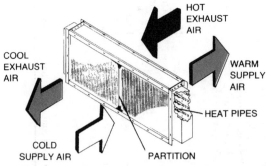

Figure 14.7 Heat pipe exchanger. *(Courtesy of SMACNA.)*

The heat pipe exchanger consists of about 100 short lengths of copper tubing sealed at both ends projecting into both airstreams. Within each sealed tube is a refrigerant charge and a snugly fitting porous cylindrical wick. There also is a vertical, gravity return type without the capillary wick. The liquid refrigerant migrates by capillary action to the warmer end, where it evaporates and absorbs heat. Under high pressure, the gaseous refrigerant flows back through the hollow center of the wick to the cooler end where it condenses, giving up its heat and starts a repeat cycle. Capacity can be easily controlled by merely tilting one end.

The gravity return type requires the cooler air, which is to condense the vapor into liquid, always to be at the top of the coil. This limits this unit to a single season of operation, unless damper arrangements are employed to alternate the cooler air to the top during each season.

The sensible heat effectiveness ranges from 45 to 65% and the coil pressure loss for the airstreams varies from 0.35 to 2.0 in.w.g. (87 to 500 Pa).

14.2.2.4 Run-around coil Standard HVAC hot or chilled water coils can be used for run-around systems, as illustrated in Figure 14.8. A coil in one or more duct systems transfers heat via some suitable liquid to other coils in another duct system(s). Ethylene glycol is commonly used as the transfer fluid, but there are many other commercial thermal fluids on the market.

A major advantage of the run-around coil system is that it can be used with widely separated airstreams, with the water solution being pumped from one to the other. In summer, the effectiveness can be improved by using air washers on the exhaust air coil. The system also is seasonally reversible.

Sensible heat recovery effectiveness is in the 55 to 85% range using chilled water–type coils. An attempt to further increase the effective-

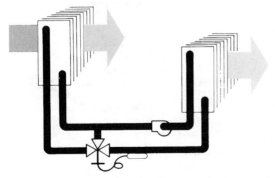

Figure 14.8 Run-around coil exchangers. *(Courtesy of SMACNA.)*

ness by adding more rows of tubing to the coils is offset by the increased air duct system pressure drop and higher fan power.

14.2.2.5 Multiple tower exchangers The multiple tower energy recovery system also permits wide separation of the supply air and exhaust air duct systems. The recovery system consists of a separate multiple contactor tower, as shown in Figure 14.9, for each of the exhaust air and supply air duct systems involved, with a sorbent liquid continuously being circulated between them. This liquid, usually a halogen salt solution, acts as the vehicle for transporting both sensible and latent heat.

The operation of the multiple tower exchange system is similar to that of the run-around coil system, except that a contactor tower containing an extended packed surface (similar to a cooling tower) is substituted for the air coils. Two pumps are required to circulate the liquid absorbent (halogen solution) allowing the system to be seasonally reversible.

The halogen solution is sprayed counterflow to the airstream through the extended packed surface. Total heat (enthalpy) transfer takes place as the solution absorbs (or rejects) heat and water from the exhaust airstream. During winter operation, care must be taken to keep the absorbent solution in a liquid form when handling low-humidity (dry) exhaust air. If the solution water concentration is reduced to 50% or lower, it can solidify and clog the spray nozzles, pump, and piping.

In addition to providing a total heat recovery effectiveness in the range of 45 to 55%, the solution spray may also act as an air washer

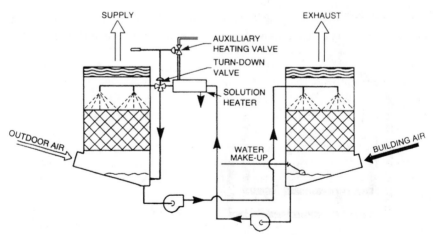

Figure 14.9 Multiple tower operation. *(Courtesy of SMACNA.)*

or scrubber. There is no contamination "carryover" between the two remote airstreams.

14.3 Recovery Device Problems

14.3.1 Pollutants

Air normally is exhausted from a space because of contaminants, carbon dioxide, and moisture. There are many contaminants that may exist within an exhaust gas, the most common of which are dust, grease, sulfur dioxide, and chlorine gas.

To remove energy from exhaust air or gases and transfer this energy to the incoming outdoor airstream, while simultaneously keeping the pollutants in the exhaust system, a large amount of surface area [approximately 0.5 ft^2 of material per cfm or (0.1 square meter per liter per second) of exchanger airflow] must be used. To obtain that amount of surface in a reasonable volume, the gases must exit and return through a series of relatively small passageways. This results in gas pressure losses and the possibility of contaminant entrapment within the confines of the energy recovery device. Many different heat exchanger designs and systems have emerged in an attempt to satisfy all of the many different requirements.

Any energy recovery device installed in an exhaust duct may be susceptible to fouling to some degree, depending upon the nature of the contaminants in the exhaust stream, and upon the dewpoint of the flowing gas. If allowed to continue, fouling will decrease the heat transfer performance. Therefore, filters are recommended unless other economical methods of cleaning the recovery device are used.

14.3.2 Condensation and freezing

Condensation and *ice* or *frost* formation results in moisture deposits on heat exchange surfaces. If entrance and exit effects are neglected, four distinct air–moisture regimes may occur as the warm airstream is cooled down from its inlet conditions to its outlet conditions. First there is a dry region with no condensation. Once the warm airstream is cooled below its dewpoint, there is a condensing region that results in wetting of the heat exchange surfaces. If heat exchange surfaces fall below freezing, the condensation will freeze. Finally, if the warm airstream temperature is reduced below 32 °F (0 °C) the condensation will form as frost. The location of these regions and rates of condensation and frosting depend on the duration or frosting conditions, airflow rates, the inlet air temperatures and humidities, heat exchanger core temperatures, the heat exchanger effectiveness, geometry, configuration and orientation, and heat transfer coefficients.

Sensible heat exchangers, which are ideally suited to applications in which heat transfer is desired but humidity transfer is not, such as swimming pools and kitchens, can benefit from the latent heat of condensation. One pound (kilogram) of moisture condensed transfers about 1050 Btu (2450 kJ) to the incoming air.

Condensation increases heat transfer rates and thus sensible effectiveness. It also can increase pressure drops significantly in heat exchangers with narrow airflow passage spacings. Frosting fouls the heat exchanger surfaces that was initially improved by the condensation. It then restricts exhaust airflows and reduces energy transfer rates. In extreme cases, the exhaust airflow (and supply airflow in the case of heat wheels) can be completely blocked.

Frosting and icing occur at lower temperatures in enthalpy exchangers than in sensible heat exchangers. In rotary enthalpy exchangers utilizing chemical absorbents, condensation may cause the absorbents to deliquesce, with permanent damage to the exchanger.

An increase in the pressure drop across a heat exchanger may indicate the onset of frosting or icing. Frosting and icing can be prevented by preheating the supply air or reducing the heat exchanger effectiveness such as reducing heat wheel speed, tilting heat pipes, or bypassing part of the supply airflow around the heat exchanger. Alternatively, a method of periodically defrosting the heat exchanger may be employed. A number of effective defrost strategies have been developed for residential air-to-air heat exchangers that also may be applied to commercial installations.

14.3.3 Fouling

Fouling refers to an accumulation of dust or condensates on heat exchanger surfaces. Fouling increases the resistance to airflow and generally decreases heat transfer coefficients, thus reducing heat exchanger performance. This increased resistance increases fan power requirements and may reduce airflow volumes.

A pressure drop increase across the heat exchanger core can be used as an indication of fouling and, with experience, may be used to establish cleaning schedules. Heat exchanger surfaces must be kept clean if system performance is to be maximized.

14.3.4 Corrosion

Moderate *corrosion* generally occurs over time in HVAC system devices, roughening metal surfaces. Severe corrosion reduces overall heat transfer coefficients and can result in increased cross leakage between airstreams because of perforation or mechanical failure.

14.3.5 Cross contamination

Cross contamination or mixing between supply airstreams and exhaust airstreams may occur in air-to-air heat exchangers. It may be a significant problem if exhaust gases are toxic or odorous. Cross contamination or leakage varies with heat exchanger type and design, airstream static pressure differences, and the physical condition of the heat exchanger.

14.3.6 Filtration

Filters are recommended in both the supply and exhaust airstreams to reduce fouling and the frequency of cleaning. Exhaust air filters are especially important if the contaminants are sticky or greasy or if particulates can plug airflow passages in the exchangers. Supply air filters eliminate insects, leaves, and other foreign materials, protecting both the heat exchanger and HVAC equipment. Snow or frost can cause severe supply air blockage problems.

15

Pumps and System Curves

15.1 Pump and System Pressures

The pressures in hydronic systems are somewhat different from those in air systems. There are similar friction losses for straight sections of pipe, the same dynamic losses caused by turbulence for fittings, and pressure drops through coils, chillers, heat exchangers, and valves. However, hydronic piping systems have other heads that must be added to or subtracted from the total head output of the pump(s).

The purpose of a pump for HVAC work is to establish fluid flow and produce sufficient pressure to overcome the resistance of a system and the system components at the design flow rate.

15.1.1 Pump heads

When working with pumps, the word *head* is often used to define *pressure*. The definitions of common head terms are noted below even though some may be defined again under other discussions:

15.1.1.1 Friction head *Friction head* is the pressure in feet (pascals or meters) of the liquid pumped that represents system resistance that must be overcome.

15.1.1.2 Velocity head *Velocity head* is the pressure needed to accelerate the liquid being pumped. (For practical purposes, the velocity head is insignificant and usually can be ignored in HVAC hydronic system calculations.)

15.1.1.3 Static suction lift *Static suction lift* is the distance in feet (meters) between the pump centerline and the source of liquid below the pump centerline.

15.1.1.4 Suction lift *Suction lift* is the combination of static suction lift and friction head in the suction piping when the source of liquid is below the pump centerline.

15.1.1.5 Suction head *Suction head* is commonly the positive pressure on the pump inlet when the source of liquid supply is above the pump centerline.

15.1.1.6 Static suction head *Static suction head* is the positive vertical height in feet (meters) from the pump centerline to the top of the level of the liquid source.

15.1.1.7 Dynamic suction lift *Dynamic suction lift* is the sum of suction lift and velocity head at the pump suction when the source is below the pump centerline.

15.1.1.8 Dynamic suction head *Dynamic suction head* is positive static suction head minus friction head and minus velocity head.

15.1.1.9 Dynamic discharge head *Dynamic discharge head* is static discharge head plus friction head plus velocity head.

15.1.1.10 Total dynamic head *Total dynamic head* is dynamic discharge head (static discharge head, plus friction head, plus velocity head) plus dynamic suction lift, or dynamic discharge head minus dynamic suction head.

15.1.2 Static (head) pressure

Static pressure or *static head* is the pressure resulting from the weight of a column of liquid; the higher the column, the greater the static pressure. In *closed systems,* static pressure on the discharge side of the pump is balanced by the static pressure on the suction side. However, static heads must be taken into account in *open systems.*

For example, the pump in Figure 15.1(A) pumps water from the tank to a higher elevation. The pump must develop enough pressure to overcome the friction losses in the pipe lines and fittings and to lift the water to the higher elevation. The pressure required for the lift is numerically equal to the height in feet (meters) of the liquid being pumped. In Figure 15.1(A), this is the difference between the two

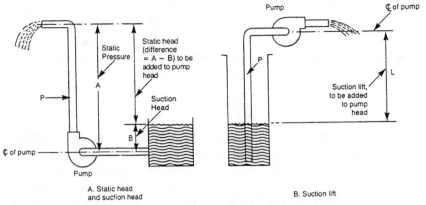

A. Static head
and suction head

B. Suction lift

Figure 15.1 Illustrations of static head, suction head, and suction lift.

levels, equal to A minus B, or the static head (static pressure difference) expressed in feet (meters) of water.

15.1.3   Suction head

Another related term illustrated in Figure 15.1(A) is *suction head*,
which exists when the source of supply is above the center line of
the pump. In this example, there is suction head because of the distance B.

Figure 15.1(B) shows a pump raising water from a reservoir to a
higher level. The pump must develop enough pressure to overcome
friction loses in the piping and it must develop pressure for the lift,
again equal to the distance L in feet (meters) of the liquid being
pumped.

In Figure 15.1(A), the static head is a pressure above atmospheric
pressure, and there is no theoretical limit to the height of the lift; it
can be as great as the pressure that the pump is capable of developing.

15.1.4   Suction lift

In Figure 15.1(B), the partial vacuum created by the pump suction
allows atmospheric pressure to push the liquid up to the pump. The
maximum *suction lift* theoretically possible would be the pressure of
the atmosphere of 14.7 psi or 34 ft of water (101.325 kPa or 10.32 m).
For ordinary HVAC system pumps, the maximum suction lift is considered to be about 26 to 30 ft (7.9 to 9.1 m). The suction lift will be
reduced by the friction loss in the suction pipe, and it may be further
reduced if the liquid being pumped should boil or vaporize because of

the partial vacuum. Such an effect occurs when pumping hot water. If too great a suction lift is attempted, some of the hot water will flash to steam, either reducing or eliminating the quantity of liquid water that can be pumped. This phenomenon, called *cavitation*, often sounds like marbles have been dropped into the pump.

15.1.5 Net positive suction head (NPSH)

To eliminate the problem of cavitation, it is necessary to maintain a minimum suction pressure at the inlet side of the pump. The actual value in feet of water (kPa or m w.g.) of internal pump losses depends on the pump size and design, and the volume of water being pumped. This must be determined by the pump manufacturer and is given by numerical values of *net positive suction head*, abbreviated NPSH. *Required NPSH,* sometimes designated NPSHR, can be considered to be the amount of pressure, in excess of the vapor pressure, required to overcome internal pump losses and to keep water flowing into the pump.

For a given pump, the required NPSH increases as capacity increases. Each system, as a result of design and physical limitations, will produce an *available NPSH*, sometimes designated NPSHA. When available NPSH is greater than the required NPSH, the problems of air release, vaporization, and cavitation will not arise.

The required NPSH value for a specific pump, available from the manufacturer or catalogue data, varies with flow and head. For any pump, the full range of values for each impeller size and operating speed is expressed as a curve (Figure 15.2).

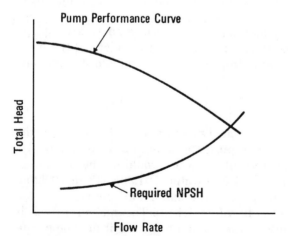

Figure 15.2 Typical required NPSH curve.

The TAB technician must remember, however, that for satisfactory pump operation, NPSHA must always exceed the NPSHR. If it does not, bubbles and pockets of vapor will form in the pump. The results will be reduction in capacity, loss of efficiency, noise, vibration, and cavitation. **The net positive suction head (NPSH) of a pump increases as the flow rate increases** (see Figure 15.2).

Equation 15.1

$$NPSHA = P_a \pm P_s + \frac{V^2}{2g} - P_{vp}$$

Where: NPSHA = net positive suction head available
(ft•w.g. or m w.g.)

P_a = atmospheric pressure at elevation of installation (use 34 ft•w.g. or use 10.32 m w.g.)

P_s = pressure at pump centerline (ft•w.g. or m w.g.)

$\frac{V^2}{2g}$ = velocity head at point P_s (ft•w.g. or m w.g.)

P_{vp} = absolute vapor pressure at pumping temperature (ft•w.g. or m w.g.)

g = gravity acceleration (32.2 ft/s² or 9.81 m/s²)

15.1.6   Friction pressure losses

Friction is the resistance to a moving fluid caused by the surface of the pipe walls. The primary purpose of a pump is to produce a designated volume of fluid at a pressure equal to the frictional resistance of the system plus the pressure losses of the other components of the system. For hydronic systems, the straight lengths of pipe losses are in terms of feet of head (water) per 100 ft of pipe (0.098 kPa/m). The pressure losses of fittings and valves normally are added in terms of equivalent length of pipe in feet (meters), based on the calculated straight pipe losses.

15.2   HVAC Pumps

The pumps used in HVAC systems fall into two major categories, *positive displacement pumps* and *centrifugal pumps*, with the centrifugal pumps being more widely used.

15.2.1   Positive displacement pumps

The usual types of positive displacement pumps in HVAC systems are the *piston pump* (reciprocating), the *rotary pump*, and the *screw pump*.

One of the characteristics that is common to all positive displacement pumps is their ability to overcome excessive pressures.

The pump curve for the positive displacement pumps is nearly linear in a vertical direction; that is, at a given speed it will pump the same volume against whatever pressure is connected at the outlet, usually from zero to over 200 psi (0 to over 1380 kPa). Many of these pumps are constructed with a built-in spring-loaded relief valve that will automatically bypass liquid internally from the discharge side to the suction side in the event of an excessive pressure buildup. This prevents damage to the pump or to the pump packing or seals. CAUTION: **Never close off the discharge valve of a positive displacement pump in an attempt to measure "shut-off pressure," because serious damage could result.** Large oil transfer pumps are a good example of a common application of rotary positive displacement pumps in HVAC work.

15.2.2 Centrifugal pumps

Centrifugal pumps are constructed with less stringent tolerances than positive displacement pumps. There is more fluid slippage, which means that as the pressures go up, more fluid slips past the impeller and less fluid is delivered to the outlet of the pump. Staging of the impellers is one of the ways in which centrifugal pumps overcome their inability to pump against high heads. Pumping through two or more stages multiplies the head capacity. A variety of drives is available, including direct drives and variable-speed drives.

15.2.3 Pump drives

The most common pump drive uses flexible couplings between the motor and the pump assembly. The alignment or misalignment of the coupling does not effect the performance of the pump or the speed at which it is driven. However, a misaligned coupling will have an amazingly short life as the elastic interface begins to break down. This sometimes allows the speed of the pump to vary and affects its output.

15.2.4 Pump connections

15.2.4.1 Single suction   *Single suction pumps* are usually constructed with the inlet at the end of the impeller and shaft with the casing arranged so that the discharge may be rotated to any position allowed by the bolt configuration.

15.2.4.2 Double suction   *Double suction pumps* do not have the flexibility of rotation, the suction and discharge being fixed and usually below the shaft centerline, each at a slightly different elevation. In either case, the suction connection is normally one or two pipe sizes larger than the discharge connection.

15.2.5   Pump rotation

Rotation of any pump is fixed by the configuration and type of vanes, and the suction and discharge connections. An arrow to indicate proper direction is often cast directly into the casing metal. In addition to proper position of the pump in the piping, rotation is also dependent on the motor or driver rotation. Rotation of motor and pump must be tested before operation. However, **pumps with mechanical seals must not be run dry, even for "bumping" to determine rotation.**

15.2.6   Pump packing and seals

15.2.6.1   Packed pumps   *Packed pumps* have a specified number of rings of packing, selected from the many types available, slipped over the shaft without undue pressure. The packing gland is installed last, with considerable care that it is not cocked. When the pump is started, the packing gland is tightened *evenly around* until one drop of liquid per second is leaking, or until the manufacturer's recommendations are met.

15.2.6.2   Mechanical seals   *Mechanical seals* are devices used in place of the packing to retain the system liquid within the pump at the shaft penetration. There are two major parts to a mechanical seal, a stationary disc and a rotating disc held in contact by a spring arrangement. One disc may be ceramic and the other carbon. Running causes a wearing-in process to maintain a complete seal. Foreign matter between the faces will destroy this seal. **Running or "bumping" a pump without water in the system will damage mechanical seals.**

15.2.6.3   Seal failure   Whereas a leaky packed pump may normally allow operation until shutdown for convenience, a seal failure may cause a complete blowout and require immediate system shutdown until repaired.

15.3   Pump Curves

Pump curves are similar to fan curves except that most pumps are directly connected to their motors, so the pump speed (rpm) remains almost constant. However, pump impellers can be changed or machined down to a specific size. Otherwise system flows, pressures, horsepowers, and efficiencies on the fan curves and pump curves are read in the same manner.

15.3.1 Efficiency

In Figures 15.3 and 15.4, the efficiency curve is a measure of the promised pump output versus the output of a theoretically perfect pump that would use 100% of its energy. Many design engineers attempt to pick pumps at peak efficiency. Pump efficiency has little value to the TAB technician. If the actual pump horsepower or wattage draw is higher than the pump specified, the TAB technician may close balancing valves to decrease the flow and lower the pump power requirements. However, the efficiency probably would be worse.

15.3.2 Power use

According to Figure 15.3 (U.S.), the pump horsepowers indicate that an operating point to the left of that line will require no more than the listed amount of horsepower: 7½, 10, 15, 20, etc. Because of the variable slopes of the curves, the pump with a 7 in. impeller will never exceed 10 HP, even though it seems to come close to it. When an 8½

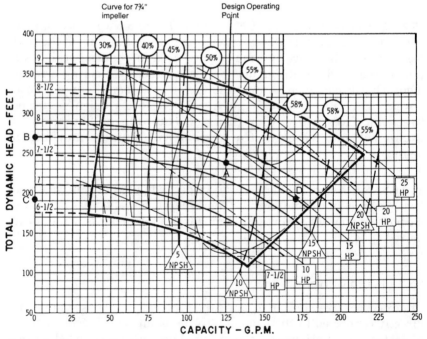

Figure 15.3 Pump performance curve as typically furnished by pump manufacturers (U.S.).

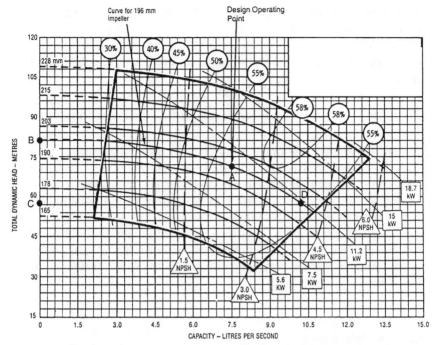

Figure 15.4 Pump performance curve as typically furnished by pump manufacturers (metric).

in. impeller is used, the 15 HP and the 20 HP curves are both crossed. For this reason caution is needed, because increasing the flow can exceed the limits of the pump motor provided for the application. This depends not only on the variety of pumps available, but also on how the system fits the specific pump curve. (The same discussion applies to Figure 15.4 using metric terms.)

15.3.3 Amperage

When the TAB technician fully opens the discharge valve to determine the maximum amperage draw, there can be many instances in which the pump motor will be overloaded. If the pump motor starter heater coils are oversized, there is a chance that the motor could be damaged before the amperage measurements are taken. **Check the size of the heater coils before starting equipment for testing**.

15.3.4 Pump curve analysis

Even though a pump submittal data sheet, which contains pump curves similar to Figures 15.3 and 15.4, is in hand for the pump to be tested, it is prudent to compare actual test measurements against the submitted pump curves.

The following procedure outlines performance testing of a centrifugal pump, including pump curve analysis used for the determination of flow:

1. With the pump(s) off, observe and record the system static pressure.

2. With the pump(s) running, *fully* close (slowly) the service (or balancing) valve located on the discharge of the pump. Record the pump discharge and suction pressures at the pump gauge taps located on, or as close as possible to, the pump volute housing. Pressure readings should be taken with one test gauge located at the same elevation (a good practice is to use a gauge manifold located at the pump base). Record the motor current and voltage. When all data are recorded, slowly open the discharge valve to the *full open position.*

15.3.5 No-flow testing

Testing of pump operating pressures with the discharge valve fully closed is known as *no-flow testing.* No-flow testing is used to determine the actual impeller diameter and pump operating curve by establishing the pump operating head at a zero gpm (L/s) flow rate. To determine the pump no-flow head, plot the intersection of the no-flow head and the zero gpm (L/s) line of the pump curve provided by the manufacturer and compare. If the no-flow head is approximately correct, continue with full-flow testing. If the no-flow head falls in between pump impeller sizes, the impeller may have been cut down in size such as shown in Figures 15.3 and 15.4 at point B.

Extreme care should be taken with no-flow testing since an error in gauge reading, calculation of head, or interpolation of curve may result in confusion and additional TAB work.

15.3.6 Full-flow testing

Full-flow testing is similar in procedure to *no-flow testing.* With the discharge valve in the full open position, record the pump suction and discharge pressures, and motor current and voltage. Plot the "operating point" intersection of the full-flow head and the established pump capacity curve to determine the actual pump flow rate.

15.4 Pump/System Curves

15.4.1 Closed system curves

The hydronic system curve is simply a plot of the change in energy head resulting from a flow change in a fixed piping circuit on a pump curve or set of pump curves (Figure 15.5). System curve construction methods differ between open and closed piping circuits.

From the pipe size and design flow rate, a calculated energy head pressure drop is determined. It should be particularly noted that system static height is of no importance in determining energy head pressure drop in a closed system. This is because the static heights of the supply and return legs are in balance; the energy head required to raise water to the top of the supply riser is balanced by the energy head regain as water flows down the return riser.

In a closed system, the system pressure can be regulated or limited by the pressure relief valve (safety valve) and the automatic water makeup valve (line pressure regulator valve). The pressure in various parts of the system will vary from top (less pressure) to bottom (more pressure) because of the static head, whether the pump(s) are running or not.

When the pump(s) are started, the discharge pressure gauge and the suction pressure gauge (which had similar readings) draw apart, with the discharge pressure gauge going to a higher reading and the suction pressure gauge going to a lower reading. If the makeup water valve is connected to the suction side of the pump, whenever the pump goes on, the pressure decrease created by the pump may cause the system automatic fill valve to add water and build the pressure back up to the pressure setting. Then when the pump is shut off, the pres-

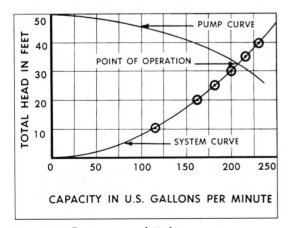

Figure 15.5 System curve plotted on a pump curve.

sure on the suction side will increase to a higher "pump off" reading, unless the pressure already was higher than the pressure setting from an earlier cycle.

15.4.2 Open system curves

This same process takes place in open systems, where the fill valve is regulating the water level in a sump or basin, that is connected directly to the suction side of the pump. When the pump draws down the water level on startup, the makeup water flows until the original water level is achieved. When the pump stops, the excess water drains back into the sump, raising the level on the suction side above its neutral position, and sometimes causing the sump to overflow.

In plotting the system curve for an open system, the statics of the system must be analyzed in addition to the friction loss. The different static conditions are illustrated in Figure 15.6.

For example, a typical cooling tower application is illustrated in Figure 15.7. In this system, the pump is drawing water from the tower sump and discharging it through the condenser to the tower nozzles, at a 10 ft (3 m) higher elevation than the sump level.

Total friction loss (suction and discharge piping, condenser, nozzles, etc.) is 30 ft (9 m) at a design flow rate of 200 gpm (12.6 L/s); the change in piping pressure drop for a change in water flow rates is determined and plotted to develop a system curve.

This system curve *cannot* be applied directly to the pump curve and the intersection taken as the accurate pumping point for the open

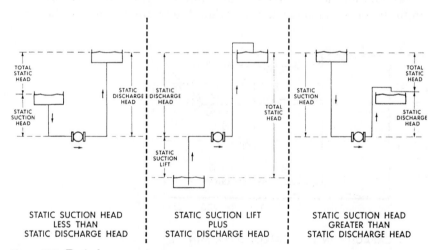

Figure 15.6 Typical open systems.

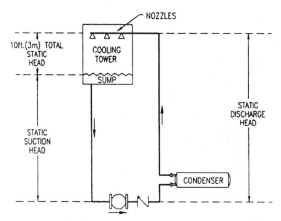

Figure 15.7 Cooling tower application.

system. A false evaluation using this criteria, but without evaluating the static height of the tower, is shown in Figure 15.8.

The illustration is false because the pump must also provide the necessary energy to raise water from the tower sump to the spray nozzles. In this case, the pump must raise each pound (kilogram) of water a distance of 10 ft (3 m) in height, or it must provide 10 ft (3 m) of energy head because of the static difference in height between the water levels.

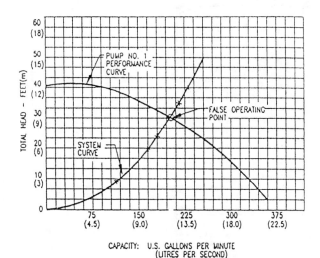

Figure 15.8 System curve for open circuit—False operating point.

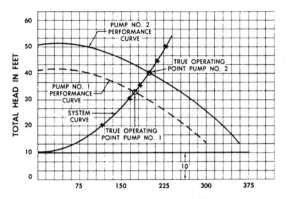

CAPACITY IN U.S. GALLONS PER MINUTE

Figure 15.9 System curve for open circuit—True operating point.

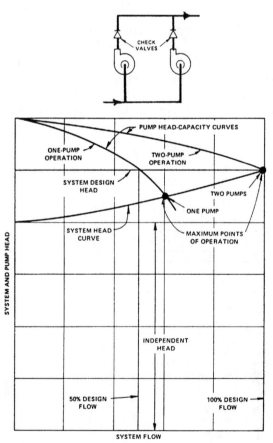

Figure 15.10 Pump and system curves for parallel piping.

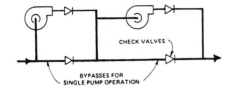

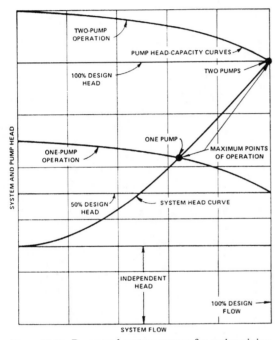

Figure 15.11 Pump and system curves for series piping.

The static difference of 10 ft (3 m) must be added to the piping pressure drop to provide total required head for each of the gpm points previously noted. The correct procedure for plotting a system curve for the piping circuit shown in Figure 15.7 is illustrated in Figure 15.9.

15.4.3 Multiple pump curves

Figure 15.10 shows two pumps piped in parallel, with a system head curve as well as the head capacity curves for single pump and two pump operation. Figure 15.11 illustrates two pumps piped in series with bypasses for single pump operation, and it indicates the use of series pumping on a hydronic system with a system head curve consisting of a large amount of system friction.

16

Pump and Hydronic Equations

16.1 Pump Laws

Pump laws apply equally to open and closed systems. These laws show only pump–system relationships and cannot be used to calculate pump curves. They can be used to calculate only system curves, so maybe they should have been called *system curve laws*. If data for one point of a system curve are known, the pump laws can be utilized to increase or decrease the flow up and down that system curve. As the flow is changed, pressure and horsepower will change. So the laws, which seemingly depend upon the pump, really govern only the system curve. If impeller diameters are changed, the new flow characteristics move the operating point along the system curve to a new operating point. See Chapter 15 for additional information.

When the system curve is changed (such as by adjusting balancing valves or cleaning strainers), the operating point is moved up or down along the pump curve to the new system curve. In practice, it is much easier to change the system curve by adjusting a balancing valve than it is to make changes to the pump, creating a new pump curve. When trouble occurs and adjusting balancing valves is not sufficient, the pump must be involved. The system curve also will change over a period of time because of pipe corrosion.

16.1.1 Affinity laws

16.1.1.1 Flow. Flow (volume) varies *directly* as the change in speed or impeller diameter.

16.1.1.2 Head. Head varies as the *square* of the change in speed or impeller diameter.

16.1.1.3 Power. Brake power varies as the *cube* of the change in speed or impeller diameter.

16.1.2 Pump equations (based on affinity laws)

Equation 16.1

$$\frac{Q_2}{Q_1} = \frac{\text{rpm}_2}{\text{rpm}_1}$$

Equation 16.2

$$\frac{Q_2}{Q_1} = \frac{D_2}{D_1}$$

Equation 16.3

$$\frac{H_2}{H_1} = \left(\frac{\text{rpm}_2}{\text{rpm}_1}\right)^2 = \left(\frac{Q_2}{Q_1}\right)^2$$

Equation 16.4

$$\frac{H_2}{H_1} = \left(\frac{D_2}{D_1}\right)^2$$

Equation 16.5

$$\frac{BP_2}{BP_1}\left(\frac{\text{rpm}_1}{\text{rpm}_1}\right)^3 = \left(\frac{Q_2}{Q_1}\right)^3$$

Equation 16.6

$$\frac{BP_2}{BP_1} = \left(\frac{D_2}{D_1}\right)^3$$

Where: Q = flow (gpm or L/s)
  rpm = revolutions per minute
   D = impeller diameter (in. or mm)
   H = head (ft.w.g. or m w.g. or Pa)
  BP = brake power (HP or kW)

16.1.3 Use of pump equations

Example 16.1 (U.S.) Specifications call for a pump with a 7.5 in. impeller to produce 360 gpm at 36 ft.w.g. and 5.0 BHP. TAB measurements indicate 247 gpm at 48 ft.w.g. and 4.2 BHP. Calculate the required head, impeller size, and BHP to obtain full flow.

Solution Using Equation 16.2:

$$D_2 = D_1 \frac{Q_2}{Q_1} = 7.5 \left(\frac{360}{247} \right)$$

$$D_2 = 10.93 \text{ in. impeller size}$$

Using Equation 16.3:

$$H_2 = H_1 \left(\frac{Q_2}{Q_1} \right)^2 = 48 \left(\frac{360}{247} \right)^2$$

$$H_2 = 101.97 \text{ ft.w.g. Head}$$

Using Equation 16.5:

$$BP_2 = BHP_1 \left(\frac{Q_2}{Q_1} \right)^3 = 4.2 \left(\frac{360}{247} \right)^3$$

$$BP_2 = 13.0 \text{ BHP}$$

Example 16.1 (Metric) Specifications call for a pump with a 190 mm impeller to produce 22.7 L/s at 108 kPa and 3.8 kW brake power. TAB measurements indicate 15.6 L/s at 144 kPa and 3.15 kW brake power. Calculate the required head, impeller size, and brake power to obtain full flow.

Solution Using Equation 16.2:

$$D_2 = D_1 \frac{Q_2}{Q_1} = 190 \left(\frac{22.7}{15.6} \right)$$

$$D_2 = 276.5 \text{ mm impeller size}$$

Using Equation 16.3:

$$H_2 = H_1 \left(\frac{Q_2}{Q_1} \right) = 144 \left(\frac{22.7}{15.6} \right)^2$$

$$H_2 = 304.9 \text{ kPa head}$$

Using Equation 16.5:

$$BP_2 = BP_1 \left(\frac{Q_2}{Q_1} \right)^3 = 3.15 \left(\frac{22.7}{15.6} \right)^3$$

$$BP_2 = 9.70 \text{ kW brake power}$$

16.2 Pump Power Equations

Equation 16.7 (U.S.)   **Equation 16.7 (Metric)**

$$WP = \frac{Q \times H \times G}{3960}$$

$$WP \text{ (kW)} = 9.81 \times \text{m}^3/\text{s} \times H \text{ (m)} \times G$$

or
$$WP \ (\text{W}) = \frac{\text{L/s} \times H \ (\text{Pa}) \times G}{1002}$$

Equation 16.8 (U.S.)

$$\text{BHP} = \frac{WP}{E_p} = \frac{Q \times H \times G}{3960 \times E_p}$$

Equation 16.8 (Metric)

$$BP = \frac{WP}{E_p}$$

Equation 16.9 (U.S.)

$$E_p = \frac{WP \times 100}{\text{BHP}} \ (\text{in percent})$$

Equation 16.9 (Metric)

$$E_p = \frac{WP \times 100}{BP} \ (\text{in percent})$$

Where: Q = flow (gpm)
H = head (ft.w.g. or m or Pa)
G = specific gravity = 1.0 for water
WP = water power (HP or W or kW)
BP = brake power (W or kW)
BHP = brake horsepower (HP)
E_p = pump efficiency (as a decimal)

The pump efficiency must be provided by the manufacturer. Where the efficiency is not known, a good rule-of-thumb is to allow 70% (0.7) as the efficiency factor.

Example 16.2 (U.S.) A pump is delivering 450 gpm of water at a 90 ft.head. Estimate the motor brake horsepower.

Solution Using Equation 16.8:

$$\text{BHP} = \frac{Q \times H \times G}{3960 \times E_p} = \frac{450 \times 90 \times 1}{3960 \times 0.7}$$

$$\text{BHP} = 14.6 \ \text{HP}$$

Example 16.2 (Metric) A pump is delivering 28.4 L/s of water at a 270 kPa head. Estimate the motor brake power in kilowatts.

Solution Using Equation 16.8:

$$BP \ (\text{W}) = \frac{WP}{E_p} = \frac{\text{L/s} \times H \times G}{1002 \times E_p}$$

$$BP \ (\text{W}) = \frac{28.4 \times (270 \times 1000) \times 1}{1002 \times 0.7} = 10{,}932 \ \text{W}$$

$$BP = 10.93 \ \text{kW}$$

Example 16.3 (U.S.) If the pump in Example 16.2 were pumping a fluid with a specific gravity (G) of 0.92, estimate the motor brake horsepower.

Solution

$$\text{BHP} = \frac{450 \times 90 \times 0.92}{3960 \times 0.7} = 13.44 \text{ HP}$$

Example 16.3 (Metric) If the pump in Example 16.2 was pumping a fluid with a specific gravity (G) of 0.92, estimate the motor brake power in kilowatts.

Solution

$$BP \text{ (W)} = \frac{28.4 \times (270 \times 1000) \times 0.92}{1002 \times 0.7}$$

$$BP \text{ (W)} = 10{,}058 \text{ W}$$

$$BP = 10.06 \text{ kW}$$

16.3 Hydronic Equations

Equation 16.8 (U.S.)

$$Q = 500 \times \text{gpm} \times \Delta t$$

Where: gpm = gallons per minute
Δt = temperature differential (°F)
Q = heat flow (Btu/h)

Equation 16.8 (Metric)

$$Q = 4.19 \times \text{L/s} \times \Delta t$$

Where: L/s = liters per second
Δt = temperature differential (°C)
Q = heat flow (kW)

Equation 16.9 (U.S.)	**Equation 16.9 (Metric)**
$$\frac{\Delta P_2}{\Delta P_1} = \left(\frac{\text{gpm}_2}{\text{gpm}_1}\right)^2$$	$$\frac{\Delta P_2}{\Delta P_1} = \left(\frac{\text{L/s}_2}{\text{L/s}_1}\right)^2 = \left(\frac{\text{m}^3\text{/s}_2}{\text{m}^3\text{/s}_1}\right)^2$$

Where: ΔP = pressure differential (psi or ft.w.g.)
gpm = gallons per minute

Where: ΔP = pressure differential (kPa)
L/s = liters per second
m³/s = cubic meters per second

Equation 16.10 (U.S.)	**Equation 16.10 (Metric)**
$$\Delta P = \left(\frac{Q}{C_v}\right)^2$$	$$\Delta P = \left(\frac{Q}{K_v}\right)^2$$

Where: ΔP = pressure differential (psi)
Q = flow rate (gpm)
C_v = valve constant or flow coefficient

Where: ΔP = pressure differential (kPa)
Q = flow rate (L/s)
K_v = valve constant or flow coefficient

Equation 16.11 (U.S.)

$$P = F/A$$

Equation 16.11 (Metric)

$$P = F/A$$

Where: P = pressure (psi)
F = force (lb)
A = area (in.^{2})

Where: P = pressure (Pa)
F = force (newtons, N)
A = area (m^2)

NOTE: 1 Pa = 1 N/m^2

Equation 16.12 (U.S.)

$$H = \frac{f L V^2}{2gD}$$

Equation 16.12 (Metric)

$$h = \frac{f L V^2}{2gD}$$

Where: h = head loss (ft)
f = friction factor
L = length of pipe (ft)
V = velocity (fps)
g = gravity (32.2 ft/s^2)
D = internal diameter (ft)

Where: h = head loss (m)
f = friction factor
L = length of pipe (m)
V = velocity (m/s)
g = gravity (9.81 m/s^2)
D = internal diameter (m)

16.4   Hydronic Equivalents

- One liter water = 1 kilogram
- Specific heat (C_p) water = 4190 J/kg • °C (at 20 °C)
- Specific heat (C_p) water vapor = 1886 J/kg • °C (at 20 °C)
- One meter water = 9.807 kPa
- One millimeter mercury = 133.33 kPa
- One cubic meter (m^3) = 1000 liters (L)
- One inch water = 249 pascals (Pa)
- Atmospheric pressure (1 Bar) = 101,325 pascals (Std.)
- One pound per square inch (psi) = 6.898 kilopascals (kPa)
- One foot water = 0.3048 meter water = 2.988 kilopascals (kPa)

16.5   Power Equations

Equation 16.13

$$BP = \frac{\text{Amps}_R - (\text{Amps}_{NL} \times 0.5)}{\text{Amps}_{FL} - (\text{Amps}_{NL} \times 0.5)} \times P_{NP}$$

Equation 16.14

$$\text{Amps}_R = \frac{\text{Amps}_{FL} \times \text{Volts}_{NP}}{\text{Volts}_R}$$

Where: BP = brake power (HP or kW)
Amps_R = running amps
Amps_{NL} = no-load amps
Amps_{FL} = full-load amps
P_{NP} = nameplate power (HP or kW)
Volts_R = running volts
Volts_{NP} = nameplate volts

17

Flow Measurements of Hydronic Systems

17.1 Hydronic Piping Systems

17.1.1 Classifications

Generally, the most economical distribution system layout has piping mains that are run by the shortest and most convenient route to the terminal equipment having the largest flow rate requirements. Branch or secondary circuits are connected to these mains. Hydronic systems may be divided into two classifications: small systems or branch circuits, and main distribution systems.

17.1.1.1 Smaller systems. The smaller systems use:

1. Series loop systems
2. One-pipe systems
3. Two-pipe, reverse return systems
4. Two-pipe, direct return systems

17.1.1.2 Larger systems. Larger systems use:

1. Two-pipe, reverse return systems
2. Two-pipe, direct return systems
3. Three-pipe systems
4. Four-pipe systems

17.1.2 Series loop systems

A *series loop system* is a continuous run of pipe or tube from a supply connection to a return connection. Terminal units are a part of the loop. Figure 17.1 shows a system of two series loops on a supply and return main (*split series loop*). The water temperature drops progressively as each radiator transfers heat to the air, the amount of drop depending on the radiator output and the water flow rate.

17.1.3 One-pipe systems (diverting fitting)

One-pipe systems (Figure 17.2) use a single loop main. For each terminal unit, a supply and a return tee is installed on the same main. One or both of the two tees is a special fitting that creates a pressure drop in the main flow to divert a portion of the flow to the unit. One

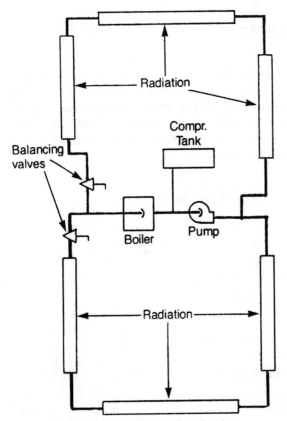

Figure 17.1 Series loop systems—Two circuits.

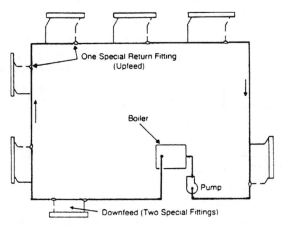

Figure 17.2 One-pipe system.

(return) diverting tee is usually sufficient to feed units above the main. Two special fittings usually are required for down-feed units to overcome thermal head.

17.1.4 Two-pipe systems

Two-pipe systems (Figures 17.3 and 17.4) may be *direct return* or *reverse return*. In the direct return system, return main flow direction is opposite the supply main flow; return water from each unit takes the shortest path back to the boiler. In the reverse return system, the return main flow is in the same direction as the supply main flow. After the last unit is supplied, the return main is connected directly to the boiler.

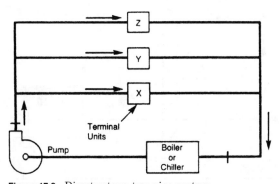

Figure 17.3 Direct return two-pipe system.

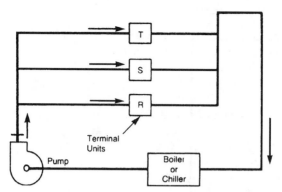

Figure 17.4 Reverse return two-pipe systems.

17.1.5 Three-pipe systems

The three-pipe system (Figure 17.5) satisfies variations in load by providing independent sources of heating and cooling to the room unit in the form of constant temperature primary and secondary chilled and hot water.

The HVAC unit contains a single secondary water coil. A three-way valve at the inlet of the coil admits the water from either the hot or the cold water supply, as required. The water leaving the coil is carried in a common pipe to either the secondary cooling or the heating equipment. The usual room control for three-pipe systems is a special three-way modulating valve that modulates either the hot or the cold water in sequence but does not mix the streams.

During the period between seasons, if both hot and cold secondary water is available, any unit can be operated within a wide capacity range from maximum cooling to maximum heating within the limits set by the temperature of the secondary chilled or hot water.

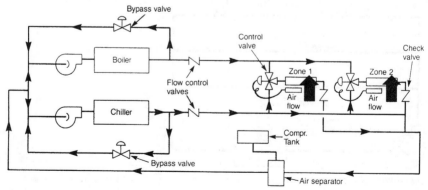

Figure 17.5 Three-pipe systems.

17.1.6 Four-pipe systems

Four-pipe systems for induction, fan coil, or radiant panel systems derive their name from the four pipes to each terminal unit. These four pipes consist of a cold water supply, a cold water return, a warm water supply, and a warm water return. The four-pipe system satisfies variation in cooling and heating to the room unit in the form of constant temperature primary air, secondary chilled water, and secondary hot water.

The terminal units usually are provided with two independent water coils, one served by hot water, the other by cold water. The primary air is cold and remains at the same temperature year round. During peak cooling and heating, the four-pipe system performs in a manner similar to the two-pipe system, with essentially the same operating characteristics. During the period between seasons, any unit can be operated at any capacity level from maximum cooling to maximum heating, if both cold water and warm water are being circulated. Any unit can be operated at or between these extremes without regard to the operation of any other unit.

Another unit and control configuration uses a single secondary water coil and three-way valves located at the inlet and leaving side of the coil. They admit water from either the hot or the cold water supply as required and divert it to the appropriate return pipe. This arrangement requires a special three-way modulating valve that controls the hot or cold water selectively and proportionally but does not mix the streams. The valve at the coil outlet is a two-position valve open to either the hot or cold water return.

17.1.7 Combination piping circuits

Figure 17.6 illustrates a *primary circuit* and two *secondary circuits*. As pipe lengths and the number of units vary, and as circuit types are combined, basic names for piping systems become meaningless. Flow, temperatures, and heads must be determined by the TAB technician for each circuit and then for the complete system.

17.2 System Flow Methods

17.2.1 Constant-volume systems

A *constant-volume* piping system is where the required system flow rate does not change during normal system operations. Constant-volume systems may be classified as *straight-through systems* or *three-way valve systems*.

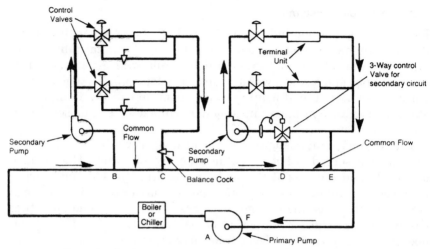

Figure 17.6   Example of primary and secondary pumping circuits.

17.2.1.1   Straight-through systems.   A *straight-through system* does not contain any automatic temperature control valves in the piping circuit. Flow through the source, outlets, and piping remains constant. Temperature control is accomplished by the modulation of system water temperature or the use of dampers at terminal units.

17.2.1.2   Three-way valve systems.   Constant system flow also may be achieved with the use of three-way automatic temperature control valves at the terminal units. Three-way or bypass valves allow circulation around a terminal unit when heat transfer is not required. System water temperatures may remain constant with systems utilizing three-way control valves allowing closer temperature control than is found in a straight-through system. Two types of three-way valves are used, *diverting valves* and *mixing valves.*

A *diverting valve* (Figure 17.7 is primarily used in two-position control applications (open and closed) where the flow is required to be diverted from one pipe to another. The valve therefore has one inlet and two outlets. In the case of terminal units, the valve diverts supply water from the coil to the bypass.

Similar to a diverting valve in appearance, a *mixing valve* has three ports, but this is where the similarity ends. Mixing valves are primarily used where proportional, or modulating, control is required. The valve contains two inlets and one outlet (Figure 17.8), allowing the mixture of two fluids. In the case of terminal units, the valve allows the mixing of bypass and coil return flow to the common return

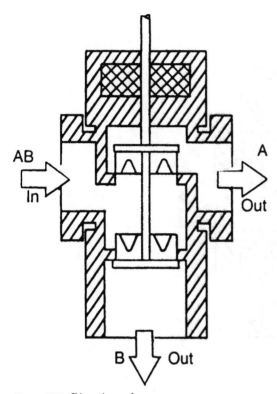

Figure 17.7 Diverting valve.

piping of the system. **It is not uncommon to find these valves installed incorrectly!**

17.2.2 Variable-volume systems

A *variable-volume piping system* is one in which the required system flow rates varies during normal system operation. Variable-volume systems have been created with the use of two-way automatic temperature control valves at terminal units or by varying the speed of the pump(s).

Two methods, *differential pressure control valves* and *variable-speed pumping*, commonly are used to maintain the minimum flow rates and pressures required to maintain satisfactory system operation.

17.2.2.1 Two-way valves. *Two-way control valves* may operate either two-position or proportionally. Both methods reduce the flow rate through a terminal unit when heat transfer is not required or is re-

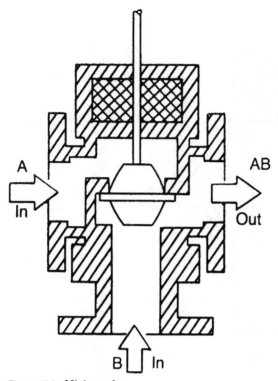

Figure 17.8 Mixing valve.

duced. The use of two-way valves at system terminal units results in varied system flow rates and pressures during normal operation.

17.2.2.2 Differential pressure control valve. A *differential pressure control valve* normally may be found in smaller variable-volume systems. As two-way control valves at terminal units close, system differential pressure increases and is monitored by a differential pressure controller. The controller modulates the differential pressure control valve to maintain a system differential pressure satisfactory for pump and primary heat exchange equipment operation.

17.2.2.3 Variable-speed pumps. Similar in control to a differential pressure control valve, the differential pressure controller resets the pump speed, reducing system differential pressure. This control application is utilized in larger variable-volume applications because of higher installation costs, but results in lower operating costs than with the use of a differential pressure control valve. Pump flows and heads may be calculated using the pump equations in Chapter 16.

17.3 Equipment Balancing and Piping Connections

17.3.1 Pump location

Pump location varies with the size and type of system. Figures 17.1, 17.3, and 17.4 illustrate pumps located in the supply main from the boiler or chiller, whereas Figures 17.2, 17.5, and 17.6 show the pumps in the return piping. A pump in the boiler return is acceptable for small systems when pump head is low, such as a 12 ft.(36 kPa) head or less, the compression tank is connected to the boiler (or a nearby main), and the highest piping and radiation is maintained at a static pressure greater than full pump head. These conditions apply to most residential or small systems.

When pump head is equal to or greater than the difference between the boiler automatic fill valve and relief valve discharge pressures, or when the highest piping or radiation can be at a static pressure less than total pump head, the pump must be located on the supply side of the boiler, with the compression tank at the pump inlet, as illustrated in Figure 17.9. This assures that pump cycling will not cause excessive pressure variations in the boiler and will not cause subatmospheric pressure at topmost system points to produce air leakage into the system. Pump cavitation is prevented by locating a properly sized compression tank near the pump inlet.

17.3.2 Compression (expansion) tanks

As a hydraulic device, the compression (expansion) tank serves as the reference pressure point in the system. Where the tank connects to the piping, the pressure equals the pressure of the air in the tank plus

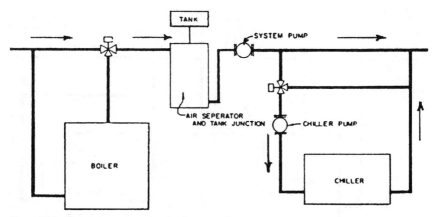

Figure 17.9 Correct compression tank connection.

or minus any fluid pressure resulting from the elevation or head difference between the tank liquid surface and the pipe.

Therefore, the location of the compression (expansion) tank connection in relation to the system pump(s) is critical. A properly located pump(s) will pump away from the junction of the tank with the system piping as in Figure 17.9. This location ensures an increase in system operating pressure over and above that set by the tank and the pressure reducing valve (makeup water line). If the pump discharges into the tank junction with the system piping (Figure 17.10), there is a decrease in system pressure during pump operation. When there is a high pump head, the system pressure can be subatmospheric causing air to be induced into the system. This could lead to air-bound terminal units, reduced flow, pump damage, and increased system deterioration (caused by new oxygen being brought into the system).

17.3.3 Pump balancing

1. Before the TAB work starts on a hydronic system, verify that the system has been flushed, cleaned, and vented of air; that valves open or a "normal" position; and that the TAB work on the air systems is completed.

2. With the pump(s) off, observe and record system static pressure at the pump(s).

3. Place the systems into operation again, check that all air has been vented from the piping systems, and allow flow conditions to stabilize. Verify that the system compression tank(s) and automatic water fill valve are operating properly. Check pump rotation.

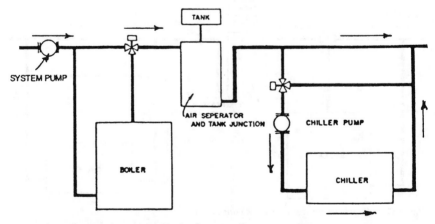

Figure 17.10 Incorrect compression tank connection.

4. Record the operating voltage and amperage of the pump(s) and compare these with nameplate ratings and thermal overload heater ratings. Verify the speed of each pump.

5. If flow meters or calibrated balancing valves are installed, which allow the flow rate of the pump circuit(s) to be measured, perform the necessary work and record the data.

6. With the pump(s) running, slowly close the balancing cock fully in pump discharge piping and record the discharge and suction pressures at the pump gauge connections. **Do not fully close any valves in the discharge piping of a positive displacement pump. Severe damage may occur.**

7. Using shutoff head, determine and verify each actual pump operating curve and the size of each impeller and compare with the submittal data curves. If the test point falls approximately on the design curve, proceed to the next step; if not, plot a new curve parallel with other curves on the chart, from zero flow to maximum flow. Make sure the test readings have been taken correctly before plotting a new curve. Preferably one gauge should be used to read differential pressure. It is important that gauge readings should be corrected to centerline elevation of the pump.

8. Open the discharge balancing cock slowly to the fully open position; record the discharge pressure, suction pressure, and total head. Using the total head, read the system water flow from the corrected pump curve established in step 7. Verify the data with that from flow meters and/or calibrated balancing valves if used.

9. If the total head is higher than the design total head, the water flow will be lower than designed. If the total head is less than design, water flow will be greater, in which case the pump discharge pressure should be increased by partially closing the pump discharge balancing cock until the system water flow is approximately 110% of design. Record the pressures and the water flow. Check pump motor voltage and amperage and record. This data should still be within the motor nameplate ratings. Start any secondary system pumps and readjust the balancing cock in the primary circuit pump discharge piping if necessary. Again record all readings.

10. Additional adjustments to the pump may be necessary during system balancing and after the balancing is complete. Record all final measurements.

17.3.4 Flow-measurement devices

If orifice plates, venturi meters, or other flow-measuring or control devices have been provided in the piping system branches, an initial recording of the flow distribution throughout the system should be made without any adjustments being made. After studying the system, adjust the distribution branches or risers to achieve balanced circuits as outlined above. Vent air from low-flow circuits and check strainers for debris. Then proceed with the balancing of terminal units on each branch. See Section 17.4 for measuring fluid flow through orifice plates.

17.3.5 Balancing HVAC units (three-way valves)

Where there is a thermostatically controlled three-way valve in series with balancing valves, and assuming that the HVAC unit coil is under full load with a stable control loop regulating the flow, additional closing of the balancing valve X in Figure 17.11 will cause a corresponding increase in flow through the temperature control valve. Temperature control engineers say that the balancing valve is *within the control loop*. This means that if the balancing valve is partially closed to balance the main system, the control valve will open to decrease the pressure drop by the exact same amount, thus reestablishing an identical flow.

However, notice that the valve positions have been changed. The flow rate has now been established but the balancing valve has less flow and the control valve has more than before. This illustrates two things: The first is that all control valves must be fully open during balancing work and remain nonoperational; the second is that a wider open balancing valve can allow a control valve to cover a wider range.

Since the intent is not to decrease the operating scale of the control valve, balancing valve X should be closed as little as possible to achieve the correct flow rates. The TAB technician should perform the balancing with full flow being established; the three-way valve open to the A position, balancing valve X fully open; and the measurements made with the gauge connected to points 1 and 6 (this is the pressure drop through the valve and coil assemblies). The automatic control valve should then be changed to position B and gauges 1 and 6 reread.

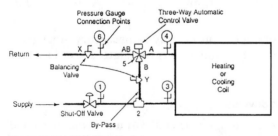

Figure 17.11 HVAC unit piping.

It may be expected (if balancing valve Y is fully open) that the pressure loss would decrease, that is, the pressure of the three-way valve and the open bypass line. The balancing valve Y is then closed until the gauge readings of 1 and 6 are identical to those in the beginning, that is, the valve–coil assembly. This is the position that balancing valve Y should be in.

The flow is of no concern (whether it is higher or lower). In fact, after the balancing (since the pressure drops in both paths are equal), the flows will be adjusted equally in both paths. The valve is reset to position A and valve X set, based on the gauge readings at 3 and 4, for the appropriate pressure drop or flow rates. It is generally recognized that the losses between 1 and 3 are too small to be significant. For this reason the gauge at 3 often is eliminated and the accuracy of the reading from 1 to 4 is accepted. This method assures that valves X and Y are not set for too low of a flow. The control valve then can be placed in proper operation and the TAB technician can be assured that the control valve has the full range in which to operate.

17.3.6 Balancing cooling tower systems

1. With the system off, confirm that the water level in the tower basin is at the correct level and that the piping system has been cleaned and flushed. On towers with variable-pitch fan blades, verify that the setting of the blades is correct for the test conditions.

2. With pump(s) off, observe and record the system static pressure at the pump(s).

3. Place the system into operation and allow the flow conditions to stabilize. Check the operation of the water makeup valve and blowdown.

4. Record the operating voltage and amperage of all fan and pump motors and compare these with nameplate ratings and thermal overload heater ratings.

5. Verify the speed and rotation of each pump.

6. With the pump(s) running, slowly close the balancing cock in each pump discharge line and record shutoff discharge and suction pressures at the pump gauge connections. **Do not use this method if a positive displacement pump(s) is used**. Follow the pump balancing procedures in Section 17.3.3.

7. Establish a uniform water distribution within the tower, where possible, and check for clogged outlets or spray nozzles. Check for vortex conditions at the tower condenser water suction connection.

8. Record the inlet and outlet pressures of the condenser(s) and check against the manufacturer's design pressure difference.

9. When a three-way control valve is used in the condenser water piping at the tower, measure the pressure difference with full water flow going both through the tower and/or through the bypass line. Set the bypass line balancing cock to maintain a constant pressure at the pump discharge with the control valve in either position.

10. Start the tower fan(s) and check rotation, gear box, belts, and sheave alignment. Measure and record fan motor amperes, voltage, phase, and speed.

11. Take inlet and outlet air dry bulb and wet bulb air temperature readings. Take test readings continually with a minimum of time lapse between readings. Note wind velocity and direction of the time of the test.

12. Take the inlet air temperature readings between 3 and 5 ft. (0.9 and 1.5 m) from the tower at all inlets. These readings shall be taken halfway between the base and the top of the inlet and then averaged.

13. If the cooling tower has a ducted inlet or outlet, make a Pitot tube traverse of the duct to verify the airflow.

14. If verification of the HVAC refrigeration equipment data is included in the TAB specifications, have the refrigeration system started. Verify the head and suction pressures and compare with design. After operation stabilizes under a normal cooling load, measure and record the condenser water inlet and outlet temperatures. Observe and record the percentage of load on the compressor where possible.

15. After setting the three-way control valve (to control head pressure) in the condenser water line, verify and record that it operates to maintain the correct head pressure by varying the flow at the tower. On units that have a fan cycling control, verify that the fan cycles to maintain design condenser water temperature. If fan inlet or outlet damper controls are used, verify that the dampers modulate to maintain the design condenser water temperature leaving the tower.

16. Make another complete set of pressure, voltage, and ampere readings at the pump(s). If the pump(s) capacity have fallen below design flow, open the balancing cock(s) at the pump discharge to bring flow within 5 to 10% of the design reading, if possible.

17. Make final measurements of all pump, fan, and equipment data and record on the TAB report forms.

18. After all balancing work has been completed and the system is operating within ± 10% of design flow, mark or score all balancing cocks, gauges, and thermometers at final set points and/or range of operation.

19. Verify the action of all water flow safety and shutdown controls.

20. Prepare all TAB report forms and submit as required.

17.4   Measuring with Flow Devices

17.4.1   Orifice plates

Depending on the geometry of the orifice, the discharge coefficient will vary. The discharge coefficient, designated K, will be supplied by the manufacturer; however, the more common sharp-edged and square-edged orifices will be found to have a coefficient, $K = 0.61$. Also, d/D will usually be ⅓ or less, d being the diameter of the orifice and D being the diameter of the pipe containing the orifice.

Equation 17.1 (U.S.)	Equation 17.1 (Metric)
$Q = 19.636 \times d^2 \times K \sqrt{\Delta h}$	$Q = \dfrac{d^2 \times K \sqrt{\Delta h}}{900}$

Where: Q = flow (gpm or L/s)
  d = orifice diameter (in. or mm)
  Δh = head differential (ft.w.g. or kPa)
  K = correction factor (usually 0.61)

If d/D_1 (see Figures 17.12 and 17.13) is greater than 0.3, then multiply Q by the following:

d/D_1	Factor × Q
0.4	1.014
0.5	1.033
0.6	1.070

17.4.2   Venturis

For any venturi tube (Figure 17.14) in which $d = ⅓D_1$, the following applies:

Equation 17.2 (U.S.)	Equation 17.2 (Metric)
$Q = 19.17 \times d^2 \sqrt{\Delta h}$	$Q = \dfrac{d^2 \sqrt{\Delta h}}{922}$

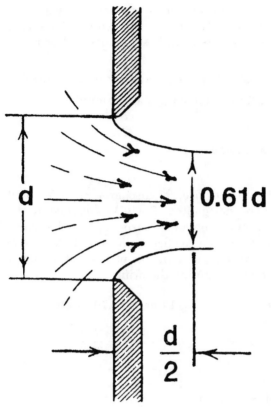

Figure 17.12 Contraction at an orifice.

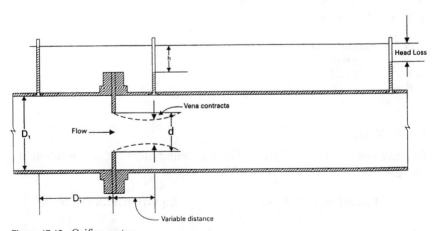

Figure 17.13 Orifice meter.

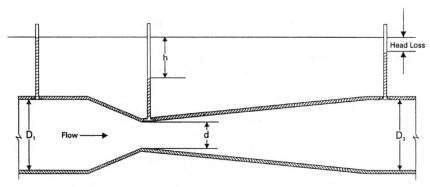

Figure 17.14 Venturi meter.

Where: Q = flow (gpm or L/s)
d = venturi; throat diameter (in. or mm)
Δh = head differential (ft.w.g. or kPa)

The manufacturer of the orifice plate or venturi tube usually provides these devices with appropriate pressure taps and instructions. Figures 17.12, 17.13, and 17.14 illustrate the calculation of d and D_1 for orifice and venturi meters.

17.4.3 Control valves

Control valves with known flow coefficients also may be used to measure flow:

<table>
<tr><td align="center">**Equation 17.3 (U.S.)**</td><td align="center">**Equation 17.3 (Metric)**</td></tr>
<tr><td align="center">$$Q = C_v \sqrt{\frac{\Delta P}{G}}$$</td><td align="center">$$Q = K_v \sqrt{\frac{\Delta P}{G}}$$</td></tr>
</table>

Where: Q = flow rate (gpm or L/s)
C_v = valve constant (U.S.)
K_v = valve constant (metric)
ΔP = pressure differential (psi or kPa)
G = specific gravity (use 1.0 for water)

Example 17.1 (U.S.) The pressure drop across a water system control valve with a C_v = 10 is 37 ft.w.g. Find the flow.

Solution

$$37 \text{ ft.w.g.}/2.31 = 16.0 \text{ psi}$$

$$Q = C_v\sqrt{\frac{\Delta P}{G}} = 10\sqrt{\frac{16}{1}} = 40 \text{ gpm}$$

Example 17.1 (Metric) The pressure drop across a water system control valve with a $K_v = 0.24$ is 11.1 m of water. Find the flow.

Solution

$$11.1 \text{ m} \times 9.807 = 108.86 \text{ kPa}$$

$$Q = K_v\sqrt{\frac{\Delta P}{G}} = 0.24\sqrt{\frac{108.86}{1}} = 2.50 \text{ L/s}$$

17.5   HVAC Coil Alternate Method for TAB

The usual method of balancing water flow at HVAC unit coils is to take the pressure drop across the inlet and outlet of the coil and check the drop against the manufacturer's published data; adjusting the water flow to the design data. But there are times on existing jobs, where the published data for a coil is not available. And even on new work, where the data are available, gauges and gauge cocks are often omitted, so there is no ready method of determining the pressure drop. In such cases the water flow across the coil is set by the *wet bulb temperature difference* or *total heat* method.

As stated earlier, the air systems should be balanced before to hydronic balancing. Assuming that airflow is at design, the cooling system total heat flow may be found by using Equation 11.11 from Chapter 11: $Q = 4.5 \times \text{cfm} \times \Delta h$ ($Q = 1.20 \times \text{L/s} \times \Delta h$).

The enthalpy or total heat difference (Δh) is found by plotting the entering and leaving wet bulb (WB) temperatures of the cooling coil on a psychrometric chart (see Chapter 12). Then use Equation 11.12 to calculate the coil chilled water flow ($Q = 500 \times \text{gpm} \times \Delta t$ or $Q = 4190 \times \text{L/s} \times \Delta t$).

Plotting enthalpy on the psychrometric chart, particularly in the field, is no simple matter. Field conditions, at best, are not conducive to paperwork, and the hairline scales of enthalpy saturation are difficult to see clearly under average lighting conditions, let alone a meagerly lit equipment room. Tables 17.1 (U.S.) and 17.2 (metric) simplify the enthalpy calculations and eliminate the use of the psychrometric chart for this task.

Example 17.2 (U.S.) Ten thousand cfm of air enters a cooling coil at 69 °F WB and leaves at 56 °F WB. The chilled water temperature measurements are 45 °F entering and 55 °F leaving. Find the water flow volume through the coil.

TABLE 17.1 Enthalpy—Btu/h per cfm for Various Wet Bulb
Temperature Differentials

Leaving WB temp (°F)	Entering wet bulb temperature (°F)								
	80	78	76	74	72	70	68	66	64
45	117	108	98.6	90.0	81.8	74.0	66.5	59.3	52.5
46	115	105	96.3	87.8	79.5	71.7	64.2	57.0	50.2
47	113	103	94.0	85.4	77.2	69.3	61.8	54.7	47.8
48	110	101	91.6	83.0	74.8	67.0	59.4	52.3	45.5
49	108	98.2	89.2	80.6	72.4	64.5	57.0	49.9	43.0
50	105	95.8	86.7	78.1	69.9	62.1	54.5	47.4	40.5
51	103	93.2	84.2	75.6	67.4	59.5	52.0	44.9	38.0
52	100	90.6	81.6	73.0	64.8	56.9	49.4	42.3	35.4
53	97.5	88.0	79.0	70.4	62.1	54.3	46.8	39.6	32.8
54	94.8	85.3	76.3	67.7	59.4	51.6	44.1	36.9	30.1
55	92.1	82.6	73.6	65.0	56.7	48.9	41.4	34.2	27.4
56	89.3	79.8	70.8	62.2	54.0	46.1	38.6	31.5	24.6
57	86.4	77.0	67.9	59.3	51.1	43.2	35.7	28.6	21.7
58	83.6	74.1	65.0	56.4	48.2	40.4	32.9	25.7	18.9
59	80.6	71.1	62.1	53.5	45.2	37.4	29.9	22.7	15.9
60	77.5	68.0	59.0	50.4	42.2	34.3	26.8	19.7	12.8
61	74.4	64.9	55.9	47.3	39.1	31.2	23.7	16.6	9.7
62	71.3	61.8	52.7	44.1	35.9	28.1	20.6	13.4	6.6
63	68.0	58.5	49.5	40.9	32.7	24.8	17.4	10.2	3.3
64	64.7	55.2	46.2	37.6	29.3	21.5	14.0	6.8	—
65	61.3	51.8	42.8	34.2	26.0	18.1	10.6	3.5	—
66	57.9	48.4	39.3	30.7	22.5	14.7	7.2	—	—
67	54.3	44.8	35.8	27.2	18.9	11.1	3.6	—	—
68	50.7	41.2	32.2	23.6	15.3	7.5	—	—	—

Solution From a psychrometric chart, h for 69 °F WB is 33.3 Btu/lb and h for 56 °F WB is 23.9 Btu/lb; then $\Delta h = 33.3 - 23.9 = 9.4$. Using Equation 11.11:

$$Q = 4.5 \times 10,000 \times 9.4 = 423,000 \text{ Btu/h}$$

Using Equation 11.12:

$$\text{gpm} = \frac{Q}{500 \times \Delta t} = \frac{423,000}{500 \times 10°} = 84.6$$

alternate solution From Table 17.1, the Btu/h/cfm for 69 °F WB entering air and 56 °F WB leaving air is 42.3. 10,000 cfm $\times$ 42.3 = 423,000 Btu/h. The balance of the solution is the same as above using Equation 11.12.

Example 17.2 (Metric) Five thousand L/s of air enters a cooling coil at 20 °C WB and leaves at 13 °C WB. The chilled water temperature measurements are 7 °C entering and 12 °C leaving. Find the water flow volume through the coil.

Solution From a metric psychrometric chart, h for 20 °C WB is 57.5 kJ/kg and h for 13 °C WB is 36.7 kJ/kg; then $\Delta h = 57.5 - 36.7 = 20.8$. Using Equation 11.11:

TABLE 17.2 Enthalpy—Watts per L/s for Various Wet Bulb Temperature Differentials

Leaving WB temp (°C)	Entering wet bulb temperature (°C)									
	27	26	25	24	23	22	21	20	19	18
7	75.1	69.7	64.7	59.8	55.0	50.3	46.1	41.9	37.8	34.0
8	72.5	67.2	62.2	57.2	52.4	47.8	43.6	39.4	35.3	31.4
9	69.4	64.4	59.4	54.5	49.7	45.0	40.8	36.6	32.5	29.7
10	67.0	61.6	56.5	51.6	46.8	42.1	37.9	33.7	29.6	25.8
11	64.3	58.9	53.9	49.0	44.2	39.5	35.3	31.1	27.0	23.2
12	61.3	55.9	50.9	46.0	41.2	36.5	32.3	28.1	24.0	20.2
13	58.1	52.7	47.8	42.8	38.0	33.4	29.2	25.0	20.9	17.0
14	55.0	49.6	44.6	39.7	34.9	30.2	26.0	21.8	17.8	13.9
15	51.7	46.3	41.3	36.4	31.6	26.9	22.7	18.5	14.4	10.6
16	48.2	43.0	37.9	33.0	28.2	23.5	19.3	15.1	11.0	7.2
17	44.6	39.4	34.3	29.4	24.6	19.9	15.7	11.5	7.4	3.6
18	41.0	35.8	30.7	25.8	21.0	16.3	12.1	7.9	3.8	—
19	37.3	32.0	26.9	22.0	17.2	12.5	8.3	4.1	—	—
20	33.2	27.7	22.8	17.9	13.1	8.4	4.2	—	—	—
21	29.0	23.6	18.6	13.7	8.9	4.2	—	—	—	—

$$Q = 1.20 \times \text{L/s} \times \Delta h = 1.20 \times 5000 \times 20.8 = 124{,}800 \text{ W}$$

Using Equation 11.12:

$$\text{L/s} = \frac{Q}{4190 \times \Delta t} = \frac{124{,}800}{4190 \times 5°} = 5.96$$

alternate solution From Table 17.2, the W/L/s for 20 °C WB entering air and 13 °C WB leaving air is 25.0. 5000 L/s × 25.0 = 125,000 W. The balance of the solution is the same as above using Equation 11.11.

18

Recording TAB Data

18.1 Report Form Data

18.1.1 Accuracy of data

To record and interpret the testing, adjusting, and balancing (TAB) data properly, it is necessary to use a complete set of well-designed TAB report forms. Such forms are an essential part of the test report and an important part of the project's history. They are the only method of properly enforcing the specifications and ensuring reliable feedback of empirical data. For the HVAC design engineer or system designer, the data can be essential. If all field data and measurements are reliably assembled and presented, the design engineer will know the job was installed in accord with the design. But even more important, with complete and accurate report data, the engineer can verify the design. The HVAC design engineer can review and analyze system design operation, accumulate a wealth of practical knowledge from it, and understand what may need to be changed in future specifications.

18.1.2 Report forms

ASHRAE Standard 111, *Practices for Measurement Testing, Adjusting, and Balancing of Building Heating, Ventilating, Air-Conditioning, and Refrigeration Systems,* does not have sample report forms, but it does have information and data needed to develop forms for specific equipment reports.

Good TAB report forms organize the report data in such a manner that they facilitate the test and balance function itself and guide the technicians through their tasks in a logical succession of steps, avoiding omissions and errors and simplifying calculations. As part of the operating and maintenance instructions, which are turned over to the owner with the job, the TAB reports become a valuable part of the permanent record of the system operating conditions and serve as a handy reference for the lifetime of the job.

Accuracy in preparing the final TAB report forms is important for several reasons:

1. The reports provide a permanent record of actual system operating conditions after the last adjustments have been made.
2. They confirm that the prescribed TAB procedures have been executed.
3. They serve as a reliable reference that can be used by the owner for system maintenance.
4. They provide the system designer with an aid for diagnosing any problem areas.

18.1.3   NEBB TAB report forms

NEBB TAB report forms shown in most NEBB publications are copyrighted. They are available for purchase and use only by those firms certified by NEBB. CAUTION: **Firms not certified by NEBB that reproduce any NEBB forms or use preprinted forms from NEBB programs for any purpose, will be subject to copyright infringement actions by NEBB.** Generic TAB forms used in examples in this chapter are for instruction and guidance only, and they may be copied.

18.1.4   Preparing TAB reports

There are several things to remember when preparing TAB report forms:

1. A TAB report must be understandable simply by review, so all potential questions regarding the report must be answered by notes or schematics.
2. The TAB report must be complete. Any "loose ends" will only result in costly return trips to the project to obtain missed data.
3. All spaces on a TAB report should have some notation. If entries are not required, then N/A for "not applicable" or a dash should be inserted in the space.
4. Computer forms should be "custom designed" to suit the equipment data and the required TAB procedures.
5. All TAB forms should include project identification, HVAC system and/or unit location, the data gathering technician's signature or

initials, date of TAB work, and remarks, which include system deficiencies and items outside design tolerances.

18.1.5 TAB report submittal

The TAB firm should prepare bound copies of a *professional looking* TAB report as required by the TAB specifications to include, but not be limited to, the following:

1. Title page with project name and address; TAB firm name, address and phone number; date; TAB supervisor's name and signature who approved the report.

2. Table of contents page.

3. A listing of all systems (air and hydronic) balanced, with those systems highlighted that were found to be performing outside of design tolerances. The review sheet should also list any system deficiencies (defective thermostats, loose dampers, etc.) noted during the TAB procedures.

4. System schematic diagrams should be furnished for all multiple outlet air and hydronic systems (see sample air system in Figure 18.1).

5. All Pitot tube traverse report forms should be included along with calculations of average velocity from velocity pressure conversions, airflow volume, static pressure, and ambient conditions.

6. Final equipment test report forms with design data, intermediate test data when required, and final test data.

7. Include under "Remarks," an explanation of variances of final data from design values.

8. A listing of all test instruments and calibration dates. List all of the instruments that have been used on the project during the course of the TAB work on an "Instrument Calibration Report" form. This includes flow-measuring hoods and other related devices.

18.2 Use of TAB Report Forms

The following brief explanations are for suggested TAB forms of major HVAC equipment. Report forms should include airflow data, hydronic data, electrical data, motor and/or drive data, and accessory data (such as filters) as they apply to the equipment. All items should be numbered consistent with the contract documents. All unnumbered terminal units or devices, such as grilles and diffusers, should be numbered and shown on a schematic drawing (see Figure 18.1). All forms

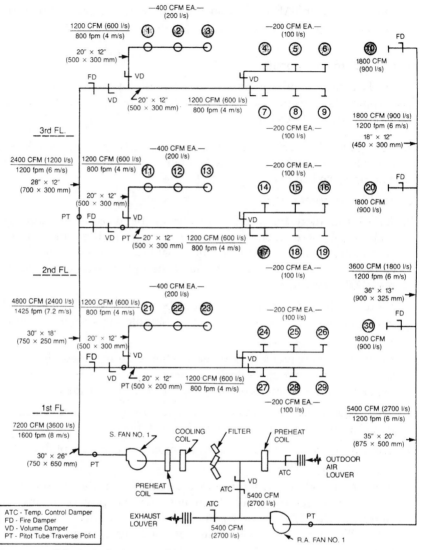

Figure 18.1 Schematic duct system layout.

should contain design data and actual "from the job" motor and equipment data.

18.2.1 Air apparatus test report

The performance or air handling apparatus with heating and/or cooling coils is to be reported on this report form. In addition, there should be space for other information that will be of benefit to the design engineer, the maintenance personnel engineer and the TAB firm. Motor voltage and amperage for three-phase motors should be reported for all three legs (T_1, T_2, T_3). If the design engineer did not specify a design quantity for any item in the test data section, place an X in the slot for the design quantity and record the actual quantity. However, if the equipment manufacturer has furnished ratings, enter them in the design columns. If motor ratings differ from design, provide an explanation at the bottom of the page under "Remarks." Figure 18.2 shows a sample metric form.

Be sure all pertinent data are included on the test report forms. If test data cannot be obtained or are not applicable, indicate such on the report forms. Indicate on the test report or on a separate fan test report form how the actual airflow was obtained (Pitot tube traverse, total of outlet airflows, or a combination).

18.2.2 Apparatus coil test report

This report form is to be used for recording performance of chilled water, hot water, steam, or direct expansion (DX) coils, and for run-around heat recovery systems. The performance of several coils or run-around systems may be shown on the same sheet (Figure 18.3).

18.2.3 Gas- or oil-fired apparatus test report

Data for gas- or oil-fired devices, such as unit heaters, duct furnaces, etc., should be recorded on this report form. This report is not intended to be used in lieu of a factory startup equipment report, but may be used as a supplement. All available design data should be reported. "HP/RPM, F.L. AMPS/S.F. (Service Factor), Drive Data" information should be included to apply to the burner motor, burner fan motor, unit air fan motor, etc., depending on the application or equipment. However, be sure to designate the motor of the recorded data.

18.2.4 Electric coil or duct heater test report

A separate report form should be used for electric furnaces or for electric coils installed in built-up units or in branch ducts. Show the manufacturers recommended minimum airflow velocity.

18.2.5 Fan test report

A fan test report form (see Figure 18.4) may be used with supply air, return air, or exhaust air fans. Since housings for various types of fans may have many different shapes and arrangements, not all entry blanks will be needed for testing a particular fan. The performance of

AIR APPARATUS TEST REPORT

PROJECT **Fawnbrook College** SYSTEM/UNIT **A/C C-1**
LOCATION **Bldg C   Fan room**

UNIT DATA	
Make/Model No.	Carrier 39E
Type/Size	MB / B32
Serial Number	1 X 10361
Arr./Class	Draw thru / II
Discharge	BH
Make Sheave	Browning
Sheave Diam/Bore	506/50 mm
No. Belts/make/size	3/BFG/B1700
No. Filters/type/size	14 RM 600×375

MOTOR DATA	
Make/Frame	G.E. 254T
H.P./RPM (W)	11.2 kW/1440
Volts/Phase/Hertz	415/3/50
F.L. Amps/S.F.	20.8/7.6 - 1.15
Make Sheave	Browning
Sheave Diam/Bore	200/44 mm
Sheave ₵ Distance	462 mm ₵
Sheave Oper. Diam.	

TEST DATA	DESIGN	ACTUAL
Total CFM (l/s)	8360	8420
Total S.P. Pa	825	630
Fan RPM	775	670
External S.P.		
Motor Volts $T_1 T_2, T_2 T_3, T_3 T_1$	415	420/420/440
Motor Amps $T_1 T_2 T_3$		19/18.7/18.9
Outside Air CFM (l/s)	675	700
Return Air CFM (l/s)	7685	7720

TEST DATA	DESIGN	ACTUAL
Discharge S.P. Pa		450
Suction S.P. Pa		180
Reheat Coil △ S.P.		
Cooling Coil △ S.P. Pa		150
Preheat Coil △ S.P.		
Filters △ S.P.		
Vortex Damp. Position		
Out. Air Damp. Position		8% open
Ret. Air Damp. Position		Max.

REMARKS:

TEST DATE **2/10/96** READINGS BY

Figure 18.2 Sample AHU report form (metric).

APPARATUS COIL
TEST REPORT

PROJECT____Fawnbrook College____

COIL DATA	COIL NO.		COIL NO.		COIL NO.		COIL NO.	
System Number	C-1		C-1					
Location	AHU C-1		AHU C-2					
Coil Type	W		W					
No. Rows-Fins/In.	4/0.55		6/0.55					
Manufacturer	Carrier		Carrier					
Model Number	39 EB 32		39EB26					
Face Area, Sq. Ft. (m²)	2.94		2.43					
TEST DATA	DESIGN	ACTUAL	DESIGN	ACTUAL	DESIGN	ACTUAL	DESIGN	ACTUAL
Air Qty., CFM (l/s)	7520	7600	6660	6710				
Air Vel., FPM (m/s)	2.36	2.39	2.52	2.54				
Press. Drop, In.w.g (Pa)	150	153	250	252				
Out. Air DB/WB								
Ret. Air DB/WB								
Ent. Air DB/WB °C	25.2°/18.0		25.0/18.3		28.6/20.4		23.9/20.0	
Lvg. Air DB/WB °C	12.2/11.1		13.3/12.8		10.1/10.0		10.0/9.4	
Air △ T °C	13.0	11.7	18.5	13.9				
Water Flow, GPM (l/s)	4.23	4.04	8.58	9.47				
Press. Drop, PSI (kPa)	19.3	17.3	16.6	20.0				
Ent. Water Temp. °C	5.6	5.6	5.6	5.6				
Lvg. Water Temp. °C	14.9	12.8	11.7	11.1				
Water △ T °C	9.3	7.2	6.1	5.5				
Exp. Valve/Refrig.								
Refrig. Suction Press.								
Refrig. Suction Temp.								
Inlet Steam Press.								

REMARKS:

TEST DATE__2/10/96__ READINGS BY_____

Figure 18.3 Sample coil report form (metric).

up to three fans may be reported on the sample form shown in Figure 18.4.

18.2.6 Rectangular duct traverse report

This report form should be used as a work sheet for recording the results of a Pitot tube traverse in a rectangular duct. It is recommended that the velocity pressures be recorded in each of the spaces provided and converted to velocities in an adjacent space at a later time. The velocities shall be averaged (not the velocity pressures).

Information for both rectangular and round duct traverses may be found in Chapter 7.

18.2.7 Round traverse report

Record the results of a Pitot tube traverse in a round duct on this work sheet–type report form (see Figure 18.5). Spaces shown are for velocity pressures and velocities taken at points across two diameters of the duct. Instructions for making the traverse may be found in Chapter 7.

18.2.8 Air outlet test reports

A sample air outlet test report form is shown in Figure 18.6. Because these report forms can be used as both work sheets and final report forms, the TAB technician is encouraged to record all readings on these test report forms. However, it is not necessary to record preliminary velocity readings on the final forms unless specified by the design engineer. If more than one set of preliminary readings is necessary or required, the data can be entered in the blank column between "Preliminary" and "Final." The "outlet number" refers to numbers similar to those assigned on the schematic layout of Figure 18.1. "Minimum" columns for the recommended minimum settings of VAV boxes should be added to the forms when needed.

Confirm that the air outlet test reports are complete with all applicable k or A_k factors and terminal device sizes. If flow measuring hoods are utilized for outlet readings, indicate their use in the remarks column (and on the "Instrument Calibration Report").

If the final adjusted airflow of any outlet varies by more than $\pm 10\%$ from the design airflow, a note should be placed in the "remarks section" indicating the amount of variance. The "remarks section" at the bottom of the sheet should be used to provide known or potential reasons for such deviation.

FAN TEST REPORT

PROJECT _Fawnbrook College_

FAN DATA	FAN NO. C - 1	FAN NO.	FAN NO.
Location	Fan room		
Service	Exhaust		
Manufacturer	Trane		
Model Number	BCL 1225		
Serial Number	19620-AA		
Type/Class			
Motor Make/Style	GE		
Motor H.P./RPM/Frame (W)	0.56kW/1440		
Volts/Phase/Hertz	415/3/50		
F.L. Amps/S.F.	1.6/N.A.		
Motor Sheave Make/Model	Browning		
Motor Sheave Diam./Bore	100/15 mm		
Fan Sheave Make	Browning		
Fan Sheave Diam./Bore	78/25 mm		
No. Belts/Make/Size	1/Gyr/1000L		
Sheave ₵ Distance	250mm ₵		

TEST DATA	DESIGN	ACTUAL	DESIGN	ACTUAL	DESIGN	ACTUAL
CFM (l/s)	720	700				
Fan RPM	1449	1450				
S.P. In/Out						
Total S.P. _Pa_	220	227				
Voltage $^{T_1 \cdot T_2}_{T_3 \cdot T_1}$ $^{T_2 \cdot T_3}$	415	420/420/420				
Amperage T_1 T_2 T_3	1.6	1.2/1.2/1.1				

REMARKS:

TEST DATE _2/12/96_ READINGS BY _JB_

Figure 18.4 Sample fan report form (metric).

ROUND DUCT TRAVERSE REPORT

PROJECT _____ SYSTEM/UNIT _____

LOCATION/ZONE _____ SERVICE _____

ALTITUDE _____ DENSITY _____ CORR. FACTOR _____

DUCT			REQUIRED		ACTUAL	
S.P. _____ AIR TEMP _____ °F(°C)			SCFM (sl/s) _____		SCFM (sl/s) _____	
SIZE _____ AREA _____		FPM (m/s) _____	CFM (l/s) _____	FPM (m/s) _____	CFM (l/s) _____	

(SEE REVERSE SIDE FOR INSTRUCTIONS)

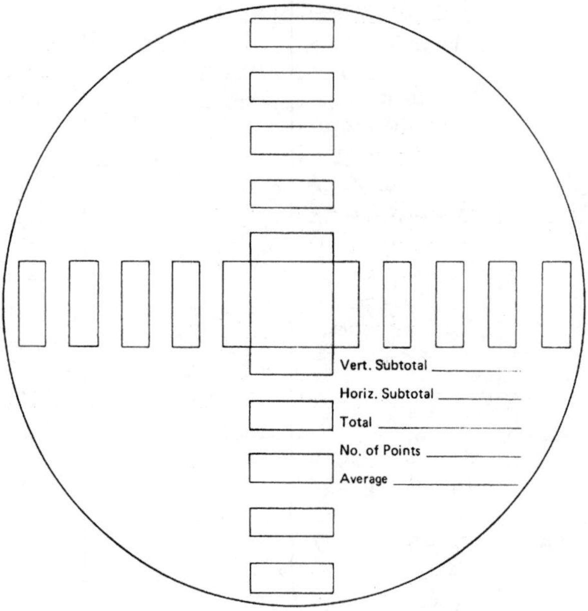

Vert. Subtotal _____

Horiz. Subtotal _____

Total _____

No. of Points _____

Average _____

REMARKS:

TEST DATE _____ READINGS BY _____

PAGE _____ OF _____

Figure 18.5 Sample round duct traverse report form.

AIR OUTLET
TEST REPORT

PROJECT _Fawnbrook College_ SYSTEM _C - 4_

OUTLET MANUFACTURER _Titus_ TEST APPARATUS _Alnor Vel. & flow hood_

AREA SERVED	OUTLET				DESIGN		PRELIMINARY				FINAL		PERCENT OF DESIGN
	NO.	TYPE	SIZE (mm)	AK	VEL m/s	AIRFLOW L/s	VEL m/s	AIRFLOW L/s			VEL m/s	AIRFLOW L/s	
Box C-1													
	1	SL	1200	29.4	3.4	100	3.5	103	\		3.5	103	
	2	"	"	"	"	"	3.3	97	\		3.3	97	
	3	"	"	"	"	"	3.7	109	\		3.7	109	
	4	"	"	"	"	"	3.6	106	\		3.7	109	
Box C-2													
	1	AL	300	45	4.0	180	4.2	189	\		4.1	185	
	2	"	"	"	"	"	4.7	212	\		4.3	194	
	3	"	450	100	3.6	360	3.0	300	\		3.5	350	
	4	"	"	"	"	"	3.3	330	\		3.5	350	
	5	"	"	"	"	"	3.7	370	\		3.6	360	
Box C-3													
	1	AS	600	180	4.2	785	3.5	702	\		4.1	738	
	2	"	"	"	4.2	"	3.8	684	\		3.9	702	
	3	"	"	"	4.2	"	4.5	810	)		4.3	774	
	4	"	"	"	4.7	846	4.7	846			4.5	810	*

REMARKS: **✱ Damper full open**

TEST DATE _2/14/96_ READINGS BY _[signature]_

PAGE _45_ OF _67_

Figure 18.6 Sample air outlet test report form.

18.2.9   Terminal unit coil check

This report form is used as a work sheet to check the water coil of terminal units. Any of the methods for determining water flow or heat transfer should be indicated on the test report form.

18.2.10   Packaged chiller test report

Use this report form as a check sheet to record the control settings and the entering and leaving conditions at the chiller. Since TAB firms normally are not responsible for startup or the proper operation of the machine, this form does not attempt to indicate the performance or efficiency of the machine except as may be determined by the design engineer from the data contained therein.

The TAB report form or the equipment manufacturer's form should be substantially completed and verified by the manufacturers' representatives and/or the installing contractor *before* the HVAC distribution systems are balanced. Temperature and pressure readings of the chiller unit evaporator and condenser should be entered during the TAB procedures. The pressures should be measured by refrigeration technicians furnished by the installing contractor.

18.2.11   Package HVAC unit test report

Test data from package units of all types may be recorded on this report form, with most of the data being furnished and verified by the installing contractor. If the unit has components other than the evaporator fan, DX coil, compressor, and condenser fan(s), use separate test report forms such as:

A form for water or steam coils

A form for direct fired heaters

A form for electric coils

A form for return air fans

18.2.12   Compressor test report

The same comments apply to this report form as for the packaged chiller test report form (see 18.2.10). This form may also be used to have the installing contractor record data for the refrigerant side of unitary systems, "bare" compressors, separate air-cooled condensers, or separate water-cooled condensers.

18.2.13   Cooling tower or evaporative condenser test report

This report form should be substantially completed and verified by the installing contractor before the system is balanced. A "pump data" section should be used for the recirculating pump in evaporative condensers. For the system pump used with cooling towers, use a separate pump test report form.

18.2.14   Heat exchanger or converter test report

This report form should be designed to record final conditions for steam or hot water heat exchangers.

18.2.15   Pump test report

Final data for each pump performance should be recorded on this form. The actual impeller diameter entry is that indicated by plotting the head curve or by actual field measurement where possible. Net positive suction head (NPSH) is an important item for pumps in open circuits and for pumps handling fluids at elevated temperatures (see Chapter 15).

Confirm that all test data have been properly entered on the test report form. Attach the manufacturer's pump capacity curves, with the actual pump operating point plotted, to the test report form when available. Note how the actual pump flow rate was determined (flow meter, pump curve, etc.)

18.2.16   Balance valve or flow meter test report

This report form should be used for recording data from balance valves or flow meters in hydronic systems.

18.2.17   Boiler test report

This report form may be used by the installing contractor to verify substantially the data on this test report, particularly when factory startup services are involved. A flue gas analysis is beyond the scope of TAB procedures, but data could be added in the "Remarks" section if available and required by the design engineer.

18.2.18   Instrument calibration report

This report form should be used for recording the application and date of the most recent calibration test or calibration for each instrument used in the testing, adjusting, and balancing work.

19

Basic Electrical Data

19.1 Basic Electricity

19.1.1 Current flow

It is important to know the definitions of *volt, amp,* and *ohm.* Notice the similarities of electrical systems to water and air systems; each has pressure, flow, and resistances, only in electrical systems it is a flow of electrons rather than water or air. *Direct current* (DC) flows very much as a fluid in an open hydronic system (in one end of a circuit and out the other end). *Alternating current* (AC), when viewed as a hydronic system, would reverse its flow in the conduit 120 times per second, and flow in the same direction 60 times per second (60 Hertz or cycles per second) [50 times per second (50 Hz) in some countries].

Equation 19.1 (Ohm's Law)

$$E = IR \qquad \text{or} \qquad I = E/R$$

Equation 19.2

$$P = EI$$

Where: E = volts (V)
I = amperes (A) (current flow)
R = ohms (Ω) (resistance)
P = watts (W) (power)

From Equation 19.1, one can see that when the voltage (E) is constant, an increase in resistance (R) will reduce the current flow (I) or vice versa, similar to hydronic flow. However, unlike the water pipes, the wire will make an attempt to handle the flow; and in extreme instances, it will become red hot and glow (as in electric resistance heaters). This is the reason that fuses and circuit breakers are used to protect the wiring system and equipment from being overloaded.

19.1.2 Resistance

Parallel electrical circuits resemble HVAC terminal units piped in parallel circuits. Using a simple circuit with two units and one pump, it is known that the water flow will split in accordance with the resistance across each unit. If both resistances are the same, the flow will split 50–50.

Equation 19.3 (Parallel Circuits)

$$1/R_t = 1/R_1 + 1/R_2 + 1/R_3 + 1/R_n$$

Where: R_t = total system resistance (ohms)

R_n = individual resistances (ohms)

Equation 19.3 states mathematically that the parallel current flow will work similar to hydronic flow, with the circuit with the highest resistance receiving the lowest flow.

Equation 19.4 (Series Circuits)

$$R_t = R_1 + R_2 + R_3 + R_n$$

Where: R_t = total system resistance (Ohms)

R_n = individual resistances (Ohms)

Resistances are added together with electrical circuits in series as in hydronic circuits. As more resistances are added, the flow becomes less and less.

The electrical diagram in Figure 19.1 is similar to a piping circuit with a pump at E, two chillers in series at R_1 and R_2, seven terminal units piped in parallel, and a strainer, valve, etc., piped in series in

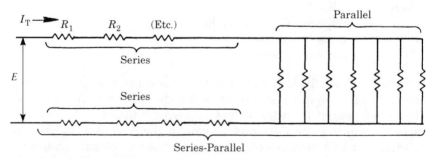

Figure 19.1 Series–parallel circuit.

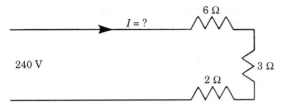

Figure 19.2

the pump suction piping. This shows the similarity of electrical calculations to those for hydronic and air systems, where resistances in series are added and those in parallel are combined.

Example 19.1   Resistors of 6 ohms, 3 ohms, and 2 ohms are connnected across a 240 volt line. Find the current flow in amps if the three are connected in series (Figure 19.2).

Solution   Using Equation 19.4:

$$R_t = R_1 + R_2 + R_3$$

$$R_t = 6\ \Omega + 3\ \Omega + 2\ \Omega = 11\ \Omega$$

Using Equation 19.1:

$$I = \frac{E}{R}$$

$$I = \frac{240\ \text{V}}{11\ \Omega} = 21.8\ \text{amps}$$

Example 19.2   If the three resistors in Example 19.1 are connected in parallel, find the current flow (Figure 19.3):

Solution   Using Equation 19.3:

$$1/R_t = 1/R_1 + 1/R_2 + 1/R_3$$

$$1/R_t = 1/6 + 1/3 + 1/2 = 1$$

$$R_t = 1\ \Omega$$

Using Equation 19.1:

$$I = \frac{E}{R} = \frac{240\ \text{V}}{1\ \Omega} = 240\ \text{amps}$$

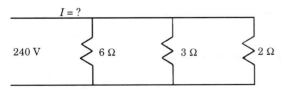

Figure 19.3

Example 19.3 If the 3 Ω and 6 Ω resistors are in parallel with each other, but in series with the 2 Ω resistor (from Example 19.1), find the current flow (Figure 19.4):

Solution Solving the parallel circuit first:

$$1/R_t = 1/6 + 1/3 = 1/2$$

$$R_t = 2\ \Omega$$

Solving the series circuit (R_t from above now becomes R_1):

$$R_t = R_1 + R_2 = 2\ \Omega + 2\ \Omega = 4\ \Omega$$

$$I = \frac{240\text{ V}}{4\ \Omega} = 60.0\text{ amps}$$

Example 19.4 Using the information from Example 19.1, (a) how much power is delivered to the circuit, and (b) how much power is delivered to the 3 ohm resistor?

Solution Using Equation 19.2:
a. $P = EI$
   $P = 240\text{ V} \times 21.8\text{ A} = 5232\text{ W}$
b. $P = EI$ and $E = IR$, so $P = I^2R$ (the amperage remains the same)
   $P = (21.8\text{ A})^2 \times 3\ \Omega = 1426\text{ W}$

19.1.3 Voltage

A *measured voltage* may not be exactly one of the values of the voltages indicated. Voltages can vary, and in normal situations a variation of ±10% does not adversely affect equipment operation. When voltage readings are taken with a volt meter there is no apparent way to tell the difference between 220 volt single-phase circuits and 220 volt three-phase circuits. When measurements are taken, it is found that voltages do vary somewhat; that three-phase circuits are usually 220 volt, and that single-phase circuits are usually 230 volt. However, **phasing cannot be determined just by voltage readings**.

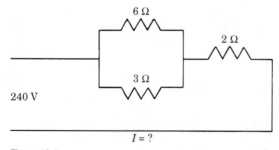

240 V

$I = ?$

Figure 19.4

19.2   Electric Power

19.2.1   Electric service

The standard electrical service in the Unites States is alternating current (AC) at 60 hertz (Hz), formerly known as cycles per second (cps). The two most common types of service normally used with HVAC systems are single-phase and three-phase circuits.

Single-phase, three-wire circuits are used for most residential services as well as for light commercial services. For heavier commercial or industrial systems, or where even larger motors are used, three-phase, three- or four-wire service is provided.

19.2.2   Single-phase circuits

A *measured voltage* may not be exactly one of the values of voltages indicated in Figure 19.5. Voltages can vary, and in normal situations a variation of ± 10% does not adversely affect equipment operation.

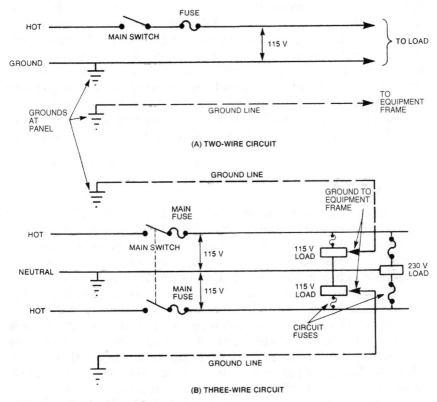

Figure 19.5   Single-phase AC service.

The basic 115 V two-wire circuit shown in Figure 19.5(A) is very common. There is a potential or *pressure* of 115 V between the hot wire and the neutral or ground. This service is obsolete but is still found in some areas.

The *neutral* or *ground wire* is another matter. The neutral normally has no voltage potential. Theoretically, if the neutral contacts a pipe or a person, nothing will happen. The neutral is connected to the generator. The word *theoretically* is used because in actual field conditions, stray currents can find their way into the neutral and then it can become dangerous. A neutral should be treated with the same respect as a known hot wire.

Figure 19.5(B) shows another common single-phase (1ϕ) circuit, which serves homes or buildings that require greater power with items such as ranges, clothes driers, and central air conditioners. This circuit represents the type of three-wire service normally entering a modern residence. Two of the three wires are "hot" wires and one is "neutral." The voltage potential between either of the hot wires and the neutral is the same 115 V previously discussed. There are actually three circuits, two separate 115 V circuits (from each of the hot wires to the neutral) and a 230 V circuit (between the two hot wires). The neutral in a 230 V connection serves as a ground only for safety and is not connected as part of the power circuit. It is connected to the frame of the machine, to carry off any stray currents or short circuits resulting from failures. **Ground or neutral wires are never switched or fused**. The main advantage of the 230 V, two–hot wire circuit is that it allows each of the hot wires to carry half of the current flow. Therefore, twice the current will be handled by the same wire sizes.

19.2.3   Three-phase circuits

The three-phase (3ϕ) concept is somewhat more difficult to understand. In the case of the single-phase, three-wire circuit, two different electrical pulses are being sent down two different hot wires. After one starts, the second starts 1/120th of a second later. These pulses continue indefinitely at the same frequency and having the same *phase relationship* between the two wires. This can be thought of as plus 115 V and minus 115 V between the hot wires and the neutral wire.

Applying the same analysis to three-phase circuits, the 3ϕ generator sends a pulse down an additional (third) wire. There are now three combination voltage–current pulses going down each of three wires. After the first set of pulses start in the first wire, the second wire set of pulses starts 1/180th of a second later, and the third wire set of pulses starts 1/180th of a second after that. Each set of pulses will be

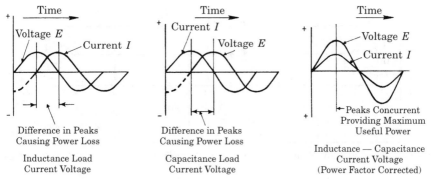

Figure 19.6 Current and voltage–time curves and power factor.

1/60th of a second long and each of the three wires will be out of phase with each other by one third of a pulse.

In large buildings, the use of many large motors running under light loads and other electrical loads can cause these pulse sets to get out of phase (Figure 19.6). This causes what is known as a *low power factor*. Utility companies penalize the user for this low power factor by increasing the charge per kilowatt of electrical power used. TAB technicians should not get involved in power factor corrective actions, but they should be aware of the problem.

Three-phase circuits are shown in Figures 19.7, 19.8, and 19.9. Three-phase circuits will have three hot wires with equal voltages, usually 208, 230 or 460 V between each "leg." A fourth neutral wire may be included to provide single-phase 110, 120, or 277 V circuits for lighting and small power applications.

When voltage readings are taken with a volt meter, there is no apparent way to tell the difference between 220 V single-phase circuits and 220 V three-phase circuits. When measurements are taken, it is found that voltages do vary somewhat; that three-phase circuits usually measure 220 V, and that single-phase circuits usually measure 230 V. However, **the type of current phase cannot be determined from measured voltage readings**.

The voltage readings between any two of the three "hot legs" of three-phase equipment should essentially be the same. Variations ex-

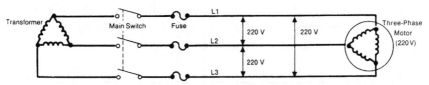

Figure 19.7 220 V three-wire delta three-phase circuit (415 V in Australia).

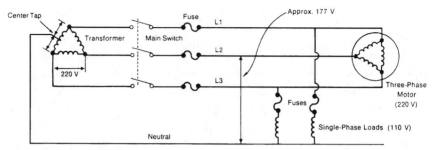

Figure 19.8 220 (415) V three-wire delta three-phase circuit with 110 (240) V single-phase supply.

ceeding ±2% could damage equipment and should be reported to the proper authorities (after the equipment is turned off) so that corrections can be made.

19.3   Electric Wiring

19.3.1   Wire

Wire used in electric circuits can be made from many different materials. When electricity (electrons) flows easily in the material, it is called a *good conductor*. If it is difficult to pass an electric current through a material, it is known as an *insulator*.

Some good conductors of electricity are copper, silver, and aluminum. Most wire is made from copper because it is a good conductor and can be purchased at a reasonable price. Silver is a better conductor of electricity than copper, but its higher cost usually prevents it from being used as ordinary wire.

Copper wire is used the most, whereas aluminum wire may be used for high-voltage lines that carry electricity long distances from power plants to users and for some local wiring. The wiring used in HVAC work and in the service connections to equipment can be a single solid

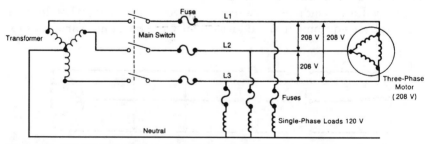

Figure 19.9 120/208 V four-wire wye circuit (240/415 V in Australia).

wire or an encased wire made up of many strands of smaller wires. Stranded wiring normally is more flexible.

Wire with braided covering, called *shielded cable*, is used extensively in control and communications work. A *coaxial cable* uses two conductors, one surrounding the other. The outer braided conductor shields magnetic fields from the inner conductor.

19.3.2 Wire sizes

The size of electrical wires is given by a number ranging from 18 to 2000. This numbering system is known as the American Wire Gauge (AWG), which gives specific information about the various sizes of wire.

Sizes from 18 to size 0, the larger the wire number, the smaller the diameter of the wire; then the opposite is true from size 00 to size 2000. It is important that the proper size of wire be selected for the job.

Round wire is measured in *circular mils*. A mil is a one thousandth of an inch (0.001 in.). A circular mil is the cross-sectional area of a wire that has a diameter of 1 mil (1 mil = 25 μm).

Circular mils are easier to visualize through an example. Five (5) squared is equal to 25. Therefore, a wire 5 mils in diameter would have a cross-sectional area that would contain 25 circular mils. Another wire has a diameter of 8 mils. Its area then would be 64 circular mils.

There is a good reason for converting the diameter of wire into circular mils. Working with the cross-sectional area of the wire makes it possible to select the right size of wire to carry safely the prescribed number of amps in the circuit.

19.4 Transformers

Look at the transformer diagram illustrated in Figure 19.10. For the voltage reduction indicated, the number of turns of wire shown on the primary side should be twice the number of turns shown on the secondary side. Voltage is "transformed" by the transformer stepping down the voltage to one half the original voltage (one half of the wire turns). If the primary and secondary connections were swapped, this same transformer could step up the voltage from 440 V to 880 V. The ballasts in fluorescent lights in buildings step up the voltages near 2500 V, the required voltage to produce light in the tubes.

19.4.1 Center taps

The function of the center tap of the transformer is illustrated in Figure 19.11. If a 220 V difference exists between the legs of the second-

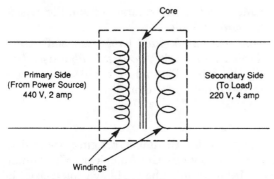

Core

Primary Side
(From Power Source)
440 V, 2 amp

Secondary Side
(To Load)
220 V, 4 amp

Windings

Figure 19.10 Basic transformer.

ary side (half of the wire turns of the primary side), it is logical that a 110 V difference would exist between one leg and a center tap (a fourth of the wire turns of the primary side). Most single-phase residential transformers have high voltages on the primary side, but the secondary voltages use a "center tap" (the ground) to furnish two 110 V circuits along with the 220 V power (Figure 19.11). This size transformer, when used with overhead services, looks like a large can and usually is attached to a pole near the residence. It can supply power to several residences or buildings, or just to a single building. Transformers for use with underground power services usually are found in square metal boxes sitting in yards.

19.4.2 Large transformers

Larger transformers are ground mounted and, if outside, are usually in a large metal housing or a wire fence enclosure. These transformers can be either single phase or three phase.

19.4.3 Small transformers

Many small transformers, such as 110 V/24 V or 220 V/24 V, are used in control circuits for HVAC equipment. When working in electric

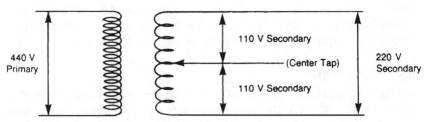

440 V
Primary

110 V Secondary

(Center Tap)

110 V Secondary

220 V
Secondary

Figure 19.11 Transformer with tapped secondary.

panels, TAB technicians must be aware that transformers often are powered from external electrical circuits and may be "hot," even though the main power switch has been pulled for the equipment.

19.5 Motors

19.5.1 Types of motors

Motors used on HVAC equipment are designed for alternating current, except in rare cases. Most small motors will use single-phase current, whereas the larger motors will use three-phase current. However, some rural areas have only single-phase current. There are many different motor speeds, but approximately 1800 rpm and 3600 rpm are the most common. The actual speed of the motor will vary with the load imposed. *Split-phase, capacitor start, synchronous, induction, shaded-pole*—all are part of the many different types of motors that the TAB technician will need to know (Table 19.1). The characteristics of each is important for troubleshooting, as the wrong type of motor often is used. The TAB team also should be aware of special applications, such as *explosion proof* or *totally enclosed* motors.

19.5.2 Rotation direction

Motors rotating in the wrong direction are a common occurrence when a new system is first started. The normal TAB procedures deal with this situation, as correct motor rotation is vital to the performance of the unit. The direction of the motor usually is changed in three-phase motors by switching any two of the three phase power-wires. In single-phase motors, the change of direction is accomplished by switching two of the internal motor leads that connect to the motor line terminal lugs.

A *word of caution*: Certain fans and most pumps will develop measurable pressures and some fluid flow when the rotation is incorrect. Rotation arrows can be found on many types of equipment. Correct rotation is obvious on some units. Flow and amperage readings also can be used to determine whether something is amiss. Whenever a piece of equipment does not perform as specified and the current flow is much lower than design, rotation is one item to be checked.

19.5.3 Motor power equations

Single-Phase Circuits

Equation 19.5 (U.S.)

$$\text{BHP} = \frac{I \times E \times PF \times \textit{Eff}}{746}$$

Equation 19.5 (Metric)

$$BP = \frac{I \times E \times PF \times \textit{Eff}}{1000}$$

TABLE 19.1 AC Motor Characteristics

Motor type	HP rating	Speed characteristics	Full voltage		Remarks
			Starting torque	Starting current	
			Polyphase		
Squirrel-cage induction	Small to large	Constant and multispeed	High to normal	Low to normal	Most widely used for constant-speed service
Wound-rotor	All	Constant or variable	High	Low	For applications requiring high starting torque and low starting current, or limited variation in speed control
Synchronous	Medium to large	Strictly constant	Normal to low	Low to normal	For constant-speed service and where power factor correction is required
			Single-phase		
Capacitor-start, induction-run	Small[a]	Constant	High	Normal	General purpose
Capacitor-start, capacitor-run	Small[a]	Constant	High	Low	High efficiency
Split-phase	Fractional	Constant	Normal	Normal	Least expensive of higher starting torque types, general purpose
Permanent split-capacitor	Fractional and small integral	Constant or adjustable varying	Low	Normal	Quiet, efficient; low running current; poor starting torque
Shaded-pole	Fractional	Constant or adjustable varying	Low	—	Inexpensive; poor starting torque; least efficient; high running current

[a]Up to 7.5 HP (5.6 kW).

Three-Phase Circuits

| **Equation 19.6 (U.S.)** | **Equation 19.6 (Metric)** |

$$\text{BHP} = \frac{I \times E \times PF \times \textit{Eff} \times 1.73}{746}$$

$$BP = \frac{I \times E \times PF \times \textit{Eff} \times 1.73}{1000}$$

Where: BHP = brake horsepower (HP)
BP = brake power (kW)
I = amps
E = volts
PF = power factor (decimal value)
$\textit{Eff}$ = efficiency (decimal value)

In Equations 19.5 and 19.6, the power factor and efficiency values must be used to obtain the actual motor brake horsepower. Because these values usually are difficult to obtain, a reasonable estimate can be used. The normal range of both is between 80 and 90%. Therefore, 80% might be used for one value and 90% for the other value to obtain a brake horsepower estimate.

Motor brake power is calculated to verify that the proper size motor has been installed, i.e., that the installed motor is not overloaded and is operating within its service factor. It also is used to determine that the pump or fan is operating with the required efficiencies. The system designer usually has specified the total amount of power or energy that may be consumed to perform a specific function.

The following equations can be used to obtain an accurate (but not exact) brake power by measuring motor amperages and voltages under no load and full load conditions.

Equation 19.7

$$\text{Amps}_R = \frac{\text{Amps}_{FL} \times \text{Volts}_{NP}}{\text{Volts}_R}$$

Equation 19.8

$$BP = \frac{\text{Amps}_R - (\text{Amps}_{NL} \times 0.5)}{\text{Amps}_{FL} - (\text{Amps}_{NL} \times 0.5)} \times P_{NP}$$

Where: BP = Brake power (HP or kW)
Amps_R = Running amps
Amps_{NL} = No-load amps
Amps_{FL} = Full-load amps
P_{NP} = Nameplate power (HP or kW)
Volts_R = Running volts
Volts_{NP} = Nameplate volts

Example 19.5 Find the brake horsepower (kW) of a motor rated at 10 HP (7.5 kW) that is drawing (each leg) 22, 20, and 21 A at 211 V (assume PF and Eff)

Solution (U.S. and Metric) Using Equations 19.6:

$$BHP = \frac{I \times E \times PF \times Eff \times 1.73}{746}$$

$$BHP = \frac{21 \, (Ave.) \times 211 \times 0.8 \times 0.9 \times 1.73}{746}$$

$$BHP = 7.4 \text{ HP}$$

$$BP = \frac{I \times E \times PF \times Eff \times 1.73}{1000}$$

$$BP = \frac{21 \, (Ave.) \times 211 \times 0.8 \times 0.9 \times 1.73}{1000}$$

$$BP = 5.52 \text{ kW}$$

Example 19.6 A 10 HP (7.5 kW) 208 V motor is rated at 24 A full load. It is drawing (each leg) 19, 19, and 20 A at 210 V. Find the BHP (kW) if the no-load amps measures an average of 8.5 A.

Solution Using Equation 19.8:

$$BP = \frac{19.33 - (8.5 \times 0.5)}{24.0 - (8.5 \times 0.5)} \times 10 \text{ (U.S.)}$$

$$BP = \frac{15.08}{19.75} \times 10 = 7.64 \text{ HP (U.S.)}$$

$$BP = \frac{15.08}{19.75} \times 7.5 = 5.73 \text{ kW (metric)}$$

19.6   Motor Controls

19.6.1   Motor overload devices

A simple on–off toggle switch, a safety switch, or an individual circuit breaker in an electrical power panel is *not* an overload protection device for a motor. Many "ordinary looking" toggle switches *do* contain overload protection for smaller single-phase motors. Many small motors do have built-in overload protection and do not need additional protection. The circuit breaker provides overload protection only for the wiring circuits, but not for any connected motor(s).

The electric current to a motor must be switched off and on to stop and start the motor (manually or automatically). The switching device is commonly called a *motor starter*. This is not to be confused with a "safety switch," which is a device that *must* be placed in the "off" position before any work is done on a motor or electrical equipment. This prevents the motor from accidently starting from remote control devices.

If starting current passes through a motor for any length of time, which is considerably greater than the full load rating, the windings will be overheated and damage may occur to the insulation, resulting in a burned up motor. Motors must be provided with protection to prevent this. Many smaller single-phase motors often have built-in protection. A device is included in the motor that senses the overload and breaks the circuit and stops the motor. A manual reset button usually is provided to restart the motor, or it may reset automatically. Most larger single-phase and three-phase motors must have external protection.

The starting current "inrush" may be from 300 to 1000% greater than the full-load amperage (FLA) rating for the motor. It will depend on the type motor, the load, and how long it takes to get up to speed. The starter has to allow for the starting current inrush without tripping out and still provide protection against exceeding the motor FLA during continuous service.

19.6.2 Motor starters

There are a larger number of different types of starters, each with various advantages and limitations. In most cases, a specific type of starter is required by a particular type of motor. For example, a full voltage *magnetic starter* (Figure 19.12) usually is used with an induction motor. *Reduced voltage* or *reduced current* starters, although more expensive than a magnetic starter, often must be used with larger motors to prevent disruption (by producing large drops in line voltage) of marginally adequate power services. There are two basic types of push-button stations used with magnetic starters, the *maintained contact* station and the *momentary contact* station. The important point to remember is that after an interruption of the motor current with

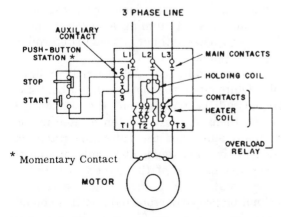

Figure 19.12 Three-wire magnetic starter and control switch.

the momentary contact station, the motor will not restart until the "start" button is pushed.

Motors are protected by putting a small electric heater in the starter in series with the phase line and locating it by a bimetallic, heat-sensitive switch. Since the heaters will take a short time to heat up the bimetallic switch, the motor will have time to get started and up to speed. These *heater coils* are available in different sizes or amperage ratings. They should be matched to the rating of the motor being protected. If the motor is overloaded, the heater coils will remain hot and the bimetallic switch will shortly trip out, breaking the circuit to the magnetic holding coil and shutting down the motor.

When fuses are used with motors, they must be dual-element fuses. These are the only type that allow for large inrush currents without blowing out.

19.6.3 Heater coil sizing

Heater coil sizing is affected by the ambient or surrounding temperature at the starter location. If this surrounding temperature is appreciably higher than the motor operating temperature, one size larger heater may be required. A motor operating right at full-load amperage may go into the allowable *service factor* range occasionally because of low line voltage or intermittent loads. To prevent nuisance tripping, heater coils one size larger may be required. TAB technicians should not replace heater coils with oversize ones unless directed to by the persons responsible for the equipment. **Never jump out or bypass heater coils or fuses in order to run motors.** With no protection you may be held responsible for any malfunctions or damage that may occur.

On larger motors, the starter may have provisions for *reduced voltage starting*. This allows a motor to start running at a lower speed before full voltage is applied. An adjustable timer is used to determine how long the reduced voltage is applied.

The problem most likely encountered by the TAB technician will be improperly sized heater coils. Quite often, the motor will not be drawing its full-load amperage but will be tripping out the starter. All heater coils have a number on them. A chart is often located in the starter cover that tells how many amps various heaters are rated for. The TAB technician should verify that the heater coils are large enough to handle the full-load amperage of the motor. If not, the proper heater will have to be installed. Of course, if the heater coils are too large, the motor will not be properly protected and this should be corrected also.

19.7   Taking Electrical Measurements

19.7.1   Motor starter measurements

The motor starter or the safety switch is the main source of access to motor terminal leads for measurement of voltage and amperage. The starter also can contain holding coils, auxiliary contacts, control transformers, and a push-button station or a "hand–off–automatic" selector switch. This last item is useful in troubleshooting. If the switch is turned to the "hand" position, the motor should run if each phase line is hot, unless there is trouble in the motor. This is because the "hand" portion of the switch bypasses the various controls in this circuit. The "automatic" portion of the switch is connected to the circuit containing auxiliary devices such as thermostats, safety lockouts and other external switches used to control or turn off the motor. If the motor runs on "hand", but not on "automatic," one of the control or safety interlocks usually is open.

19.7.2   Safety

Emphasizing safety, the TAB technician can find many different voltages around starters and starter combinations (Figure 19.13). If a remote room thermostat was added in series with the push-button station of the A unit, the 110 V control circuit might have been required to be 24 V. It is not good practice to use line voltages (110 V or higher) for control circuits, but to save money, 240 V control circuits are not uncommon.

If line phase terminals need to be changed to correct for polyphase motor rotation, the changes can be made either in the starter or at the motor terminal connections. Amperage and voltage readings must be taken "live" in the vicinity of voltages up to 480 V. **Extreme care must be exercised when using the test instrument inside a "live" electrical junction box or starter**. One wrong move could cause serious burns or electrocution.

19.7.3   Amp/volt readings

19.7.3.1   Constant-speed motors. To take amperage readings (see Figures 19.12 or 19.13), the clamp-on ammeter would be placed around line L1, then line L2, then line L3 (one line at a time). The amperages, which are recorded, seldom have the same values, but should be within 10% of one another. The two voltage test probes are placed across the different pairs: L1/L2, L1/L3, and L2/L3. Again, the voltages should be recorded and the values should be within 10% of each

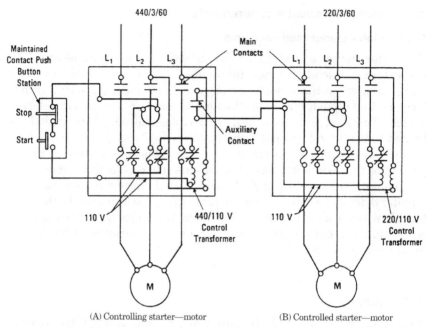

(A) Controlling starter—motor   (B) Controlled starter—motor

Figure 19.13 Interlocked starters with control transformers.

other. When the voltages or amperages are not within 10% of one another, there is an indication of trouble. If the voltage and amperage readings of one phase line is zero, *single phasing* could be occurring, and the equipment must be shut down *immediately*.

Single phasing results when one phase of a polyphase circuit is open (no current). Motors can continue to run under this condition, with a lower power output and possible overheating. Motors usually will not start under this condition but will produce a loud hum.

19.7.3.2 Part-winding motors.   On part-winding, three-phase motors, all six motor leads must be measured, i.e., amps from lines L1 and L4 added, lines L2 and L5 added, and lines L3 and L6 added to obtain the total load.

19.7.3.3 Variable frequency drives. The solid-state components of variable-frequency drives (VFD) and some newer controls can cause distorted amperage readings because the current flow occurs in short pulses instead of the normal sine wave. The average clamp-on volt-ammeter typically will read low. A true RMS (root mean square) type of instrument is required to take these amperage readings.

19.7.4   Root mean square (RMS)

RMS (root mean square) comes from a mathematical equation that calculates the effective or heating value of any alternating current (AC) wave shape. In electrical terms, the alternating current RMS value is equivalent to the DC heating value of a particular wave form, either voltage or current.

For example, if a resistive heating element in an electric furnace is rated at 15 kW of heat at 240 V AC RMS, then the same amount of heat from 240 V of direct current would be obtained.

Electrical power systems components, such as fuses, bus bars, conductors, and thermal elements of circuit breakers, are rated in RMS current because their main limitation has to do with heat dissipation.

If a volt-ammeter is labeled and specified to respond to the true-RMS value of current, it means that the internal circuit calculates the heating value according to the RMS equation. This method will give the correct heating value regardless of the current wave shape.

Some low-cost volt-ammeters, which do not have true-RMS circuitry, use a shortcut method to find the RMS value. These meters are specified to be "average responding–RMS indicating." These meters capture the rectified average of an AC waveform and multiply the number by 1.1 to calculate the RMS value. The value they display is not a true value but is based on an assumption about the wave shape. The average-responding method works for pure sine waves, but can lead to large reading errors when a waveform is distorted by nonlinear loads, such as adjustable-speed drives or computers. Figure 19.14 gives some examples of the way the two different types of meters respond to different wave shapes.

Multimeter Type	Response to sine wave	Response to square wave	Response to single phase diode rectifier	Response to 3 Ø diode rectifier
Average responding	Correct	10% high	40% low	5–30% low
True RMS	Correct	Correct	Correct	Correct

Figure 19.14   A comparison of average-responding and true RMS units.

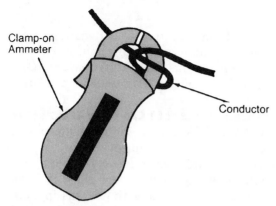

Clamp-on
Ammeter

Conductor

Figure 19.15 Looped conductor doubles ammeter reading.

19.7.5 Volt-ammeter

An accurate clamp-on volt-ammeter, preferable a true-RMS unit, is the electrical testing instrument required by the TAB technician. These devices have several scales for different ranges of volts and amps. Amperage is measured by clamping the probes around a *single* conductor. Figure 19.15 shows a method that doubles the amperage reading when there is a low current flow. This allows a more accurate reading (if there is enough slack in the wire), but the amperage reading *must* be divided by two.

Most TAB report forms require that electrical measurements be supplemented by motor speed (rpm) measurements. See Chapter 8 for data on rotation-measuring instruments.

20

HVAC Control Systems

20.1 HVAC System Control Basics

The proper operation of automatic temperature control (ATC) systems lies with the temperature control contractor and/or the installing mechanical contractor, and it is not the responsibility of the TAB technician. However, TAB technicians should now realize how important good automatic control systems and ATC valves and dampers are to good HVAC system operation. A TAB technician must know how changes in control settings can affect the operation of a unit or of a whole system. Without this knowledge, TAB technicians may not be able to accomplish testing and balancing work, and troubleshooting work may become an impossibility.

For example, using the three-pipe system with terminal units shown in Figure 20.1, each terminal unit is controlled by a three-way "blending" valve. To balance the chilled-water side of this system, all three-way valves must be opened to the chilled-water supply main. These valve position settings then become important to the TAB technician. One valve in a system left indexed to the hot water main could invalidate the entire set of readings, causing the work to be done again. This is but one example to show that a good working knowledge of automatic temperature control systems is necessary for TAB technicians.

Automatic temperature controls are used in HVAC systems to maintain design conditions within an occupied space. Maintaining design conditions involves the control of temperatures, humidities, and duct and space pressures. Other automatic controls provide for the safe and economical operation of the entire HVAC system.

20.1.1 Types of ATC systems

Control systems are classified by the source of power. Four basic types of control systems are used:

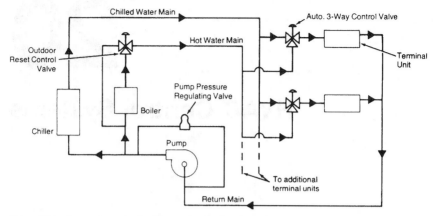

Figure 20.1 Arrangement of a three-pipe hydronic system.

1. Electric control
2. Electronic and/or direct digital control (DDC)
3. Pneumatic control
4. Self-contained controls

20.1.1.1 Electric controls. Electric controls use low-voltage and line voltage current to modulate control devices such as dampers or valves. Electric controls usually are two position and are used in smaller ATC applications. Larger and more complex systems normally use pneumatic, electronic, or DDC control systems.

20.1.1.2 Electronic controls. Electronic control systems use electricity as a source of power. Small electric currents are used to transport signals to amplifiers where the signal is strengthened and used to operate a control device similar to those used in electric systems. Direct digital control (DDC) and energy management control (EMC) systems fall under this category.

20.1.1.3 Pneumatic controls. Compressed air is used as the power source of pneumatic control systems. This system provides proportional, or modulating control of diaphragm- or piston-type control devices.

20.1.1.4 Self-contained controls. Self-contained controls differ from all of the previous control systems in that they do not use an external source of power. Power to control is developed internally, or from pressures within the system. Condenser water regulating valves and thermostatic expansion valves are examples of self-contained controls.

20.1.1.5 Control systems mix. Even though there are four distinct classifications of control systems, it is possible and likely that a project may utilize a combination of the above systems to satisfy the scope of control requirements.

20.1.2 Control loops

No matter which type of control system is used, or whether the purpose is to control cooling, heating, humidity, or pressure, all control applications must involve a fundamental *control loop*. A control loop consists of four components:

1. A controller (thermostat, receiver–controller)

2. A controlled device (valve, damper)

3. A controlled agent (steam, hot or chilled water)

4. A sensing device (transmitter, bimetal strip)

For example, a duct temperature sensing device such as a remote bulb, monitors the temperature of a supply air duct and sends a signal to the controller.

The controller monitors the signal as sent by the sensing device and reacts by either opening or closing a controlled device, such as a steam valve. As a result of more or less of the controlled agent (steam in a heating coil), the action of the controlled device creates a change in the air temperature of the duct and causes the sensing device to change the signal once again to the controller.

The operation of the control loop is never ending during normal operation of the HVAC system.

20.1.3 Types of control action

Controllers (such as thermostats, humidistats, and receiver–controllers) have two possible sets of control actions:

1. Modulating or two position

2. Direct or reverse acting

20.1.3.1 Modulating or two position. *Modulating control* (also called proportional control) is obtained when the control signal sent by the controller to the controlled device is constantly changing in small increments to increase or decrease the capacity of a terminal unit gradually to the load conditions. *Two-position control*, which also can be on–off, may only assume two positions: fully open or fully closed. Two-position controllers are normally used with equipment that may only operate in an on–off application. Equipment such as gas valves or small hermetic air conditioning compressors would fall under this category.

20.1.3.2 Direct or reverse acting. Pneumatic controllers may increase or decrease the output (branch) control pressure with changes in space conditions monitored by the sensing element. A *direct acting* controller will increase the output (branch) control pressure as the controlled variable (temperature, humidity, pressure) increases. A *reverse acting* controller increases control pressure as the controlled variable decreases. The action of the pneumatic controller must be properly matched with the control device or the control loop will produce unexpected results or those opposite of that desired.

The position of a controlled device when deenergized is considered the *normal position*. Control devices (such as valves or dampers) are either *normally open* (NO) or *normally closed* (NC). Some electric devices also contain switches that are normally open or normally closed until moved to the opposite position by a controller.

20.1.3.3 Safety controls. As stated earlier, control systems maintain design conditions within a space but also provide for the safe and economical operation of the HVAC system. The devices discussed up to now are considered operating controls. The other type of control that the TAB technician must be aware of is the *safety* or *limit control*. Safety or limit controls are used to provide safe system operation and may interrupt the operating controls at any given time to ensure safe system operation. Examples of safety controls are freezestats, firestats, flow switches, smoke detectors, and refrigeration high–low pressure cutouts.

20.2 ATC System Components

20.2.1 Control diagrams

In a typical job specification, there are general descriptions of various types of control applications, which the automatic temperature control (ATC) contractor translates into a set of drawings called *control diagrams*. These control diagrams, after approval by the system designer, are used by the ATC contractor for control system installation in coordination with the HVAC system contractor. The data found in these diagrams are extremely important to the TAB technician. These diagrams frequently are the only description of how the HVAC systems will operate.

20.2.2   Troubleshooting with ATC diagrams

Control system diagrams also can be used to assist the TAB technician in troubleshooting. For example, it is found that when the "hand–off–automatic" switch of a fan motor starter is in the "automatic" position, the fan will not run. But the fan will run when the switch is in the "hand" position. This may indicate that some type of automatic temperature control device or safety switch is preventing the fan from running. A review of the system control diagrams and the sequence of operation indicates that there are auxiliary contracts and a low-limit safety control in the fan control circuit. The two devices are inspected to verify if one or both is responsible for the "open" in the control circuit.

The TAB technician needs to have enough knowledge about different types of control systems and to be able to read control diagrams so that an improperly used or an improperly located device can be found before a system is damaged or before time is wasted in an attempt to do the TAB work under fluctuating conditions.

20.2.3   Control relationships

All control system sensors and controllers are generally *linear* (which means "straight line"). Figure 20.2 indicates the error induced by nonlinear control devices such as dampers and valves. Linear control in pneumatic systems may translate to 1° of temperature change from 1 psi (6.9 kPa) of air pressure change. One pound per square inch (6.9 kPa) of fluid pressure change on the discharge side of a pump also could result in a 2 or 3 psi (13.8 to 20.7 kPa) of control pressure change. These systems are linear as long as each increment of "controlled variable" produces the same increment of "signal." The system would be nonlinear if different amounts of signals emanated from a

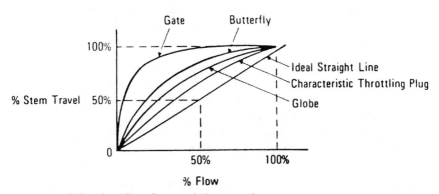

Figure 20.2   Valve throttling characteristic comparison.

fixed increment of the controlled variable. For example, a system is nonlinear if at 68 °F (20 °C) a 1° change produces a 1 psi (6.9 kPa) control signal; but at 95 °F (35 °C) a 1° change produces a 2 psi (13.8 kPa) control signal.

Most actuators are linear; that is, if an actuator has a signal range of 10 psi (69 kPa) from fully open to fully closed, a 5 psi (34.5 kPa) signal will cause a 50% travel. However, if the actuator device is used on a valve or damper, an actuator change of 50% will seldom change the fluid flow by the same 50%. From Figure 20.2 it can be seen that a 50% stem travel of a gate valve from wide open will have little effect on the fluid flow.

20.2.4 ATC valves

The pressure drop across an ATC valve is proportional to the square of the fluid flow rate. This relationship is indicated by the general curve shown in Figure 20.2, which can apply to most systems, although the numbers may vary. The nonlinearity of the controlling device is shown by this curve. To minimize the resulting control inaccuracies, the controller and the controlled device must be carefully matched so that an average linearity is achieved. This cannot be done across the entire range of the device; therefore, the devices are matched for a "normal operating range," which is a matter of judgment of the system designer or the ATC Contractor.

Equation 20.1 (U.S.)   **Equation 20.1 (Metric)**

$$\Delta P = \left(\frac{Q}{C_v}\right)^2$$ $$\Delta P = \left(\frac{Q}{K_v}\right)^2$$

Equation 20.2 (U.S.)   **Equation 20.2 (Metric)**

$$C_v = Q\sqrt{\frac{1}{\Delta P}} = Q\sqrt{\frac{2.3}{H}}$$ $$K_v = Q\sqrt{\frac{1}{\Delta P}}$$

Where: ΔP = pressure differential (psi or kPa)
H = head loss (ft.w.g.)
Q = flow rate (gpm or L/s)
$C_v\ (K_v)$ = valve constant (U.S. or metric)

Example 20.1 (U.S.)   A control valve with a C_v of 40 has a flow rate of 70 gpm. Find the pressure drop across the valve (a) in psi and (b) in feet of water.

Solution   Using Equation 20.1:

(a) $\Delta P = \left(\dfrac{Q}{C_v}\right)^2 = \left(\dfrac{70}{40}\right)^2 = 3.06$ psi

(b) $H = \Delta P \times 2.3 = 3.06 \times 2.3$

$H = 7.04$ ft.w.g.

Example 20.1 (Metric)   A control valve with a K_v of 1.1 has a flow rate of 4.5 L/s. Find the pressure drop across the valve.

Solution   Using Equation 20.1:

$$\Delta P = \left(\frac{Q}{K_v}\right)^2 = \left(\frac{4.5}{1.1}\right)^2 = 16.74 \text{ kPa}$$

20.2.5   Valve and damper categories

ATC valves and dampers generally fall into two categories, *modulating* and *two-position* (open or closed). Modulating means that the device can assume infinite positions of control as required by the sensor.

Three-way valves also fall into two categories, *mixing valves* and *diverting valves*. A *three-way mixing valve*, Figure 20.3(A), has two inlet connections and one outlet connection, and a double-faced disc operating between two seats. It is used to proportionally mix two streams entering through the inlet connections and leaving through the common outlet.

A *three-way diverting valve,* Figure 20.3(B), has one inlet connection, two outlet connections, and two separate discs and seats. It is used to divert the flow to either of the outlets or to proportion the flow to both outlets.

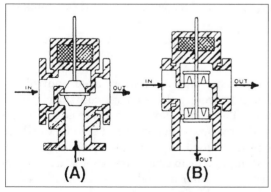

(A)   **(B)**

Figure 20.3 Typical three-way mixing and diverting valves.

Pneumatic valves and valves with spring return electric operators can be classified as *normally open* or *normally closed.*

A *normally closed* valve will assume a closed position when all operating force is removed.

A *normally open* valve will assume an open position when all operating force is removed.

20.2.6 Control system adjustment and calibration

In most control systems, with the exception of some DDC systems, controllers can be adjusted in range, set points, sensitivity, and differential. This often is done at the ATC system control panels, which, like motor starters, are the heart of the control systems. There usually is some calibration and control adjustment in sensing devices and actuators that may be located in other areas. The TAB technician is *not* authorized to make changes to the control system. It would be advantageous to everyone concerned with HVAC systems to have the ATC contractor on the job to make adjustments during the system balancing process. Initial position settings of valves and dampers can be made for the ATC contractor before the TAB work. Coordination and cooperation between the TAB firm and the ATC contractor is the key to a smoothly running, well-balanced HVAC system.

20.3 Direct Digital Controls (DDC)

20.3.1 Introduction

Direct digital controls (DDC) use microprocessors that perform control logic functions instead of conventional components such as thermostats, receiver controllers, and other electromechanical or electropneumatic devices. DDC systems can interface with pneumatic systems but tend to be mostly electronic.

20.3.2 DDC operation

All control logic is performed by microprocessor-based controllers. These controllers receive inputs from field sensors, contact closures, or other types of equipment. The controllers then process this information based upon their software programs and send signals to HVAC equipment. They can be programmed to start and stop equipment, open and close valves and dampers, or execute very sophisticated routines based upon the way they are programmed. These controllers may operate central HVAC systems and/or terminal units.

In addition to HVAC systems and units, DDC systems also can be tied into security, life safety, lighting, and other building systems.

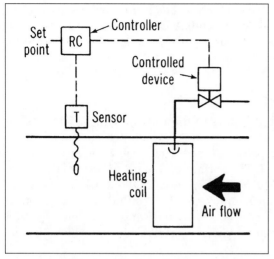

Figure 20.4 Conventional control loop. (Courtesy of Heating/Piping/Air Conditioning)

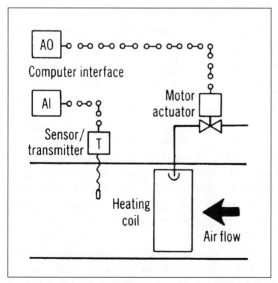

Figure 20.5 DDC control loop. (Courtesy of Heating/Piping/Air Conditioning)

A conventional control loop (electric, electronic, or pneumatic) is shown in Figure 20.4 for a heating coil. A DDC system does the same thing (Figure 20.5) except the computer or microprocessor takes the place of the receiver or controller.

20.3.3 Microprocessor inputs

There are two basic types of inputs to the microprocessor. The first is the *analog input* (AI), which is a variable input such as temperature, humidity, pressure, static pressure, water or airflow, or any variable medium that can be sensed. The second is the *digital input* (DI) or, as called in some literature, the *binary input* (BI). DI and BI are the same and show an on or off, open or closed, flow or no-flow, etc., condition. The on or off signal can be very rapid and is called a pulsed signal, which would simulate a varying rate such as heat flow (kW or MBtu/h).

The digital input (DI) is easily converted to a signal that the computer can understand, because it speaks the same language as the computer, i.e., digital or on or off. The analog input (AI) must be converted from the analog form (i.e., a needle or pointer on a gauge) to a digital form (exact numbers to an assigned decimal accuracy). This is done through an analog-to-digital converter. Assuming that the sensor is accurate, the value of a temperature AI to the DDC system can be accurate to 0.1 °F (0.05 °C).

Once the computer has digitized the very close approximation of the analog value, it can compare it to the set point it retains in memory (in software) to provide a digital output (make a decision) to operate a controlled device (hot water valve) to assure that the actual sensed analog value (temperature) is the same as its set point.

20.3.4 Microprocessor outputs

To convert a digital percentage value to a variable voltage, current or pneumatic air pressure that can automatically operate a hot water valve, a digital-to-analog converter is used to furnish an *analog output* (AO). When examined very closely, the AO is a very finely stepped digital signal. For example, 65.0% of full flow might indicate an analog or current output of 14.40 mA within a signal range of 4 to 20 mA, which is hardly detectable. In some systems where the output is directly to a pneumatic system, the output would be referred to as a *pneumatic output* (PO).

The other type of output, the *digital output* (DO) or in some systems the *binary output* (BO), is easily converted from the computer's digital signal to the on or off, open or closed, etc., signal needed to start or stop a motor or open and close a valve.

20.3.5 DDC example

Using the DDC control loop example in Figure 20.5, for the duct supply air temperature (SAT) there will be one AI, and for the hot water
valve there will be one AO, each wired to the DDC panel. Set point of
the SAT on the heating coil is maintained by the DDC microprocessor,
which continually monitors the SAT, compares the AI SAT sensor
value to the set point, and calculates an AO (makes a decision) to
reposition the hot water valve to allow more, or less, hot water flow
such that the AI SAT sensor value comes into line with the set point.

20.4 Building and HVAC Communications

20.4.1 Protocol standards

There are many DDC systems being used for HVAC systems that cannot interface with each other. Many items of HVAC and electrical
equipment used today come factory assembled with some type of microprocessor control module. Requiring each manufacturer to be compatible with individual proprietary control systems to integrate the
control of all components is becoming increasingly expensive and slow-
ing the pace of technological advancement.

Many in the HVAC industry believe that the development of communication *protocol standards* will solve industry problems with DDC
controls, by simplifying the process by which control systems are designed and procured. A standard protocol will permit designers to
treat DDC components as interchangeable components with a standard or uniform formatting of data.

20.4.2 BACnet protocol

A proposed ASHRAE Standard SPC-135P, *BACnet—A Data Communication Protocol for Building Automation and Control Networks,* is
attempting to set an industry communication standard that is focusing
on HVAC controls. This open standard protocol, if accepted industry-
wide, would allow automation and other building component systems
to interface with each other as shown in Figure 20.6. BACnet also
would provide a building owner with the broadest set of options when
upgrading or expanding HVAC systems without replacing any of the
original DDC controls.

20.4.3 Open protocols

In lieu of BACnet, all manufacturers of automation systems and controllers could open up their proprietary protocols and give everyone
"translator books." However, as shown in Figure 20.7, the DDC system

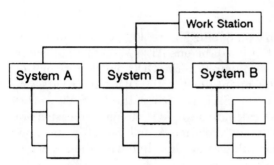

Figure 20.6 BACnet typical schematic. (Courtesy of Heating/Piping/Air Conditioning)

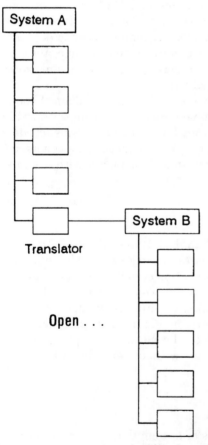

Figure 20.7 Open protocol typical schematic. (Courtesy of Heating/Piping/Air Conditioning)

is still closed, as system A always will be in control of the building owner's options when upgrading or expanding HVAC systems.

A number of open protocols are competing to set their own "industry standard" for building control communications. Building communication preprogrammed computer chips, which provide network services, can be provided to HVAC equipment manufacturers that may be compatible with most open protocols. System costs probably will decide which communications networks will survive.

The malfunction of HVAC equipment and systems because of control system networking problems can only add to the problems facing TAB technicians.

21

Cleanroom Testing

21.1 Cleanroom Basics

21.1.1 Federal standard 209E

21.1.1.1 Development. Federal Standard 209, *Airborne Particulate Cleanliness Classes in Cleanrooms and Clean Zones,* is probably the most widely referenced contamination control document in existence. Although it is nominally a U.S. government publication intended for use by federal agencies, it has been adopted as the standard for air cleanliness classification by American industry and most other countries with high-tech capabilities.

The original version of the document was released in December of 1963 as Federal Standard 209. The first revision, Federal Standard 209A, was published in August of 1966. A second revision, Federal Standard 209B, was issued in April of 1973, and Amendment Number 1 to Federal Standard 209B in May of 1976. No significant changes were made to the original document by any of these revisions.

Major changes were made to the document, which was published in October of 1987 as Federal Standard 209C. For the first time methodology for statistical analysis of the particle count data was included as well as changes to classification levels and data collection methods. The most recent version, Federal Standard 209E, was released September 1992, with significant changes including the addition of SI (metric) unit designations for cleanrooms.

21.1.1.2 Airborne particles. Airborne particulate matter can be organic or inorganic, viable or nonviable. Most contamination control problems concern the total (gross) contamination within the air, but applications exist for specific contamination control of bacteria, spores, and viruses that are contained in the air. Airborne particles range in size from 0.001 μm to several hundred micrometers (also called microns). Conditions for a clean space vary widely with industrial and research requirements. The control of airborne particulates is presented in Federal Standard 209E. Table 21.1 gives the new U.S. and Metric range of particle sizes and cleanroom class limits.

21.1.1.3 Airborne particulate cleanliness class. The level of cleanliness specified by the maximum allowable number of particles per cubic foot of air (per cubic meter of air) is shown for the cleanroom classes in Table 21.1. The class level in U.S. units is established by the maximum allowable number of particles, 0.5 μm and larger, per cubic foot. The class level in SI (metric) units is taken from the logarithm (base 10) of the maximum allowable number of particles, 0.5 μm and larger, per cubic meter.

A class 100 (class M3.5) cleanroom means that particle concentrations per cubic foot (cubic meter) may not be higher than 750 (26,500) of a 0.2 μm size, plus 300 (10,600) of a 0.3 μm size, and plus 100 (3530) of a 0.5 μm size or larger.

21.1.1.4 Cleanroom class verification. When cleanrooms are tested and certified by NEBB Certified Cleanroom Performance Testing Firms, measurements and observations of applicable environmental factors related to the cleanroom or clean zone during verification are recorded. Such factors may include, but are not limited to, air velocity, air volume change rate, room pressurization, makeup air volume, unidirectional airflow parallelism, air turbulence, air temperature, humidity or dewpoint, and room vibration. The presence of equipment and personnel activity should also be noted.

Verification of air cleanliness in cleanrooms and clean zones shall be performed in accordance with the appropriate particle counting method or methods in the NEBB *Procedural Standards for Certified Testing of Cleanrooms.*

21.1.1.5 Micrometer particle size. One micrometer is one thousandth (0.001) of a millimeter or 0.000039 in. The thickness of a human hair is approximately 80 to 100 μm and 1000 μm (1 mm) is about the thickness of a dime. The chart in Figure 21.1 shows the sizes of common particles.

Sometimes "particles per liter" is used in place of "particles per cubic meter" in the metric system. One thousand liters equals 1 m^3. For example, 10.6 particles per liter equals 10,600 particles per cubic meter.

TABLE 21.1 Airborne Particle Cleanliness Classes[a]

		Class limits									
		0.1 µm, volume units		0.2 µm, volume units		0.3 µm, volume units		0.5 µm, volume units		5 µm, volume units	
SI	English[c]	m³	ft³	m³	ft³	m³	ft³	m³	ft³	m³	ft³
M 1	1	350	9.91	75.7	2.14	30.9	0.875	10.0	0.283	—	—
M 1.5		1,240	35.0	265	7.50	106	3.00	35.3	1.00	—	—
M 2		3,500	99.1	757	21.4	309	8.75	100	2.83	—	—
M 2.5	10	12,400	350	2,650	75.0	1,060	30.0	353	10.0	—	—
M 3		35,000	991	7,570	214	3,090	87.5	1,000	28.3	—	—
M 3.5	100	—	—	26,500	750	10,600	300	3,530	100	—	—
M 4		—	—	75,700	2,140	30,900	875	10,000	283	—	—
M 4.5	1,000	—	—	—	—	—	—	35,300	1,000	247	7.00
M 5		—	—	—	—	—	—	100,000	2,830	618	17.5
M 5.5	10,000	—	—	—	—	—	—	353,000	10,000	2,470	70.0
M 6		—	—	—	—	—	—	1,000,000	28,300	6,180	175
M 6.5	100,000	—	—	—	—	—	—	3,530,000	100,000	24,700	700
M 7		—	—	—	—	—	—	10,000,000	283,000	61,800	1,750

[a]Class limits are given for each level. The limits designate specific concentrations (particles per unit volume) of airborne particles with sizes equal to and larger than the particle sizes shown. The class limits shown in Table 21.1 are defined for classification purposes only and do not necessarily represent the size distribution to be found in any particular situation.

[b]Concentration limits for intermediate classes can be calclated, approximately, from the following equations:

$$\text{Particles/ft}^3 = N_c(0.5/d)^{2.2}$$

where N_c is the numerical designation of the class based on U.S. units, and d is the particle size in micrometers, or

$$\text{Particles/m}^3 = 10^M(0.5/d)^{2.2}$$

where M is the numerical designation of the class based on SI units and d is the particle size in micrometers.

[c]For describing the classes, SI levels and units are preferred; however, U.S. units may be used.

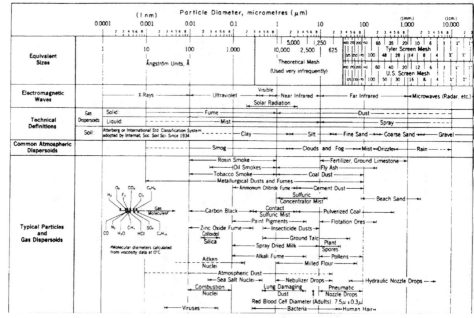

Figure 21.1 Characteristics of particles and particle dispersoids. *(Courtesy of Stanford Research Institute.)*

21.1.2 Cleanroom airflow

The types of cleanroom systems can be described in terms of combinations of the following parameters: air change rates, airflow patterns, method of filtration, and method of handling bypass or recirculated air. It is important that return air outlets be sized with correct air velocities [such as 500 to 700 fpm (2.5 to 3.5 m/s)] to ensure that the proper pressurization of the cleanroom can be maintained.

21.1.2.1 Air change rates. *Air change rates*, the number of times the total volume of a given room in cubic feet (m^3) is changed and filtered in a minute (or a hour), together with filter efficiency, are the two most important factors in providing and maintaining environmental conditions free of contaminants.

Air change rates in cleanrooms vary from as few as 30 per hour in class 100,000 (M6.5) rooms to as many as 600 to 700 per hour in class 100 (M3.5) and class 10 (M2.5) rooms (see Table 21.2). Care must be taken when testing rooms with 600 complete air changes per hour to ensure temperature and relative humidity control under such dynamic airflow conditions.

21.1.2.2 Airflow patterns. Air from the high-efficiency particulate air filters should be directed so that the cleanest air is at the most critical work areas. As contaminants are entrained, they should be conveyed

TABLE 21.2 Typical Cleanroom Air Change Rates and Velocities

Cleanroom class	Room air velocity		Airchange rates[a]	
	Feet per minute	Meters per second	Per hour	Per minute
1	90–120	0.45–0.60	720	12
10	80–100	0.50–0.50	600–720	10–12
100	80–100	0.40–0.50	600–720	10–12
1000	25–30	0.12–0.15	180–240	3–4
10,000	8–10	0.04–0.05	60–120	1–2
100,000	4–6	0.02–0.03	30	0.5

[a]Eight foot (2.4 m) ceiling height.

to less critical portions of the room for removal by the return air system.

These criteria generally result in (1) the introduction of large quantities of air at low velocities in the area of the most critical work surfaces and (2) unidirectional movement, usually downward through the room, prior to removal from the space. The choice of a specific airflow arrangement should be based on the criticality of the conditions to be maintained in the space, the size of the room, and the ratio of space occupied by critical operations to the overall room size.

21.1.2.3 Multidirectional airflow. A satisfactory arrangement for conventional or *multidirectional flow* air distribution is shown in Figure 21.2. Air is supplied through large ceiling outlets, flows generally downward, and is removed near the floor level.

Multidirectional air systems function satisfactorily for many applications. When they are supplemented by local unidirectional flow work stations, they can provide a high degree of contaminant control for critical operations.

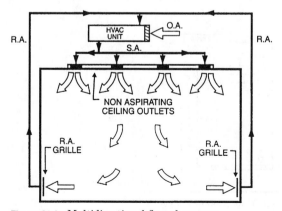

Figure 21.2 Multidirectional flow cleanroom.

21.1.2.4 Unidirectional airflow. In a *unidirectional flow* or *laminar flow* system, air is introduced evenly from filter banks from an entire surface of the room, such as the ceiling or a wall, flows at constant velocity across the room, and is removed through the entire area of an opposite surface. Unidirectional flow provides a direct, predictable path that a submicrometer size particle will follow through the cleanroom, with minimum opportunity for contaminating room components. It also captures the particles constantly generated within the room and introduced into the airstream, thereby reducing the potential for cross-contamination.

To provide good dilution and sufficient air motion to prevent settling of particles, airflow velocities of approximately 90 fpm $\pm$ 20 fpm (0.45 m/s $\pm$ 0.1 m/s) are recommended as standard design for unidirectional flow cleanrooms.

21.1.2.5 Vertical unidirectional flow cleanroom. The *vertical unidirectional or laminar flow cleanroom* (Figure 21.3) has a ceiling consisting of *high efficiency particulate air* (HEPA) filters. As the numerical class of the cleanroom gets lower, a greater percentage of the ceiling will require HEPA filters. For a class 100 (M3.5) room, almost the entire ceiling will consist of HEPA filters. Ideally, a grated or perforated floor may serve as the air return or exhaust. Air in the unidirectional room moves uniformly from the ceiling to the floor. After moving through the ceiling filters, it enters the cleanroom essentially free of all particles.

A pressurized ceiling plenum of filters, individual ducted filters, or fan module filters may provide filtered air to the clean space. Care must be taken with the pressurized ceiling plenum to diffuse the air in order to keep the velocity through each filter uniform. Each HEPA filter should be factory tested and the pressure drop recorded. Filters

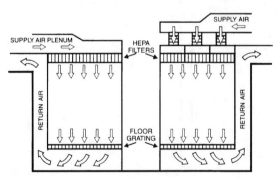

Figure 21.3 Vertical unidirectional or laminar flow cleanroom.

within the same plenum should have equal pressure drops. Airflow balancing can be accomplished by volume dampers built into the perforated floor, in the sidewall returns, or in the ducts connected to the plenums. Individual ducted filter boots should have dampers for each filter to facilitate balancing. It should be noted that when ducted filters are used, the area above the ceiling is not normally a clean area.

21.1.2.6 Horizontal unidirectional flow cleanroom. The *horizontal unidirectional or laminar flow cleanrooms* (Figure 21.4) uses the same filtration airflow technique as the vertical unidirectional system, except that the air flows from one wall of the room to the opposite wall. The supply wall consists entirely of HEPA filters supplying air at approximately 90 fpm (0.45 m/s) across the entire section of the room. The air then exits through return air devices at the opposite end of the room. As with the vertical unidirectional room, this design removes contamination generated in the space at a rate equal to the air velocity and does not allow cross-contamination perpendicular to the airflow.

In this design, the air first coming out of the filter wall is as clean as air at the ceiling in a vertical unidirectional room. The process activities can be oriented to have the most critical operations at the cleanest end of the room, with progressively less critical operations located toward the return air end of the room.

21.2   Cleanroom Air Filters

21.2.1   Rating air filters

The three operating characteristics that distinguish the various types of HVAC air filters are *efficiency, airflow resistance,* and *life or dust-holding capacity.* Efficiency measures the ability of the air filters to

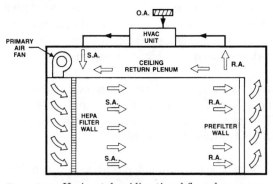

Figure 21.4 Horizontal unidirectional flow cleanroom.

remove particulate matter from an airstream. Average efficiency during the life of the filter is the most meaningful characteristic for most types and applications.

Airflow resistance (or merely *resistance*) is the static pressure drop across the filter at a given airflow rate. The term *pressure drop* is used interchangeably with resistance.

Dust-holding capacity defines the amount of a particular type of dust that an air filter can hold when it is operated at a specified airflow rate to some maximum resistance value or before its airflow is seriously reduced as a result of the collected dust.

Complete rating of air filters then requires data on efficiency, resistance, dust-holding capacity, and the effect of dust loading on efficiency and resistance.

21.2.2 Air filter tests

Air filter testing is complicated by a number of technical and practical considerations to the extent that no individual test adequately describes all filters. Ideally, performance testing of equipment should simulate the operation of the device under operating conditions and furnish performance ratings in terms of characteristics important to the equipment user. In the case of air filters, this is made difficult by the wide variations in the amount and type of particulate matter in the air being cleaned. Another complication is the difficulty of closely relating measurable performance to the specific requirements of users. Recirculated air tends to have a larger amount of lint than does outside ventilation air.

In general, three types of tests, together with certain variations, are employed to determine air filter efficiency.

21.2.2.1 Weight arrestance. A standardized synthetic dust consisting of various particle sizes is fed into the air filters and the percentage of the weight of the dust removed is determined. Under ASHRAE Standard 52.1-1992, this type of efficiency measurement is named *synthetic dust weight arrestance* to distinguish it from other efficiency values. The term often is abbreviated as *weight arrestance*.

21.2.2.2 Dust-spot efficiency. Atmospheric dust is passed into the air filter, and the discoloration effect of the cleaned air is compared with that of the incoming air. This type of measurement is named *dust-spot efficiency* under the ASHRAE Standard 52.1-1992.

21.2.2.3 Particle size. Uniformly sized aerosol particles are fed into the air filter and the percentage removed by the filter is determined. For example, particle size concentration is measured upstream and downstream of the air filter. This is referred to as *fractional removal efficiency* under proposed ASHRAE Standard 52.2P.

21.2.3 HEPA filters

The ability to obtain the level of cleanliness required in present-day cleanrooms rests solely upon the use of *high-efficiency particulate air* (HEPA) filters. The HEPA filter medium is composed of glass fibers of a variety of sizes—both lengths and diameters. They are bound together primarily by the interlacing of the fibers with the help of a binder. This medium is much more dense than the media in less efficient filters, which causes a higher pressure drop and requires greater filter area. In the HEPA filter, this is accomplished by folding and pleating the medium (Figure 21.5).

Because HEPA filters operate at such low velocity rates through the medium, the pressure drop versus airflow volume curve is linear (straight line). HEPA filters are tested and systems are designed to have the initial filter pressure drop at approximately 1 in.w.g. (250 Pa). HEPA filters reach about 90% of their maximum holding capacity at twice the initial pressure drop.

In testing and handling HEPA filters, the medium should never be touched, because it is easily damaged. Media damage also can occur if the filter frame is dropped or banged against a hard object.

21.2.4 Cleanroom filtration

Cleanroom air filtration involves several steps. The secondary air system (used for temperature and humidity control) should provide pre-

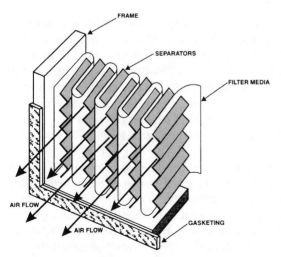

Figure 21.5 HEPA filter, isometric cutaway.

filtering and intermediate air filtering to remove larger airborne particles. Final high-efficiency particulate air (HEPA) filters, located in the primary airstream, are used to remove the smaller airborne particles. These filters can be placed in the air handling unit or in the cleanroom ceiling or wall, depending on the class of cleanroom and its use.

HEPA filters have a large increase in pressure drop as they load up with particulate matter. This will cause the cleanroom airflow volume to decrease as the operating point moves up the fan curve unless the fan speed is modulated automatically to compensate for the pressure increase.

Supply air for the cleanroom must be distributed very uniformly in the plenum above a nonducted HEPA filter ceiling to obtain uniform flow through the ceiling-mounted filters and the cleanroom itself. This also applies to the plenum of a wall-mounted HEPA filter system.

21.3   Cleanroom HVAC Systems

Cleanroom HVAC systems are somewhat different from general commercial HVAC systems. Room walls may become duct walls, rooms become pressure vessels, walls or ceilings may be used as air diffusers, and a single room or space may have three separate fan–duct systems. **A TAB technician NEVER should enter an existing cleanroom area without instructions from the facility manager**. Gowning procedures may be required, even outside the cleanroom, and procedures vary from facility to facility.

21.3.1   Room pressurization

A cleanroom facility may consist of multiple rooms with different requirements for contamination control. All rooms in a clean facility should be maintained at static pressures sufficiently higher than atmospheric to prevent infiltration by wind or other effects. Differential pressures should be maintained between the rooms sufficient to assure airflow outward progressively from the cleanest spaces to the least clean during normal operation and during periods of temporary upsets in the air balance, as when a door connecting two rooms is suddenly opened (Table 21.3).

Static pressure regulators can maintain desired room pressures by operating dampers, fan inlet vane controls, vaneaxial fans, controllable pitch-in-motion controls, or a combination of these to vary the ratio of supply air to makeup air or exhaust air. To provide control over room pressures, airflow variations should be minimized. Exhaust airflow from rooms through hoods should be maintained constant by continuous hood operation or appropriate bypasses. In many systems,

TABLE 21.3   Air Pressure Relationships

Application	Pressure differential
General	0.05 in.w.g. (12 Pa) higher than surroundings
Between cleanroom and uncontaminated section	0.05 in.w.g. (12 Pa), minimum
Between uncontaminated and semicontaminated section	0.05 in.w.g. (12 Pa)
Between semicontaminated section and locker area	0.01 in.w.g. (2.5 Pa)

door openings to the outside are protected by airlocks, and provision is made for offsetting the pressure loss variations across filters as the dust loading increases.

21.3.2   Room temperatures

Temperature controls provide stable conditions for materials and instruments and for personnel comfort. Heat loads from lighting are high but stable; personnel loads vary; the heat generated by process operations, including soldering, welding, heat treating, and heated pressure vessels, is usually high and variable.

The large quantities of air supplying the cleanroom diffuses internal heat gains such that the temperature differential, between the room entering supply air and the room air, is quite low. However, areas of concentration of heat-producing equipment and supply air patterns should be analyzed to determine resulting temperature gradients (Table 21.4). Large cleanrooms have multiple zones of temperature control because of radically different cooling requirements of the various localized areas.

21.3.3   Room humidity

In cleanrooms, humidity control is affected more by external influences (such as weather changes) than by variations in moisture gen-

TABLE 21.4   Cleanroom Temperature and Humidity

	Temperature	Humidity
Capacity range	67°–77 °F (19°–25 °C)	40–55%
Control point	72 °F (22 °C)	45%
Control tolerance		
General applications	± 2 °F (± 1.1 °C)	± 5%
Critical applications	± 0.5 °F (± 0.3 °C)	± 2%
Capacity and control response rate	2.5°–4 °F (1.4°–2.2 °C) change per hour	

eration within the space. When processes involving evaporation take place within the cleanroom, they usually are confined within ventilated enclosures. Some precision manufacturing processes require humidities lower than 35%. Precautions often are taken to control static electricity by using ionization grids and grounding straps.

Corrosion of precisely manufactured surfaces, including bearings, electrical contact surfaces, ballbearing raceways, and miniature-gear trains, occurs with above 50% relative humidity. At relative humidities much below 40%, static charges may form, attracting dust particles that later may become airborne in objectionable concentrations.

21.3.4 Makeup and exhaust airflow

Ventilation and makeup air is required for both room and unitary equipment application. In the use of suction exhaust benches, where the exhaust air is discharged out of the area through ductwork, the makeup air may be supplied from within the area or may be ducted in. If this supply air is ducted from an external source, it should be prefiltered and conditioned. If the air is supplied from within the area, the space air conditions must not be adversely affected.

Activities in the cleanroom often require the use of exhaust air equipment. Work stations emitting toxic fumes, ovens emitting heated fumes, small machinery operations, and the like, all must have exhaust air capability. Work stations use the cleanroom air as their source of makeup air; therefore, the cleanroom makeup air volume is increased by the volume of exhaust air. The exhaust air ducts for the equipment should be carefully adjusted to maintain unidirectional airflow within the cleanroom and room pressurization.

Exhaust air systems, in which contamination control is necessary, may be contained within unitary equipment or may be remote with central ducts. In addition, special pollution control systems may also be necessary before the air can be discharged to the atmosphere. As a safety precaution, exhaust air ducts for highly toxic materials should be kept at a negative pressure within the entire confines of the building. **Exhaust air filter housings NEVER should be opened by TAB technicians.**

21.4 Typical Cleanroom Systems

21.4.1 Class 10 (M2.5) cleanrooms

The system shown in Figure 21.6 is an example of a class 10 (M2.5) cleanroom with three fan–duct systems. The primary air is supplied through a pressurized plenum and filtered through ultra-low penetra-

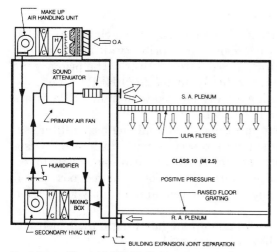

Figure 21.6   Class 10 (M2.5) cleanroom example.

tion air (ULPA) filters. A raised floor or grating provides laminar air-flow within the cleanroom area.

The primary air is supplied to the pressurized plenum by a controllable pitch, axial flow fan provided with a sound attenuator. The large volume of primary air [540 air changes per hour for a cleanroom of 10 ft height (3 m) and 90 fpm (0.45 m/s) air velocity through the ULPA filters] requires the use of vaneaxial fans.

The secondary air is provided by an air handling unit consisting of a mixing box, a coil section with cooling coil for cooling and dehumidification, a heating coil for heating and reheat, and a supply fan section with a centrifugal fan.

The outside air required for room pressurization and makeup air for the process equipment exhaust, if applicable, is supplied by a makeup air handling unit. The makeup air handling unit consists of an outside air intake, a filter section with 2 in. (50 mm) prefilters and 95% efficient bag filters, a coil section with a preheating coil with face and a bypass damper and a cooling coil, and a supply fan section with a centrifugal fan provided with a frequency inverter. The preheating coil and the cooling coil maintain a fixed leaving air temperature year round. The preconditioned outside air and the return air from the cleanroom to the secondary air handling unit are mixed in the mixing box of the secondary air handling unit.

Room pressurization is provided by the makeup air handling unit. A static pressure sensor installed in the room modulates the supply fan electric motor speed through the frequency inverter to maintain the required positive pressure.

21.4.2   Class 100 (M3.5) cleanrooms

The system shown in Figure 21.7 is an example of a class 100 (M3.5) cleanroom. The primary air is provided by a primary air handling unit and distributed through medium pressure ductwork. The primary air is supplied to the room by individually ducted final HEPA filters.

The entire cleanroom ceiling area is covered with final HEPA filters installed on the T-grid ceiling system.

The primary air handling unit consists of a mixing box, a filter section with prefilters, and a supply fan section with an internally isolated controllable pitch, axial flow fan with a builtin sound attenuator.

The secondary air handling unit consists of a mixing box, a coil section with a cooling coil and a heating coil, and a supply fan section with a centrifugal fan.

Outside air is provided by a makeup air handling unit, which has the same components as shown in Figure 21.6.

The return air from the cleanroom is transferred through floor-mounted registers into an air-tight return air plenum (or basement). The air flows through a chase area ceiling into the plenum and is ducted back to the mixing boxes of the primary and the secondary air handling units. The chase areas are used for returning the air. Branch exhaust ducts (for work station exhaust) are located in these chases and connected to the main exhaust ductwork located in the basement area. Since the return air chases are used for process equipment connections and services, the return air is prefiltered.

The cooling coil and the heating coil, in the secondary air handling unit, maintain a fixed leaving air temperature on the discharge side

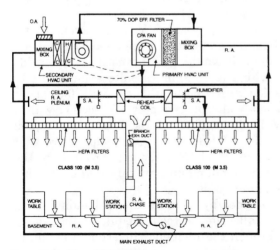

Figure 21.7 Class 100 (M3.5) cleanroom example.

of the primary supply fan. The design temperature for each room is maintained by an electric or hot water reheat coil mounted in the supply ductwork to the zone. If the relative humidity in any zone increases above the set point, the secondary air handling unit cooling coil provides more dehumidification, and at the same time the heating coil will be in the reheat mode to maintain design temperature. When the relative humidity decreases below the set point, a duct-mounted humidifier provides humidification.

21.4.3   Class 10,000 (M5.5) cleanrooms

The system shown in Figure 21.8 is an example of a Class 10,000 (M5.5) cleanroom with local class 100 (M3.5) areas.

The air handling unit consists of a mixing box, a filter section with 2 in. prefilters and 95% NBS bag filters, a coil section with a cooling coil (with an opposed blade damper installed above the cooling coil) and a heating coil, and a supply fan section with an internally isolated, airfoil, centrifugal fan, provided with an inlet vane damper. The 2 in. (50-mm) prefilters are used only for the startup of the system.

The air is distributed through medium pressure ductwork and supplied to the space by individually ducted final HEPA filters. HEPA filters tested for 99.97% efficiency are provided for the class 10,000 (M5.5) area. Canopies with HEPA filters tested for 99.99% DOP efficiency on 0.3 μm are provided above the work stations to provide a local class 100 environment.

The air from the room is transferred through low sidewall registers into return air chases. From the ceiling return plenum the air is ducted to the mixing box of the air handling unit.

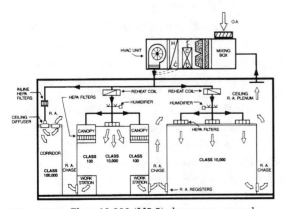

Figure 21.8   Class 10,000 (M5.5) cleanroom example.

The corridor adjacent to the cleanroom has a class 100,000 (M6.5) cleanliness level. The air is supplied through duct-mounted HEPA filters and ceiling diffusers.

Outside air is provided by a central makeup air handling unit connected to several air handling units. The preconditioned outside air and the return air from the cleanrooms is mixed in the air handling unit mixing box. A part of the mixed air is cooled and dehumidified by the cooling coil and the rest of the mixed air bypasses the cooling coil. The quantity of the air passing through the cooling coil is constant and is dependent upon the room sensible and latent loads. The concept of primary and secondary air is used with a single air handling unit.

21.5   Cleanroom Systems TAB

21.5.1   Main supply air systems

Before the cleanroom facility is tested for airflow patterns, volumes, and velocities, the quantities of airflow must be measured for the main supply air to the cleanroom and for the makeup supply air system. The reserve air handling capacities needed to accommodate a loaded filter capacity also must be determined.

It is very important that the proper testing instrumentation be selected and calibrated (if required) before the cleanroom TAB procedures are initiated. Chapters 7 and 8 have a description of basic instrumentation used for testing HVAC systems.

1. Confirm that every item affecting the airflow of the main supply air duct system is ready, such as doors being closed, ceiling tiles (supply air or return air plenums) in place, etc.

2. Confirm that all automatic control devices will not affect the testing, adjusting, and balancing (TAB) operations.

3. Establish the conditions for the maximum demand system airflow, which generally is a cooling application with "wetted" coils.

4. After verifying that all dampers are open or set, start all related systems (return, exhaust, etc.) and the system being balanced with each fan running at the design speed (rpm). Upon starting each fan, immediately check the fan motor amperage. If the amperage exceeds the nameplate full-load amperage, stop the fan to determine the cause or to make the necessary adjustments.

5. Again confirm that all related system fans serving each area within the space being balanced are operating. If they are not, pressure differences plus any infiltration or exfiltration pressures may adversely influence the balancing. Preliminary studies will

have revealed whether or not the supply air quantity exceeds the exhaust air quantity from each area. Positive and negative pressure zones should be identified at the time.

In most cleanroom applications, pressure zones will be a primary consideration. The pressure differentials may be as high as 0.25 in.w.g. (63 Pa) static pressure but normally are in the range of 0.05 to 0.10 in.w.g. (12.5 to 25 Pa). These differentials must be maintained during all airflow balancing and cleanroom certification testing. If differential pressures were allowed to vary during the TAB procedures, it would be difficult to repeat the test results, making the final results unacceptable.

6. If the cleanroom is served by a primary system for filtration and a secondary system for makeup air, room pressurization, and air conditioning, all systems should be in operation during all TAB work and cleanroom certification testing. If the fan systems have a return air system and an outside air intake, the modulation of the dampers also may adversely affect the balancing and testing procedure.

7. Determine the volume of air being moved by the supply fan at design rpm by one or more of the following methods:
   a. Pitot tube traverse of main duct or ducts leaving fan discharge.
   b. Fan curves or fan performance charts. To determine fan performance using a fan curve or performance rating chart, it is necessary to take amperage and voltage readings. In addition, a static pressure reading across the fan must be recorded. With rpm, brake horsepower (watts), and static pressure, the fan manufacturer's data sheets may be used to determine the airflow (cfm or L/s) predicted by the manufacturer. Fan performance can deviate substantially from the fan curves if a "system effect" is present or a substantial amount of extra duct fittings have been installed.
   c. Velocity readings taken across coils, filters, and/or dampers on the intake side of the fan. This must be used as an approximation only.
   d. Filter velocity profile.

8. If the supply fan volume is not within ± 10% of the design capacity at design rpm, determine the reason by reviewing all system conditions, procedures, and recorded data. Check and record the air pressure drop across filters, coils, eliminators, sound traps, duct elbows, etc., to see whether excessive loss is occurring. Particularly study duct connections and casing conditions at the fan inlet and outlet for "system effect."

9. If the measured airflow of the supply air fan, central return air fan, or central exhaust air fan varies more than 10% from design, adjust the drive of each fan to obtain the approximate required cfm (L/s). Record fan suction static pressure, fan discharge static pressure, amperage, and cfm (L/s) measurements. Confirm that the fan motor is not overloaded and the proper heater coils are installed.

10. Make a preliminary spot check of filter velocities and area pressures.

11. The HVAC systems are considered balanced in accordance with NEBB cleanroom procedural standards when the value of the air quantities is measured and found to be within 10% of the design air quantities (unless there are conditions beyond the control of the TAB firm).

21.5.2 Makeup air systems

The same TAB procedures used in Section 21.5.1, Main Supply Air Systems, should be followed for makeup air systems.

21.5.3 Reserve air handling capacity

This test is to determine the amount of excess capacity beyond design amounts available in the main supply air handling unit to compensate for increased system resistance caused by maximum allowable filter loading. Reserve air handling unit capacity is to be expressed in in. w.g. (Pa).

Determine the external static pressure of the main supply air system (see Section 21.5.1). Calculate the "reserve system capacity" by either of the following methods:

1. Plot a point on the fan curve at actual rpm and minimum acceptable system airflow. Determine the external static pressure at which this occurs. The difference between this value and the measured external static pressure is reserve capacity expressed in in. w.g. (Pa).

2. At the measured value of airflow, plot the static pressure that can be developed without exceeding the BHP (W) of the air handling unit motor. The difference between this static pressure and the measured external static pressure is the reserve capacity in in. w.g. (Pa).

21.5.4 Reporting

1. Report the measured or calculated main supply air volume and makeup air volume in total standard cfm (std. L/s), scfm per square foot (std. L/s/m^2) of work area, and air changes per hour, to the nearest 10 scfm (5 std. L/s) or 0.1 air change per hour.

2. Report the calculated reserve air handling unit capacity in static pressure to the nearest 0.01 in.w.g. (2.5 Pa).

21.6 Cleanroom Pressurization Tests

The purpose of cleanroom pressurization tests is to verify the capability of the cleanroom systems to maintain the specified pressure differential in the cleanroom. These tests should be performed after the facility has met the acceptance criteria for airflow velocity, uniformity, parallelism, and other applicable tests.

21.6.1 Test instruments

Use inclined manometer(s) or a mechanical differential pressure gauge (see Chapter 7).

21.6.2 Procedures

1. Verify that all doors of the cleanroom enclosure are closed.

2. Measure and record the pressure differential between the room and the vestibule (if present), and between the vestibule and the exterior ambient.

3. If no vestibule is present, measure and record the pressure differential between the room and the exterior ambient.

4. If the clean space is subdivided into more than one room, measure the pressure differentials between the innermost room and the next room in order. Continue until the last room (or vestibule), has been measured against the exterior ambient.

21.6.3 Reporting

Report all measured values to the nearest 0.01 in.w.g. (2.5 Pa).

21.6.4 Acceptance

Pressurization levels and acceptance values such as 0.03 to 0.05 in.w.g. (7.5 to 12.5 Pa) normally are specified by the cleanroom owner or operator.

21.7 Cleanroom Certification Testing

Two basic types of certification tests for cleanroom systems are employed by NEBB Certified Cleanroom Performance Testing Firms to evaluate a facility properly: *initial performance tests* and *operation monitoring tests*. The initial performance tests may result in corrections of problems within the system and, therefore, are normally conducted before occupancy of the facility. However, a cleanroom facility cannot be fully evaluated until it has performed under full occupancy and the manufacturing process to be performed within it is operational. The techniques for conducting initial performance tests and operation monitoring are similar.

21.7.1 Initial tests

Sources for contamination are external and internal. For both multidirectional flow and unidirectional flow cleanrooms, the major source for external contamination is through the primary air loop. Therefore, similar leak testing of the HEPA filter banks may be conducted. According to Federal Standard 209E, equipment applying light scattering principles shall be used for detection of particle sizes 0.5 μm and smaller. For particles sizes 5.0 μm and larger, microscopic counting is allowed, with the particles collected on a membrane filter through which a sample of air has been drawn. In general, the light scattering method is required for evaluation of class 10 (M2.5) and class 100 (M3.5) rooms, whereas either method may be used for class 10,000 (M5.5) or class 100,000 (M6.5) rooms.

HEPA filters, for class 1000 (M4.5) and better cleanrooms, should be tested both before installation and while in place. HEPA filters for class 10,000 (M5.5) and above may be tested after installation. Field tests for pinhole leaks are required at the following places: the filter medium, the sealant between the medium and the filter frame, the filter frame gasket, and the filter bank supporting frames. A pinhole leak at the filter bank can be extremely critical since the concentration of the leak varies inversely as the square of the pressure drop across the hole.

When the filter bank is properly sealed, other sources of external contamination must be tested. In laminar flow facilities, this consists of physical barriers such as walls, partitions, windows, doors, and the like. In conventional flow rooms, in addition to the above, all ductwork downstream of the filter bank must be tested. If the primary air loop is properly pressurized, sufficient control will be obtained.

21.7.2 Operational monitoring tests

Internal sources of contamination may be generated by the process, by service equipment, or by operating personnel. Therefore, to evaluate the performance of the cleanroom properly in controlling internal contamination, tests must be conducted at critical areas within the facility. The contamination level with a conventional flow cleanroom reaches a plateau and the level of contamination tends to equalize throughout, whereas in a laminar flow cleanroom, contamination stratifies. It is therefore critical to sample work areas properly within a laminar flow facility. This sampling should also be done with a light scattering particle counter, providing a high rate of sampling. Since this test is to evaluate a work area with respect to both external and internal contamination, a DOP challenge need not be introduced. However, the cleanrooms should be occupied and in operation.

Average velocity readings throughout the cleanroom should be taken with thermal anemometers to determine velocity gradients and air patterns. Several readings should be taken to obtain an average of each filter in the bank of a laminar flow cleanroom, since their construction produces a definite gradient and must be properly evaluated and balanced.

Other tests, such as temperature and humidity gradients, lighting levels, and sound and vibration levels, often are required in evaluating the performance of a cleanroom facility. All testing within cleanrooms should be done only by NEBB Certified Cleanroom Performance Testing Firms.

22

Sound and Vibration

22.1 Fundamentals of Sound

22.1.1 Sound Waves

Sound can be defined as vibrations transmitted through an elastic medium, similar to waves in water, that are perceived by the human ear. *Noise* can be defined as unwanted sound or sound that is disturbing.

Sound waves travel through air at 68 °F (20 °C) at 1125 fps (343 m/s). They travel through water at approximately 5000 fps (1520 m/s). The wavelength of the sound is the distance it travels during one vibration or *cycle* (Figure 22.1), which is similar to a sine wave. The period of vibration is measured in *seconds*. The number of complete cycles of vibration that occurs per unit of time is the *frequency*, measured in cycles per second or *hertz* (Hz). The relationship between frequency, speed of sound, and wavelength is shown by Equation 22.1.

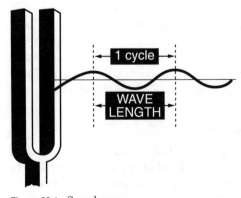

Figure 22.1 Sound wave.

Equation 22.1

$$\lambda = \frac{c}{f}$$

Where: λ = wavelength in feet (meters)
c = speed of sound, 1125 ft/s (343 m/s)
f = frequency in hertz (cycles per second)

Example 22.1   What is the wavelength of 20 Hz?

Solution

$$\lambda = 1125/20 = 56.3 \text{ ft. (U.S.)}$$

$$\lambda = 343/20 = 17.15 \text{ m (Metric)}$$

Example 22.2   What is the wavelength of 20,000 Hz?

Solution

$$\lambda = \frac{1125}{20,000} = 0.056 \text{ ft. or } 0.68 \text{ in.}$$

$$\lambda = \frac{343}{20,000} = 0.017 \text{ m or } 17 \text{ mm}$$

22.1.2   Octave bands

The audible frequency range extends from about 20 to 20,000 Hz. Within the frequency range of interest, a sound source is characterized by its sound power output in octave or $\frac{1}{3}$-octave bands, although narrower bandwidths may be appropriate for certain analyses. An *octave* is a frequency band with its upper band limit twice the frequency of its lower band limit.

Table 22.1 lists the preferred series of octave bands and the upper and lower band limit frequencies. One third–octave band center frequencies and upper and lower band limits also are listed in Table 22.1. Octave and $\frac{1}{3}$-octave bands are identified by their center frequencies, not by their upper and lower band limit frequencies. Analysis in octave bands is used for rating acoustical environments in rooms, while $\frac{1}{3}$-octave bands are used in vibration analysis and troubleshooting.

22.1.3   Relationships among sound terms

The relationship among *sound power level* (L_w), *sound pressure level* (L_p), and *sound intensity level* (L_i) must be thoroughly understood. In practice, the terms *sound pressure level* and *sound power level* are used most frequently. The definitions of these terms follow.

TABLE 22.1   Center Frequencies for Octave and $\frac{1}{3}$-Octave Band
Series

Octave bands (Hz)			$\frac{1}{3}$-Octave bands (Hz)		
Lower	Center	Upper	Lower	Center	Upper
			22.4	25	28
22.4	31.5	45	28	31.5	35.5
			35.5	40	45
			45	50	56
45	63	90	56	63	71
			71	80	90
			90	100	112
90	125	180	112	125	140
			140	160	180
			180	200	224
180	250	355	224	250	280
			280	315	355
			355	400	450
355	500	710	450	500	560
			560	630	710
			710	800	900
710	1,000	1,400	900	1,000	1,120
			1,120	1,250	1,400
			1,400	1,600	1,800
1,400	2,000	2,800	1,800	2,000	2,240
			2,240	2,500	2,800
			2,800	3,150	3,550
2,800	4,000	5,600	3,550	4,000	4,500
			4,500	5,000	5,600
			5,600	6,300	7,100
5,600	8,000	11,200	7,100	8,000	9,000
			9,000	10,000	11,200
			11,200	12,500	14,000
11,200	16,000	22,400	14,000	16,000	18,000
			18,000	20,000	22,400

22.1.3.1   Sound power level (L_w).   The fundamental characteristic of an acoustic source (fan, etc.) is its ability to radiate power. Sound power level cannot be measured directly; it must be calculated from sound pressure level measurements. The sound power level of a source (L_w) is the ratio, expressed in *decibels*, of its sound.

A considerable amount of confusion exists in the relative use of sound power level and sound pressure level. An analogy may be made in that the measurement of sound pressure level is comparable to the measurement of temperature in a room, whereas the sound power level is comparable to the cooling capacity of the equipment conditioning the room. The resulting temperature is a function of the cooling capacity of the equipment and the heat gains and losses of the room. In exactly the same way, the resulting sound pressure level would be a function of the sound power output of the equipment together with the sound reflective and sound absorptive properties of the room.

Given the total sound power output of a sound source and knowing the acoustical properties and dimensions of a room, it is possible to calculate the resulting sound pressure levels.

22.1.3.2 Sound power level of a source (L_w). The ratio, expressed in *decibels*, of a source's sound power to the reference sound power, which is 10^{-12} (or 10^{-13} watts, now obsolete). The reference power should always be stated.

22.1.3.3 Sound power of a source (W). The rate at which sound energy is radiated by the source. Without qualification, overall sound power is meant, but often sound power in a specific frequency band is indicated.

22.1.3.4 Sound pressure. Sound pressure is an alternating pressure superimposed on the barometric pressure by sound. It can be measured or expressed in several ways, such as *maximum sound pressure* or *instantaneous sound pressure*. Unless such a qualifying word is used, it is the effect of root-mean-square (RMS) pressure that is meant (see Section 19.7.4 for RMS discussion).

22.1.3.5 Sound pressure level (L_p). A measure of the air pressure change caused by a sound wave expressed on a decibel scale reference to a reference sound pressure of 2×10^{-5} Pa or 0.0002 microbar.

22.1.4 Decibel

The term *decibel* (abbreviated dB) is a combination of two words. "Bel," in honor of Alexander Graham Bell, is a dimensionless unit for the logarithmic ratio of two power qualities. "Deci," meaning "10," indicates that the decibel is one tenth of a bel. The decibel scale counts ratios in powers of 10. A small amplifier with a gain of 10 dB has multiplied the input power by a factor of 10. An amplifier with a gain of 40 dB has multiplied the power at the input by a factor of 10,000 ($10 \times 10 \times 10 \times 10$) or 10^4. Therefore the decibel is a logarithmic term. Because sound pressure and sound power both can be expressed in decibels, to avoid confusion it is always best to say, "decibels, sound power level" or "decibels, sound pressure level."

On the decibel scale, the sound power level of 0 dB does not indicate an absence of sound. Zero decibel does indicate a level of sound at a frequency of 1000 Hz, which is just barely audible to a person with normal, unimpaired hearing. A difference in sound level of 1 dB is about the smallest relative change that the average person may be able to detect paying very close attention.

22.1.5   Sound power level

Simply stated, *sound power level* (L_w) is the total acoustic power radiating from a sound source. One does *not* hear sound power, as one does not see the candle power of a 100 W incandescent light bulb. But the output from any incandescent 100 W light bulb is always the same.

If the sound radiates uniformly in all directions (Figure 22.2), the radiated power is spread equally over the surface of an imaginary sphere enclosing the source. The sound intensity, i.e., the sound power passing through a unit area of the spherical surface, is inversely proportional to the surface area of the sphere. If the radius of the sphere is doubled, the surface area increases four times and the sound intensity is reduced to one fourth.

The sound intensity itself is proportional to the sound pressure, which can be measured with a sound meter. If the total area of a surface enclosing a sound pressure amplitude on that surface is measured, it is possible to calculate the sound levels found normally in the everyday workplace.

The sound power level (L_w) in decibels is expressed by the following:

Equation 22.2

$$L_w = 10 \log_{10} \left[\frac{W}{W_{ref}} \right]$$

Where: L_w = sound power level (dB)
   W = sound power (watts)
   W_{ref} = reference sound power (10^{-12} W)

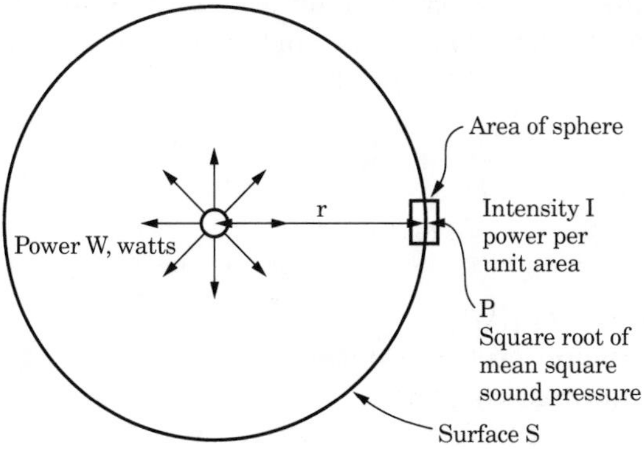

Figure 22.2   Nondirectional sound radiation.

Note that the word "level" is always used to indicate that a ratio, expressed in decibels, of two quantities proportional to power is involved.

The most recent standard reference for sound power levels has become 10^{-12} W and it is consistent with metric units. Some older texts and data give 10^{-13} W as a reference which is consistent with English units. Conversion may be as follows:

Equation 22.3

dB re 10^{-12} W = dB re 10^{-13} W − 10

22.1.6 Sound pressure level

Sound pressure level (L_p) is the acoustic pressure at a point in space where a listener's ear or a microphone of a sound level meter is located. One can hear sound pressure measured in decibels, the loudness depending on the distance from the sound source. As with the 100 W incandescent light, the greater the distance, the less sound (light) is received. However, obstructions or walls may affect the sound just as light colored walls or mirrors may reflect the light.

The amplitude of air pressure fluctuations can vary over a wide range within the audible range. The quietest sound at a frequency of 1000 Hz, which can be heard by an average person (threshold of hearing), corresponds to a pressure fluctuation amplitude of 20 micropascals (20 μPa). The loudest sound, just bearable, corresponds to about 100 Pa (Table 22.2). There is a ratio of 5,000,000:1 between the loudest and the quietest sounds that can be heard by an average person. Applying a linear scale to a measurement of sound pressure fluctuations would give large, unusable numbers. Therefore sound pressures are expressed as a logarithmic ratio of the measured sound value to a standard reference value. This standard reference value has been fixed as 20 μPa. Since acoustic intensity (or loudness) of a sound is proportional to the square of the sound pressure fluctuation, the sound pressure level is defined as:

Equation 22.4

$$L_p = 10 \log_{10} \left[\frac{P_{rms}^2}{P_{rel}^2} \right] = 20 \log_{10} \frac{P_{rms}}{P_{ref}}$$

Where: L_p = sound pressure level (Pa)
P_{rms} = RMS value of sound pressure (Pa)
P_{ref} = reference sound pressure (Pa)

The reference sound pressure has a value of 2×10^{-5} Pa (20 μPa) or 0.0002 microbar. To repeat, this is the amplitude of the sound pres-

TABLE 22.2 Typical Sound Pressures and Sound Pressure Levels

Sound source	Sound pressure (Pa)	Sound pressure level (dB)	Subjective response
Military jet takeoff at 100 ft (30 m)	200	140	Extreme danger
Artillery fire at 10 ft (3 m)	63.2	130	Extreme danger
Passenger's ramp at jet airliner (peak)	20	120	Threshold of pain
Loud rock band[a]	6.3	110	Threshold of discomfort
Platform of subway station (steel wheels)	2	100	
Unmuffled large diesel engine at 130 ft (39 m)	0.6	90	Very loud
Computer printout room[a]	0.2	80	
Freight train at 100 ft (30 m)	0.06	70	
Conversation speech at 3 ft (0.9 m)	0.02	60	Moderately loud
Window air conditioner[a]	0.006	50	
Quiet residential area	0.002	40	
Whispered conversation at 6 ft (1.6 m)	0.0006	30	Quiet
Buzzing insect at 3 ft (0.9 m)	0.0002	20	
Threshold of good hearing	0.00006	10	Silent
Threshold of excellent youthful hearing	0.00002	0	Threshold of hearing

[a]Ambient.

sure that roughly corresponds to the threshold of hearing at a frequency of 1000 Hz. Because a pascal (Pa) is equal to a newton per square meter (N/m^2), sometimes one will see $P_{ref} = 2 \times 10^{-5}$ N/m^2.

22.2 Use of Decibels

22.2.1 Weighting networks

The sound levels associated with different sound sources are usually measured as a function of frequency. They often contain sound energy over a broad band of frequencies. When it is desired to determine whether or not a noise problem exists or to obtain a measure of the overall sound level in an area, a sound level meter with weighting networks normally will be used. This type of sound level meter usually contains two weighting filters that pass all of the sound energy between the frequencies of 16 Hz and 20,000 Hz. The weighting filters are labeled A and C. Figure 22.3 shows the attenuation characteristics of these filters. Table 22.3 lists the attenuation values as a function of octave band center frequencies. The *A-weighted* filter is most often used when making overall noise measurements. The attenuation of the sound signal with an A-weighted filter at the lower frequencies corresponds to the fact the human ear is not as sensitive to sound at these lower frequencies as it is at the higher frequencies.

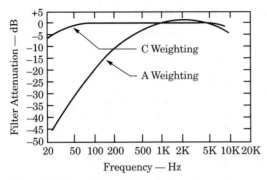

Figure 22.3 A and C weighting networks.

The weighted sound level meter gives only a single number reading for sound level. It does not give any information relative to the frequency content of the signal. However, a weighted sound level meter can be used to tell whether a noise signal contains frequencies primarily above 1000 Hz or below 1000 Hz. To do this, make sound readings using both the A and C weighting networks. Subtract the A-weighted sound pressure level in dB(A) from the C-weighted sound pressure level in dB(C). If the difference is large, the signal is primarily composed of frequencies less than 1000 Hz. If it is small, the signal is primarily composed of frequencies above 1000 Hz.

22.2.2 Octave bandpass filters

Generally, more information is desired concerning the frequency content of a sound signal than that afforded by weighting networks. Such a refinement can be achieved by using octave or third octave bandpass filters. Figure 22.4 shows a set of octave band filters. Table 22.4 lists

TABLE 22.3 Attenuation Associated with Weighting Networks

Frequency (Hz)	Curve A (dB)	Curve C (dB)
16	−56.7	−8.5
31.5	−39.4	−3.0
63	−26.2	−0.8
125	−16.1	−0.2
250	−8.9	0.0
500	−3.2	0.0
1,000	0.0	0.0
2,000	1.2	−0.2
4,000	1.0	−0.8
8,000	−1.1	−3.0
16,000	−6.6	−8.5

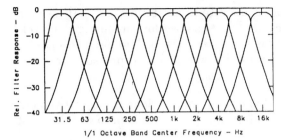

Figure 22.4 Octave band filter set.

the upper and lower band limits and the corresponding band center frequencies for the octave frequency bands from 31.5 to 16,000 Hz.

When a set of octave or third octave bandpass filters are used to analyze a sound signal, only the sound energy with frequency components contained in the frequency bandpass of each filter is allowed to pass through the respective filters. For example, only those parts of a sound signal that have frequencies between 710 Hz and 1420 Hz are allowed to pass through the 1000-Hz octave bandpass filter. All the components of the signal that have other frequencies are prevented from passing through this filter. Thus, if octave bandpass filters are used to analyze a sound signal, it is possible to determine the sound energy that is contained in the octave frequency bands from 31.5 Hz to 16,000 Hz.

22.2.3 Combining decibel levels

Because decibels are logarithmic units, normal addition or subtraction cannot be used to combine decibel pressure levels. The nomograph in

TABLE 22.4 Band Limits and Center Frequencies for Octave Frequency Bands

Lower band limit	Band center frequency	Upper band limit
22	31.5	44
44	63	88
88	125	177
177	250	355
355	500	710
710	1,000	1,420
1,420	2,000	2,840
2,840	4,000	5,680
5,680	8,000	11,360
11,360	16,000	22,720

Figure 22.5 can be used, as it is quite accurate for all HVAC system work. The calculation procedure may be used for combining the levels of two sound sources and for combining the octave band levels of a single noise source to determine the overall sound pressure or sound power level.

1. The combining of octave band data may be done in any convenient sequence. However, combining adjacent levels is probably the most error-free method.

2. The amount added from Figure 22.5 must be added to the higher level of the two levels.

3. Sound pressure levels may be combined or sound power levels may be combined (but not to each other). Both must have the values in decibels re: 0.0002 microbar.

4. Values may have decimals to one place during the combining process, but the final dB value should be rounded off to a whole number.

Example 22.3 Determine the total sound pressure level for the following sound pressure levels: $L_{p1} = 78$ dB, $L_{p2} = 83$ dB, $L_{p3} = 89$ dB

Solution Using the nomogram in Figure 22.5, 83 dB − 78 dB = 5 dB difference. Using the bottom scale, 5 dB lines up with approximately 1.2 dB on the top scale. So 1.2 dB is added to the higher pressure level (83 + 1.2 = 84.2 dB).

Next, 89 dB − 84.2 dB = 4.8 dB difference on the bottom scale lines up with about 1.2 dB, which is added to the higher number (89 + 1.2 = 90.2 dB). The total sound pressure level is 90 dB (use whole numbers).

22.2.4 Background sound

Often when measuring the sound pressure levels associated with sound from a specified sound source, sound from other sound sources may be present. When this occurs, the sound pressure levels associated with the sound from the other sources must be subtracted from the measured sound levels to obtain the correct source sound pressure

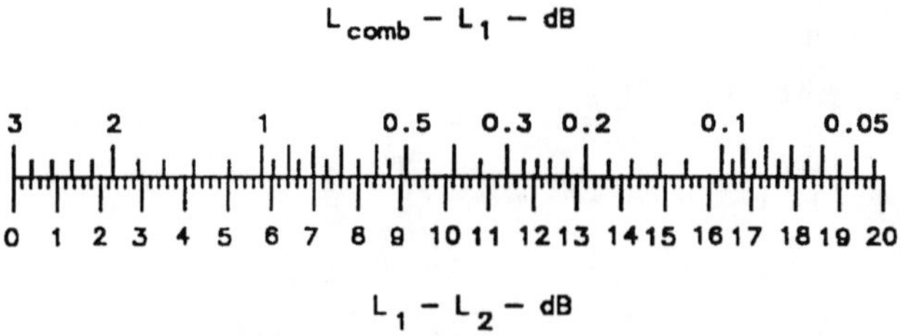

$$L_{comb} - L_1 - dB$$

$$L_1 - L_2 - dB$$

Figure 22.5 Nomogram for combining the sound levels of uncorrelated sound sources.

levels. The nomogram in Figure 22.6 can be used to accomplish this. Let $L_{p \ (source+background)}$ be the sound pressure level associated with a specified sound source in the presence of background sound from other sound sources, and let $L_{p \ (background)}$ be the background sound pressure level with the specified sound source turned off. The sound pressure level, $L_{p \ (source)}$, associated with only the specified sound source is obtained by subtracting the upper scale difference from the total sound pressure.

Example 22.4 Determine the specified source sound level when the following levels are present: Total sound level is 80 dB and Background level is 78 dB.

Solution Using the nomogram in Figure 22.6, 80 dB $-$ 78 dB = 2 dB difference. Using the bottom scale, 2 dB lines up with approximately -4.4 dB on the top scale. So -4.4 dB is subtracted from 80 dB (80 $-$ 4.4 = 75.6 dB). The specified source therefore is about 76 dB (answer rounded off to a whole number).

22.2.5 Combining octave band levels

Each octave band or third octave band segment is a part of the whole. Therefore, the addition of the sound pressure levels in all of the octave or third octave bands will give the "overall" sound pressure level as read on the "C weighting" or "flat" scale of the sound level meter.

Example 22.5 A small room in a plant contains a compressor with known octave band sound pressure levels re: 0.0002 microbar measured at a distance of 10 ft (3 m). Find the total octave band sound pressure levels in the room at 10 ft (3 m) if a fan is added. The fan data were obtained in an acoustically similar room at the same distance (use the nomogram in Figure 22.5).

Solution

Octave band center frequency (Hz)	63	125	250	500	1000	2000	4000	8000
L_p Compressor (dB)	94.0	82.0	80.0	82.0	88.0	88.0	88.0	79.0
Fan (dB)	96.0	82.0	76.0	72.0	68.0	62.0	56.0	50.0
dB difference	2.0	0.0	4.0	10.0	20.0	26.0	32.0	29.0
Add to higher L_p	2.1	3.0	1.5	0.4	0.0	0.0	0.0	0.0
Total L_p in room at 10 ft (3 m)	98.1	85.0	81.5	82.4	88.0	88.0	88.0	79.0

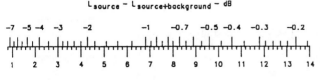

Figure 22.6 Nomogram for determining the sound pressure level of a sound source in the presence of background sound.

Example 22.6   Calculate the overall sound pressure level (dB) from the solution to Example 22.5 (use the nomogram in Figure 22.5).

Solution

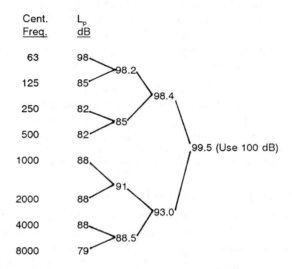

22.3   Response to Sound

22.3.1   Frequency ranges

Table 22.5 and Figure 22.7 indicate some of the significant frequency ranges associated with hearing. As can be seen from the table, the audible frequency range for a normal youth with no hearing loss is from around 16 Hz to 20,000 Hz. For a large number of adults the upper frequency limit may be around 10,000 Hz to 12,000 Hz. The *speech intelligibility range* of 200 to 6000 Hz contains those frequen-

TABLE 22.5   Significant Frequency Ranges for Hearing

Description	Frequency range (Hz)
Range of human hearing	16 to 20,000
Speech intelligibility	200 to 6000
Contains the frequencies most necessary for understanding speech	
Speech privacy range	250 to 2500
Contains speech sounds that intrude most objectionably into adjacent areas	
Typical small table radio	200 to 5000
Male voice	350
Energy output tends to peak	
Female voice	700
Energy output tends to peak	

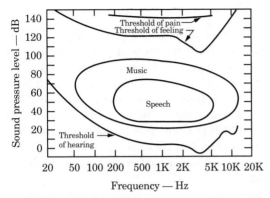

Figure 22.7 Thresholds of hearing.

cies which are most essential for understanding sentence communication. The *speech privacy range* of 250 to 2500 Hz includes those frequencies that tend to interfere most with speech when they intrude into an area from another area.

Most noise control problems require information on sound levels over the frequency range of 63 to 4000 Hz. Originally, this frequency range was 64 to 4096 Hz. The octave band center frequencies originally were 64 Hz, 128 Hz, 256 Hz, 512 Hz, 1024 Hz, 2048 Hz, and 4096 Hz. For convenience, and since the small changes in frequency are not detectable by the ear, the original octave band center frequencies were changed to those listed in Table 22.4.

22.3.2 Subjective response

Hearing can be defined as the subjective response to sound. From a mechanical standpoint, the response of the ear to sound is fairly predictable. Table 22.6 gives the relation between the mechanical characteristics of sound and the subjective response characteristics of the ear.

TABLE 22.6 Subjective Response Characteristics of the Ear

Mechanical characteristics of sound	Subjective response characteristics of the ear
Amplitude, pressure, intensity	Loudness
Frequency	Pitch, timbre
Spectral distribution of energy	Quality

22.3.3 Loudness

Loudness is the physiological response to sound pressure and intensity. It was found that the sound pressure levels of pure tones over a wide range of frequencies were judged to be equally as loud as a 1000 Hz reference tone set at a fixed sound pressure level. These "equal loudness contours" for pure tones are shown in Figure 22.8. The figure indicates that the human ear is much more sensitive to sound at high frequencies (>500 Hz) than it is to sound at low frequencies. For example, a pure tone at 100 Hz must have a sound pressure level of around 54 dB to be perceived as having the same loudness as a 40 dB pure tone at 1000 Hz.

Loudness describes the magnitude of the auditory sensation an individual experiences relative to sound; i.e., a sound is twice, half, three times, etc., as loud as a reference sound. Table 22.7 shows the subjective response of the ear to changes in sound levels. The table indicates that it is usually necessary to have a change of sound level of at least 5 dB for a change in loudness of a sound to be clearly noticeable. A 5-dB reduction in sound level is generally a good "rule of thumb" number to attempt to achieve before the cost of expensive noise control measures can be justified.

22.3.4 Pitch

Pitch is the subjective response of the ear to frequency. Even though pitch is primarily a function of frequency, it is also a function of intensity. For example, if a musical note of 200 Hz frequency is sounded at a moderate and then a high loudness level, nearly all listeners would agree that the louder sound has a lower pitch, in spite of the fact that the frequency remains unchanged.

22.3.5 Timbre

Timbre is the subjective response of the ear to sound that makes it possible to distinguish between two tones that have the same intensity and fundamental frequency but different waveforms. For example, it is easy to recognize the sound of a violin as being different from that of a trumpet, even though these two instruments are sounding the same note with equal intensity. Timbre is primarily a function of waveform, but it is also a function of intensity and frequency.

22.3.6 Quality

Quality of a sound signal refers to the frequency composition of the sound energy contained in the signal. Almost every sound source has

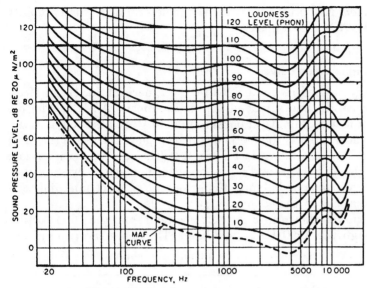

Figure 22.8 Equal loudness contours for pure tone in a free field.

its own unique characteristics that can be associated with the source. The distribution of sound energy associated with a particular sound source within the audible frequency range yields a distinct and unique character to the sound from the source.

22.4   Indoor Noise

HVAC and other types of mechanical and electrical equipment noise are often the primary types of intruding or background noise that exist in many indoor spaces. With regard to the design of HVAC and other types of mechanical and electrical systems, it is necessary to quantify and to determine the acceptability of the noise generated by these systems that intrudes into building spaces. *Room criterion* (RC) and

TABLE 22.7   Subjective Effects of Changes in Sound Levels

Change in sound level	Change in perceived loudness
3 dB	Just perceptible
5 dB	Clearly noticeable
10 dB	Twice or half as loud
20 dB	Much louder or quieter

noise criterion (NC) procedures are used to determine the acceptability of indoor HVAC-related noise.

22.4.1 Room criterion curves

The primary method recommended by the American Society of Heating, Refrigerating, and Air-Conditioning Engineers (ASHRAE) for determining the acceptability of background HVAC-related sound in unoccupied indoor areas is the method which employs the use of the *room criterion* (RC) *curves* shown in Figure 22.9. The curves, as they relate to HVAC system background noise, are based on sound pressure levels, spectrum shape or balance, tonal content of spectrum, and temporal fluctuations in the sound pressure levels.

The RC curves shown in Figure 22.9 extend from the 16 Hz octave band through the 4000 Hz octave band. These are the general fre-

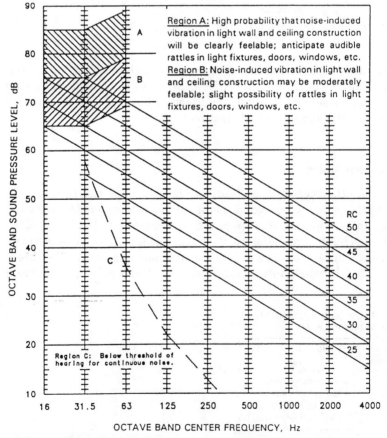

Figure 22.9 Room criterion curves.

quency limits for noise produced by HVAC systems. When determining the RC criterion based on measured octave band sound pressure levels, the lower frequency limit of the analysis is determined by the capabilities of the instrument(s) used to make the sound measurements.

22.4.2 RC curve procedures

Two parts are used in the procedures to determine the RC noise rating associated with HVAC background noise. The first is the calculation of a number that corresponds to the speech communication or masking properties of the noise. The second is designating the quality or character of the background noise.

1. Calculate the arithmetic average of the measured octave band levels in the 500 Hz, 1000 Hz and 2000 Hz octave bands. Round off to the nearest integer. This is the *RC level* of the room or space.

2. Draw a line with a −5 dB per octave slope that passes through the *calculated* RC level at 1,000 Hz. For example, if the RC level is RC 33, the line will pass through a value of 33 at the 1000 Hz octave band. This average value sometimes may not be equal to the measured value of the octave band sound pressure level of the background noise in the 1000 Hz octave band.

3. At a distance 5 dB above the −5 dB per octave slope line, draw a parallel dotted line from 16 Hz to 500 Hz (Figure 22.10). At a distance 3 dB above the −5 dB per octave slope line, draw a parallel dotted line from 500 Hz to 4000 Hz. The location of the measured octave band levels in relation to these lines will determine the subjective character of the background noise.

22.4.3 Subjective character

22.4.3.1 Neutral noise. Noise that is classified as neutral has no particular identity with frequency. It is usually bland and unobtrusive. Background noise that is neutral usually has an octave band spectrum shape similar to the RC curves in Figure 22.9. If the octave band data do not exceed the RC curve by 5 dB at frequencies of 500 Hz and below and by more than 3 dB for frequencies of 1000 Hz and above, the background sound is neutral and (N) can be placed after the calculated RC level.

22.4.3.2 Rumbly noise. Noise has a rumble from an excess of low-frequency sound energy. If any of the octave band sound pressure levels below the 500 Hz octave band are more than 5 dB above the RC curve associated with the background noise in the room, the noise

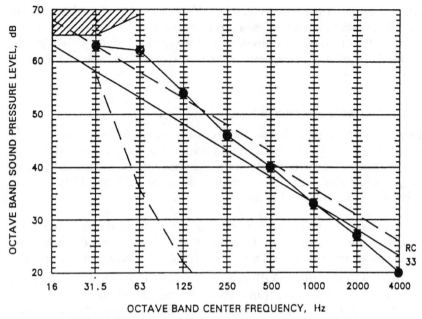

Figure 22.10   RC level for example 22.7.

will be judged to have a "rumbly" quality or character. If the background sound has a rumbly quality, place (R) after the RC level.

22.4.3.3 Tonal noise.   Noise that has a tonal character usually contains a humming, buzzing, whining, or whistling sound. When a background sound has a tonal quality, it will generally have one octave band in which the sound pressure level is noticeably higher than the other octave bands. If the background sound has a tonal character, place (T) after the RC level.

22.4.3.4 Acoustically inducted perceptible vibration.   The cross-hatched region of the RC curves in Figure 22.9 indicate the sound pressure levels in the 16 to 63 Hz octave frequency bands at which perceptible vibration in the walls and ceiling of a room can occur. These sound levels can be associated with rattles in cabinet doors, pictures, ceiling fixtures, and other furnishings in contact with walls or ceilings. If the background sound levels fall in this region, place (PV) after the RC level.

22.4.4 Noise spectrum

It is desirable to have background sound that has an octave band spectrum that has a neutral character or quality. If the noise spectrum is such that it has a rumble, hiss, or tonal character, it will generally be judged to be objectionable.

Room criterion procedures can be easily used to determine the acceptability of sound in an area based on measured sound pressure levels in the area. However, care must be exercised in using room criterion procedures for determining the acceptability of sound in an area based on calculated sound pressure levels. A balanced sound spectrum in a room is comprised of sound from the fan and duct system and sound from air flow through the air diffusers in the room. Sound from both sources must be present in the system sound calculations when using room criterion procedures to determine the acceptability of sound.

Example 22.7 The measured octave band sound pressure levels of background noise in an office area are given below:

Octave band center frequency (Hz)	31.5	63	125	250	500	1000	2000	4000
L_p (dB)	63	62	54	46	40	33	27	20

Determine the RC level and the corresponding character of the noise.

Solution The RC level is determined by obtaining the arithmetic average of the octave band sound pressure levels in the 500 Hz, 1000 Hz, and 2000 Hz octave bands, or

$$RC = \left[\frac{40 + 33 + 27}{3} \right] = 33 \text{ dB}$$

Thus, the RC level is RC 33. The measured octave band sound pressure levels for the background noise are plotted in Figure 22.10. The −5 dB octave slope line (level in 1000-Hz octave band is 33 dB) is shown in Figure 22.10. A dashed line 5 dB above the RC 33 slope line or curve for frequencies below 500 Hz and a dashed line 3 dB above the RC 33 curve for frequencies from 500 Hz and above also are shown in the figure. An examination of the figure indicates that at frequencies below the 250 Hz octave band, the octave band sound pressure levels of the background noise are 5 dB or more above the RC 33 curve. Thus, the background noise has a "rumbly" character. The octave band sound pressure levels above 500 Hz are equal to or below the RC 33 curve, so there is no problem at these frequencies. The RC rating of the background noise is RC 33(R).

22.4.5 Noise criteria curves

Noise criteria (NC) curves are shown in Figure 22.11. These curves apply to steady noise and specify the maximum noise levels permitted in each octave band for a specified NC curve. For example, if the noise requirements for an activity area call for a NC 20 rating, the sound pressure levels in all eight octave frequency bands must be less than or equal to the corresponding values for the NC 20 curve. Conversely

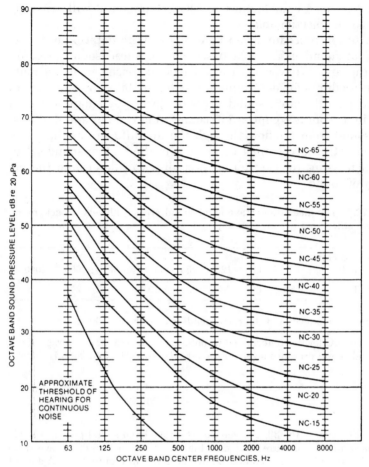

Figure 22.11 Noise criteria curves.

the NC rating of a given noise equals the highest penetration of any of the octave band sound pressure levels into the curves. If the farthest penetration falls between two curves, the NC rating is in the interpolated value between the two curves.

In the past, NC criteria curves have been used to specify acceptable HVAC background sound levels in indoor areas. However, experience has indicated when HVAC background noise is present, the use of NC levels has often resulted in a poor correlation between the calculated NC levels and an individual's subjective response to the corresponding background noise.

ASHRAE no longer recommends the use of NC criteria curves as a method for determining the acceptability of HVAC background sound levels in unoccupied indoor areas (Table 22.8). However, many HVAC design engineers still specify and use NC levels.

Example 22.8 The following octave band sound pressure levels were measured in a laboratory work area. Find the NC rating of the noise in the work area.

Octave band center frequency (Hz)	63	125	250	500	1000	2000	4000	8000
L_p (dB)	50	55	58	58	55	50	45	39

Solution Figure 22.12 shows a plot of the above data relative to the NC curves found in Figure 22.11. Since the octave band sound pressure level in the 500-Hz octave band penetrates to the NC 55 curve, the NC rating of the work area is NC 55.

TABLE 22.8 Design Guidelines for HVAC System Noise in Unoccupied Spaces[a]

Space	RC(N)[b]	Space	RC(N)[b]
Private residences, apartments, condominiums	25–35	Laboratories (with fume hoods)	
Hotels or motels		Testing or research, minimal speech communication	45–55
Individual rooms or suites	25–35	Research, extensive telephone use, speech communication	40–50
Meeting or banquet rooms	25–35		
Halls, corridors, lobbies	35–45	Group teaching	35–45
Service and support areas	35–45	Churches, mosques, synagogues with critical music programs	25–35[c]
Office buildings			
Executive and private offices	25–35	Schools	
Conference rooms	25–35	Classrooms up to 750 ft^2 (70 m^2)	40 (max)
Teleconference rooms	25 (max)		
Open plan offices	30–40	Classrooms over 750 ft^2 (70 m^2)	35 (max)
Circulation and public lobbies	40–45		
Hospitals and clinics		Lecture rooms for more than 50 (unamplified speech)	35 (max)
Private rooms	25–35		
Wards	30–40	Libraries	30–40
Operating rooms	25–35	Courtrooms	
Corridors	30–40	unamplified speech	25–35
Public areas	30–40	amplified speech	30–40
Performing arts spaces		Indoor stadiums and gynasiums	
Drama theaters	25 (max)	School and college gynasiums and natatoriums	40–50[d]
Concert and recital halls	c		
Music teaching studios	25 (max)	Large seating capacity spaces (with amplified speech)	45–55[d]
Music practice rooms	35 (max)		

[a]These values and ranges are based on judgment and experience, not on quantitative evaluations of human reactions. They represent general limits of acceptability for typical building occupancies. Higher or lower values may be appropriate and should be based on a careful analysis of economics, space use, and user needs. They are not intended to serve by themselves as a basis for a contractual requirement.

[b]When the quality of the sound in the space is important specify criteria in terms of RC(N). If the quality of the sound in the space is of secondary concern, the criteria may be specified in terms of NC criteria.

[c]An experienced acoustical consultant should be retained for guidance on acoustically critical spaces (below RC 30) and for all performing arts spaces.

[d]Spectrum levels and sound quality are of lesser importance in these spaces than overall sound levels.

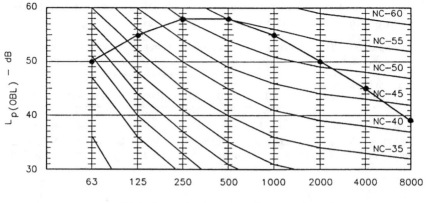

1/1 Octave Band Center Frequency — Hz

Figure 22.12 NC level for example 22.8.

22.4.6 Indoor measurements

22.4.6.1 Near field region Many sound measurement errors are made by locating the microphone of the sound level meter in the "near field region" of a sound source. The *near field region* is approximately one-fourth the distance of the sound wavelength in feet (meters) when making octave band measurements.

22.4.6.2 Wavelengths A knowledge of the wavelength of the sound also is important when determining the effectiveness of sound barriers. Sound waves radiating from a source will be reflected by objects in their paths. If the objects are large compared with the wavelength of sound, the objects will have a large effect upon the sound wave. On the other hand, if the objects are small, they will have little effect on the sound.

In general, low-frequency waves (long wavelengths) are only affected by large obstructions, attenuated by heavy sound barriers, or absorbed by thick sound absorption materials. High-frequency waves (short wavelengths) are affected by both large and small obstructions, may be attenuated by relatively light-weight sound barriers, and are absorbed by thin sound absorption materials.

Example 22.9 A compressor is radiating a pure tone of about 1000 Hz. What is the wavelength of 1000 Hz? What is the absolute minimum distance between the compressor and the wavelength location?

Solution (U.S.)

$$\lambda = \frac{c}{f} = \frac{1300}{1000} = 1.13 \text{ ft. (13.6 in.) length}$$

$$A \text{ (near field)} = \frac{\lambda}{4} = \frac{1.13}{4} = 0.28 \text{ ft. (3.36 in.)}$$

Solution (Metric)

$$\lambda = \frac{c}{f} = \frac{343}{1000} = 0.343 \text{ m length}$$

$$A \text{ (near field)} = \frac{\lambda}{4} = \frac{0.343}{4} = 0.086 \text{ m (86 mm)}$$

22.4.6.3 Discussion of Example 22.9. The closest microphone location is about 3.4 in. (86 mm) from the compressor. However, the more preferable location would probably be at least one wavelength of 13.6 in. (343 mm) from the compressor. It should be noted, however, that this applies to the measurement of the pure tone (1000 Hz) only. The compressor will generate lower frequencies and the accurate measurements of these lower frequencies will require a microphone location further away.

A rule of thumb is that the measurement should be made away from the machine at least two to three times the largest radiating dimension of the machine, or at a distance corresponding to one wavelength of the lowest frequency to be measured—whichever is the greatest.

22.4.6.4 Erroneous readings. Erroneous readings also occur when the directivity of a sound source is not adequately determined. Always be alert for unusual readings. Check equipment connections and check the sound level meter with the calibrator.

If the background noise levels are at least 8 dB(A) less than the sound pressure levels with the noise source operating, then no correction normally is required.

Operator interference or carelessness also can cause measurement errors. If a tripod for the microphone cannot be used, the microphone should be held at arms length, oriented so the meter or microphone is not directly between the sound source and the operator.

22.4.6.5 Normal sound measurements. Interior sound measurements normally require *A-weighted measurements, linear measurements,* and *octave band measurements* using a "random-incidence" microphone. A "pressure" microphone also may be used. The sound level meter must be calibrated and used according to the manufacturer's instructions.

The contract specifications should be checked as to the distance from the microphone to the noise source(s) or, if the noise source is not in the same room, the specified location where the sound pressure levels are to be measured. If this information is not contained in the contract specifications, the following locations should be used:

1. *Ceiling, wall, sill, or floor diffuser outlets:* Set tripod or hold the sound level meter so that the microphone is around 4 ft. (1.2 m) from the floor, 5 ft. (1.5 m) from the center of a diffuser, and at least 3 ft. (0.9 m) from a wall or other sound reflecting surface. If there is more than one diffuser in the room, select a diffuser location that is near the center of the room or a location in the room where the sound level is highest.

2. *Return air grilles:* Same as diffusers.

3. *Adjacent mechanical equipment room or duct pipe shaft:* Place microphone around 4 ft. (1.2 m) above the floor and at nearest desk location to the partition or floor that separates the occupied space from the mechanical equipment room. The microphone must be at least 3 ft. (0.9 m) from the nearest wall or sound reflecting surfaces (does not include floor).

4. *Air pressure reduction devices, terminal devices, air valves, single and double duct units, noise generating ductwork located above a suspended ceiling:* Place microphone around 4 ft. (1.2 m) above the floor and directly below the location where audible observation indicates noise is at a maximum.

5. *Inside mechanical equipment room:* Make a sketch of the room, locating noise generating equipment and sound measurement points. Make sound level measurements at indicated points with the microphone located around 5 ft. (1.5 m) above the floor. Key points include walls that are adjacent to noise-sensitive rooms, return air openings to open-ceiling plenums, outdoor air inlets, and equipment with exceptionally high noise levels.

22.4.6.6 Background noise measurements

1. Check for the possible intrusion of noise from sources other than that being measured in accordance with building specifications. This intruding noise is called *background noise*. The only completely effective way to determine whether background noise is influencing the measurements is to turn off the noise source being measured. When this can be done, the sound level measurements must be made at the same location. If the background sound pressure level is at least 8 dB(A) less than the sound pressure level with the noise source operating, no background noise level correction is required. If the background noise level is less than 8 dB(A)

below the noise source, use the proper procedures to correct the measured sound pressure level of the sound source for the presence of background noise.

2. If the noise source to be measured cannot be turned off, several procedures may be followed to determine the influence, if any, of the background noise on the sound level measurements.

3. If space permits, move the sound level meter very slowly away from the noise source but not closer than 3 ft. (1 m) to a wall or another noise source. If the sound levels decrease by 3 dB(A) or more every time the distance between the noise source and the microphone is doubled, and if, at the maximum distance, the reduction totals 8 dBA or more, it is probable that background noise levels are not influencing the measurements.

4. Insert the input plug of monaural headphones into the sound level meter AC output and listen for audible extraneous signals above the level of the steady noise source signal.

22.4.7   Outdoor measurements

Many sound sources can be characterized as emitting sound equally in all directions. These sound sources are referred to as *spherical sound sources* (see Figure 22.2). With respect to spherical sound sources, there are four radiation fields of interest. Two are associated with the position relative to the sound source. They are the *acoustic near field* and the *acoustic far field* (Figure 22.13).

22.4.7.1   Near field. The extent of the near field depends upon the frequency of the sound being generated, the radiation characteristics of the source, and the characteristic sound source dimensions. If the sound source is a simple spherical source, the near field can extend to a distance of around two wavelengths from the source. However, in many cases, reasonably accurate sound pressure measurements can be made, beginning at distances of a quarter wavelength from the sound source.

If the sound source is rather irregularly shaped and large and has a definite direction radiation pattern (i.e., it does not radiate evenly in all directions), the near field can extend to a distance of about two to five times the largest characteristic dimension of the source.

22.4.7.2   Far field. As the wave continues to move away from the source, the sound pressure and particle velocity approach the state where they are in phase. This region is called the *acoustic far field*.

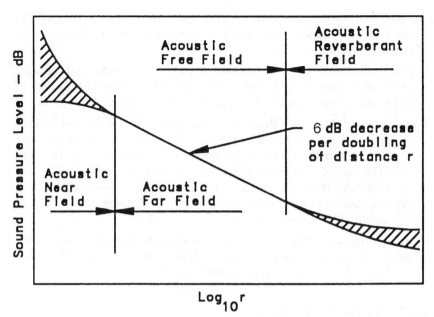

Figure 22.13 Radiation fields of a spherical sound source.

The sound pressure in the far field is inversely proportional to the distance r from the source. The relation between sound pressure and distance from the sound source is shown by Equation 22.5.

Equation 22.5

$$P_2 = \frac{P_1 r_1}{r_2}$$

Where: P = sound pressure (Pa)

r = distance from sound source (ft or m)

Every doubling of the distance from the sound source results in a halving of the sound pressure, or as will be shown later, a doubling of distance from the sound source results in a 6-dB decrease in the sound pressure level.

22.4.7.3 Free field. There are two other radiation fields that are important: the *free field* and the *reverberant field*. The *free field* is the field in which only direct radiated sound waves moving away from the sound source are present (Figure 22.13). There are no reflected sound waves present in the free field. A free field exists when a sound source is located a large distance from reflecting surfaces or when nearby surfaces are highly absorbent, such that no sound waves are reflected from the surfaces.

22.4.7.4 Reverberant field. If there are reflecting surfaces that have little or no sound absorption at a specified distance from a sound source, reflected sound waves will be generated and superimposed on the directly radiated sound waves. Thus, the sound field will consist of both directly radiated and reflected sound waves. The region in which this occurs is called *reverberant sound field.* If there are many reflected waves crisscrossing from all directions, the reverberant field is referred to as a *diffuse sound field.*

22.4.7.5 Measurements near the source. When measuring a *nondirectional noise source* with the microphone located within about 100 ft. (30 m) of the source, and with no nearby reflecting surfaces, the following relationship can be assumed:

Equation 22.6

$$L_{p2} = L_{p1} + 20 \log_{10} D_1 - 20 \log_{10} D_2$$

Where: L_{p1} = sound pressure level, dB re 0.0002 microbar (or dBA sound level), at position 1.

L_{p2} = sound pressure level, dB re 0.0002 microbar, (or dBA sound level), at position 2.

D_1 = distance (in feet or meters) from noise source, position 1.

D_2 = distance (in feet or meters) from noise source, position 2.

As stated above, under free-field conditions where the noise source is in the air far from any reflecting surfaces, the sound pressure is halved for each doubling of distance from the source. This change results in a 6 dB reduction in noise level for every doubling of distance.

If the sound power (L_w) of the noise source is known, the sound pressure level at any distance can be calculated from the following:

Equation 22.7 (U.S.)   **Equation 22.7 (Metric)**

$$L_p = L_w - 20 \log_{10} D - 0.5 \text{ dB} \qquad L_p = L_w - 20 \log_{10} D - 10.5 \text{ dB}$$

Where: L_p = sound pressure level, in dB re 0.0002 microbar.

L_w = sound power level of the source, in dB re 10^{-12} W.

D = distance in feet (meters) from the point source to the point where the sound pressure is measured.

If the noise source is *directional*, such as a mechanical equipment room louver opening in the side of a building or a fan discharge, the sound pressure level of Equation 22.7 shall be corrected for directivity as shown on the graph in Figure 22.14.

22.4.7.6 Measurements away from the source. When the noise source is 100 ft. (30 m) or further from the measuring point, the 6 dB per doubling of distance (also called the inverse square law) relationship must be corrected for air absorption, which has greater effect at high frequencies. The air absorption correction added to the results of Equations 22.6 and 22.7 is given in Table 22.9 for distances in excess of 100 ft. (30 m) from the source. These numbers should be used in place of $20 \log_{10} D$.

> **Example 22.10** Figure 22.15 shows the distance relationship between the air intake louvers of a new building and a nearby residential area. Measurements cannot be made at the residential boundary because a large topsoil storage pile blocks the line of sight between the nearest residence and the air intake louvers. Measurements can be made at a distance of 50 ft. (15 m) from the air intake louvers. The nearest residence is on the same line. The noise level (L_{p1}) at the 50 ft (15 m) location, designed as D_1, is 72 dB(A). Find the estimated noise level (L_{p2}), at the residential boundary (D_2) which is 182 ft. (55 m) from the air intake louver.
>
> **Solution** Using Equation 22.6:
>
> $$L_{p2} = L_{p1} + 20 \log_{10} D_1 - 20 \log_{10} D_2$$
>
> $$L_{p2} = 72 + 20 \log 50 - 10 \log 182$$
>
> Using logarithm tables found in Appendix G or on a calculator:
>
> $$20 \log 50 = 20 \times 1.70 = 34.0$$
>
> $$20 \log 182 - 20 \times 2.26 = 45.2$$

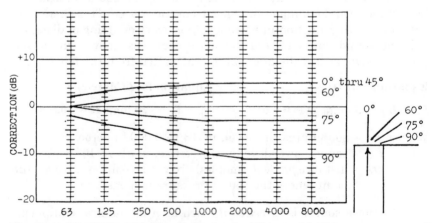

Figure 22.14 Directivity Correction Octave band midfrequencies (Hz).

TABLE 22.9   10 Log$_{10}$ D in dB at Frequency (Hz)

Distance D [ft (m)]	31–250	500	1000	2000	4000	8000
100 (30)	38	38	38	38	39	39
112 (34)	39	39	39	39	40	41
126 (38)	40	40	40	40	41	42
141 (42)	41	41	41	41	42	43
158 (47)	42	42	42	42	43	44
178 (53)	43	43	43	43	44	46
200 (60)	44	44	44	44	46	47
224 (67)	45	45	45	46	47	48
252 (76)	46	46	46	47	48	50
282 (85)	47	47	47	48	49	51
316 (95)	48	48	48	49	50	53
356 (107)	49	49	49	50	52	54
400 (120)	50	50	51	51	53	56
448 (134)	51	51	52	52	54	57
504 (151)	52	52	53	54	56	59
564 (169)	53	53	54	55	57	61
632 (190)	54	54	55	56	59	63
712 (214)	55	56	56	57	60	65
800 (240)	56	57	57	58	62	67
900 (270)	58	58	58	60	64	70
1000 (300)	58	59	59	61	66	72

$$L_{p2} = 72 + 34 - 45$$

$$L_{p2} = 61 \text{ dB(A)}$$

22.5   Vibration

Vibration is the movement of an object or group of objects moving back and forth from a state of rest. Vibration of HVAC equipment is caused by a force that is changing in its direction or its amount. The resulting

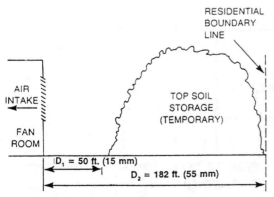

Figure 22.15   Drawing for example 22.10.

characteristics of the vibration will be determined by the manner in which the forces are generated.

The most important of these vibration characteristics are frequency, displacement, velocity, acceleration, and phase.

22.5.1   Vibration terms

22.5.1.1   Period and frequency. The amount of time required to complete one full cycle of a vibration pattern is called the *period of vibration* (Figure 22.16). If a machine completes one full cycle of vibration in 1/60th of a second, the period of vibration is said to be 1/60th of a second. The period of vibration is a simple and meaningful characteristic that often is used in vibration detection and analysis. A characteristic of equal simplicity and more meaning is vibration *frequency*.

Vibration frequency is the measure of the number of complete cycles that occur in a specified period of time, usually the number of cycles per minute (cpm). The cpm also relates to the rpm of rotating equipment, since a 1800 rpm pump may cause vibration problems at a frequency of 1800 cpm.

Frequency also may be measured in Hertz (Hz):

Equation 22.8

cpm = Hz × 60

22.5.1.2   Displacement. The total distance traveled by the vibrating object, from one extreme limit of travel to the other extreme limit of travel is referred to as the *peak-to-peak displacement* (Figure 22.16). Peak-to-peak vibration displacement is usually expressed in mils,

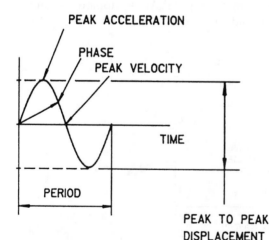

Figure 22.16   Vibration terms.

where 1 mil equals one-thousandth of an inch (0.001 in.). In metric units, the peak-to-peak vibration displacement is expressed in micrometers (μm), where 1 μm equals one thousandth of a millimeter (0.001 mm). One mil equals 25.4 μm.

22.5.1.3 Velocity. Since a vibrating object is moving, it must be moving at some speed. However, the speed of the object is changing constantly. At the top limit of the motion, the velocity is zero since the weight must come to a stop before it can go in the opposite direction. The peak velocity or greatest velocity is obtained as the object passes through the neutral position. The velocity of the motion is definitely a characteristic of the vibration, but since it is constantly changing throughout the cycle, the highest peak velocity is selected for measurement. Vibration velocity is expressed in terms of inches per second (in./s) peak [or millimeters per second (mm/s) peak].

22.5.1.4 Acceleration. As a vibrating object stops at the upper and lower limits, it must *accelerate* to pick up speed as it travels toward the other limit of travel. Vibration acceleration is another important characteristic of vibration and is the *rate of change of velocity.*

The acceleration of the object is maximum at the extreme limits of travel where the velocity is zero. As the velocity of the object increases, the acceleration decreases. At the neutral position, the velocity is maximum and the acceleration is zero. As the object passes through the neutral point, it must now *decelerate* as it approaches the other limit of travel (Figure 22.17).

Vibration acceleration is normally expressed in terms of "G" peak, where 1 G is the acceleration produced by the force of gravity at sea level—32.2 ft/s^2 (9.8 m/s^2).

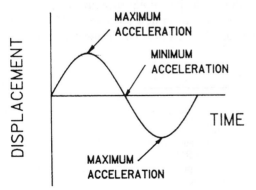

Figure 22.17 Acceleration.

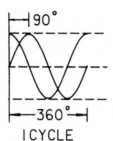

Figure 22.18 90° phase relationship.

I CYCLE

22.5.1.5 Phase

Vibration *phase* is the relationship of one vibrating object to another vibrating object at a fixed point. For example, if two objects are vibrating at the same frequency and displacement but one is at the upper limit of travel at the same instant the other is at the lower limit, the two objects are vibrating 180° *out of phase*. By plotting one complete cycle of motion of these two objects, starting at the same given instant, the points of peak displacement are separated by 180° (one complete cycle = 360°).

Figure 22.18 graphically shows two objects vibrating 90° out of phase.

Table 22.10 summarizes the characteristics of vibration.

22.5.2 Vibration isolation

22.5.2.1 HVAC equipment. The requirement for HVAC equipment isolators with a large amount of static deflection has been caused by the location of machinery on light-weight structures adjacent to critical areas. Pad materials such as rubber, cork, lead, etc., provide minimal deflections, which are adequate to isolate high-frequency noise. Since these deflections are small when compared to upper floor deflections, their use should be limited to the isolation of equipment on grade.

Steel springs can be used to obtain as much as 5 in. (125 mm) of deflection and are now widely used to isolate equipment located on

TABLE 22.10 Vibration Characteristics

Vibration characteristics	U.S. units	Metric units
Frequency	cpm	cpm
Displacement	Mils, peak-to-peak	mm, peak-to-peak
Velocity	in./s RMS	mm/s RMS
	in./s peak	mm/s peak
Acceleration	G peak	G peak
Phase	Degrees	Degrees

upper floors. Preference has been given to stable single springs that do not require housings. The equivalence of 7 to 8 in. (175 to 200 mm) of deflection can be obtained from the ever more popular air spring, which basically is a large rubber bladder designed to hold as much as 100 psi (690 kPa) of air pressure.

22.5.2.2 Theory of HVAC machinery bases. The HVAC machinery base must perform several functions:

a. Keep equipment in alignment—such as to maintain a structural tie between a fan and motor, or to provide a common base for a turbine driven compressor.

b. Provide stability for a tall machine such as an absorption machine.

c. Tie a complete package together as in the case of a long, many sectioned HVAC unit.

d. Floating concrete bases are normally used for pumps because pump bases are designed to be grouted to concrete.

e. For inertial reasons, when an increase in mass is required to resist either the imbalance of the equipment or external forces.

f. Offer resistance to external forces such as fan thrust. This is especially important for units operating above 6 in.w.g. (1500 Pa) static pressure.

22.5.3 Vibration measurements

22.5.3.1 Measurement terms. Measurement of vibration using the sound level meter and vibration integration system can be made in terms of RMS displacement, RMS velocity, and RMS acceleration, depending on which term is required. If not specified, obtain values for all three terms at all center frequency octave bands. The sound level meter, octave band analyzer, and vibration integration system will read out in decibels using the references and definitions in Table 22.11.

Conversion from decibel levels to acceleration velocity and displacement (re noted references) can be made using equations or by means of graphs, tables, or a circular slide rule often furnished with the S&V equipment.

TABLE 22.11   Vibration Definitions

Vibration level	Definition	Reference
Acceleration Level (L_a)	$L_a = 20 \log_{10} a/a_0$	Re: $a_0 = 10^{-5}$ m/s^2
Velocity Level (L_v)	$L_v = 20 \log_{10} v/v_0$	Re: $v_0 = 10^{-8}$ m/s
Displacement Level (L_d)	$L_d = 20 \log_{10} d/d_0$	RE: $d_0 = 10^{-11}$ m

22.5.3.2 Measurements. The great majority of the measurements of vibration will be to determine whether the excitation of a portion of the building structure resulting from vibration from HVAC equipment is radiating noise in excess of criteria or is the source of "feelable" and "annoying" vibration. Table 22.12 submits maximum acceleration values for acceptable vibration environments.

In normal use, accelerometers or pickups often are subjected to quite violent treatment. When dropped onto a concrete floor from hand height, an accelerometer can be subjected to a shock of many thousand G. It is necessary, therefore, to make a periodic check of the sensitivity calibration to confirm that the accelerometer has not been damaged. The most convenient means of performing a periodic calibration check is by using a battery powered, calibrated vibration source that has a small builtin shaker table that can be adjusted to vibrate at precisely 1 G.

The vibration integrator system used with sound level meters has some limitations not present in special purpose equipment that is used solely for measuring vibration. The vibration integrator system has a low-frequency cutoff of about 10 Hz. Since mechanical systems tend to have most of their vibration energy contained in the relatively narrow frequency range between 10 Hz and 1000 Hz, this cutoff is not serious if the S&V report clearly indicates the range of the measurements that have been made.

22.6 Sound and Vibration Instruments

22.6.1 Sound level meters

The American National Standards Institute (ANSI) Standards S1.4, *American National Specification for Sound Level Meters*, and S1.11, *American Standard Specification for Octave-Band and Fractional Octave-Band Analog and Digital Filters*, establishes the accuracy requirements for the measuring instruments. NEBB certified sound and vibration firms use type 1 precision instruments.

TABLE 22.12 Maximum Acceleration Values

Use of space	Time	RMS acceleration[a]
Critical areas such as hospital operating rooms	Anytime	$0.0036/\sqrt{t}$ m/s^2
Residences	Day	$0.072/\sqrt{t}$ m/s^2
	Night	$0.005/\sqrt{t}$ m/s^2
Offices	Anytime	$0.14/\sqrt{t}$ m/s^2
Factories or workshops	Anytime	$0.28/\sqrt{t}$ m/s^2

[a]t = time in seconds up to 100 s. For times longer than 100 s, use $t = 100$ s. (All values are in m/s^2 for the frequency range of 1 to 80 Hz.)

It should be emphasized that the accuracy requirements given in ANSI S1.4 refer to a single-frequency signal (pure tone) with the meter setting at 80 dB and not to the random-type noise usually encountered in field measurements. Although the instrument accuracy is ± 0.5 dB, the overall accuracy for general measurements in the field is ± 1.5 dB.

22.6.1.1 SLM requirements. At a minimum, the sound level meter (SLM) should have:

1. A, C, and linear weighing networks
2. Octave and third octave band filter sets
3. Fast, F, and slow, S, exponential time averaging
4. RMS and peak detectors
5. Overload detector
6. Peak-hold capability for measuring impulse or short-duration sound levels
7. AC and DC outputs for external recording devices
8. Accuracy: ± 0.5 dB
9. Matching calibrator for sound level meter

22.6.1.2 Microphone Diffuse-field sound measurements are generally associated with making sound measurements in an environment where sound waves can arrive at the microphone from many angles simultaneously. Most indoor sound measurements are diffuse-field measurements. The random-incidence microphone is recommended for this type of sound measurement. Pressure microphones may also be used for diffuse-field sound measurements.

22.6.1.3 SLM filters. If the sound signal being measured has no pure tones or contains no significant energy levels within narrow frequency bandpasses, the measurement of octave band levels is usually sufficient. However, if pure tones or significant narrow band levels exist, it may be necessary to measure third octave band levels.

The octave band filter set will cover the octave band center frequencies from 31.5 Hz to 16,000 Hz. The third octave band filter set will cover the third octave band center frequencies from 25 Hz to 20,000 Hz.

22.6.1.4 Windscreen. Often it is necessary to make sound measurements in areas where there is an unwanted airflow. Indoors, this can be related to sound measurements around air handling equipment and air distribution or ventilating systems. Outdoors, this can be associated with making sound measurements in the presence of wind.

A spherical light foam windscreen placed over the microphone can be used to minimize the errors resulting from air turbulence acting on the microphone. Sound measurements should not be attempted in moving air streams that have velocities significantly greater than 25 mph (11.2 m/s). Windscreens should be used for all sound measurements in moving airstreams with velocities greater than 5 mph (2.2 m/s).

22.6.2 SLM calibrators

All type 1 precision sound level meters and corresponding microphones are individually calibrated at the factory with appropriate traceability to National Institute of Standards and Technology (NIST). However, the performance of sound level meters must be checked each time they are used for sound measurements. This can be accomplished with a calibrator that applies a known sound pressure level at a fixed frequency (or at fixed frequencies) to the microphone. The calibrator should be obtained from the manufacturer of the sound level meter, and it must fit the specific microphone to which it is to be coupled. Most common calibrators fit directly over the microphone and generate a known pressure level (usually 94 dB, 114 dB, or 124 dB) within the enclosed volume around the microphone. The sound pressure level comes from the "back and forth" motion of a piston, from a small loudspeaker, or from another microphone used as a loudspeaker.

Calibrators are intended to indicate small changes that must be made to the sound level meter readout to achieve accurate sound measurements. These changes should be less than ±2 dB. If larger adjustments are required, the sound level meter and calibrator should be sent to the manufacturer for a complete checkup and recalibration. **The calibrator should be recalibrated at least once every 12 months.**

22.6.3 Vibration equipment

The same instruments used for sound measurements may also be used for vibration measurements. Equipment for vibration measurements should consist of:

1. Accelerometer or vibration pickup

2. Sound level meter with integrated or attached third octave band filter set, or a frequency analyzer

3. Calibrator

4. External recording device (optional).

22.6.3.1 Field measurements. For field measurements, the accelerometer should have integrated electronics so that the output of the accelerometer is a voltage signal proportional to acceleration. If a frequency analyzer is used that allows for a charge input, then the accelerometer can have a charge output proportional to acceleration.

If occupant comfort or vibration exposure of sensitive equipment is being investigated, the analysis frequency range is usually 1 to 100 Hz. If vibration induced noise is being investigated, the analysis frequency range may go as high as 1000 Hz. It is often difficult to obtain accurate field vibration measurements at frequencies above 1000 Hz.

If an integrated or attached third octave band filter set associated with a sound level meter is used for vibration measurements, the filter set must go down to the 1 Hz third octave band center frequency. Most frequency analyzers will have third octave band filters that range from 1 Hz to 20,000 Hz. In general, third octave band vibration measurements are sufficient for most vibration analyses.

22.6.3.2 Attaching the vibration pickup. The point at which the vibration pickup (accelerometer or transducer) should be placed is usually obvious. When it is not obvious, some exploration of the vibration pattern over a portion of the building structure or machinery is necessary to locate maximum vibration. The method of attaching the vibration pickup can seriously affect its performance.

Most vibration pickups are most sensitive to vibrations in the direction perpendicular to the largest flat surface on the pickup. Consult the manufacturer's manual for position of maximum sensitivity. The directivity of maximum motion may also be determined by experiment.

The vibration pickup or accelerometer should load the vibrating body on which it is mounted as little as possible because any additional load may change the original motion of the vibrating body. When obtaining measurements of concrete constructions or heavy machinery parts, this additional loading may have inconsequential results; however, when the vibration pickup is attached to light-gauge sheetmetal, some damping may occur.

The method of attaching the accelerometer is one of the most critical factors in obtaining accurate data. Sloppy attachment results in a reduction in the resonant frequency. The ideal way is with a threaded stud. However, this rarely is possible when measuring the vibration of a concrete slab.

A common alternative attachment method is the use of a thin layer of beeswax for sticking the accelerometer into place. However, the sur face of the concrete must be cleaned before the beeswax is applied. Because beeswax becomes soft at higher temperatures, this method is restricted to a maximum temperature near 100 °F (18 °C).

A hand-held probe with the accelerometer mounted on top is very convenient for quick-look survey work, but can give gross measuring errors because of the low overall stiffness.

22.6.4   Vibration calibration

There are two primary methods that can be used to calibrate an instrument system that is used to make vibration measurements. One involves the use of a vibration calibrator shaker calibrated to NIST standards and the other involves comparison measurements using an accelerometer calibrated to NIST standards.

22.6.4.1   Vibration calibration shaker.   A vibration calibration shaker is a small shaker that generates a known acceleration amplitude (usually 1 G) at a specified frequency. An accelerometer can be attached to this shaker and then excited at the known (usually 1 G) acceleration amplitude. The voltage output of the accelerometer can then be recorded. The accelerometer calibration is obtained by dividing the measured voltage amplitude by the known acceleration amplitude. If English units are used, the sensitivity or calibration of the accelerometer is usually given in the units of mV/G.

22.6.4.2   Accelerometer calibration.   When a calibrated referenced accelerometer is used to calibrate an accelerometer, both the reference accelerometer and the accelerometer to be calibrated are attached to a vibration shaker. The amplitude of the shaker acceleration at a specified frequency is measured with the calibrated reference accelerometer. The voltage output of the accelerometer that is being calibrated is recorded. As before, the calibration of the accelerometer (mV/g) is obtained by dividing the measured voltage amplitude (mV) by the known acceleration amplitude (G).

22.7   Duct System Problem Noises

22.7.1   Generated noise

Duct system noise that is generated by air turbulence in the duct or fittings is called *generated noise*. The maximum noise generation takes place at elbows, branch ducts, dampers, and other locations where an abrupt change of air direction occurs or where an abrupt change in air velocity takes place. Higher pressure duct systems that operated in excess of 2000 fpm (10 m/s) and 4 in.w.g. (1000 Pa) are most prone to generated noise problems.

Generated noise is predominantly low frequency (63 through 500 Hz) and transmits without noticeable loss through suspended acoustical ceilings. In general, the noise generated by airflow through elbows, branches, and other duct fittings, will depend upon:

1. The size and shape of the fitting

2. The extent of the air turbulence created

3. The airflow rate

4. The pressure drop across the fitting

22.7.2 Short branch ducts

A common problem found in all terminal outlets connected by a short branch duct to a main supply air duct is the lack of airflow or an actual reverse airflow on the upstream side of the branch duct. This phenomenon will cause the supply air to the terminal outlet to be effective in only one half to two thirds of the unit, which can double outlet velocity. The increase in sound level over that indicated in the manufacturer's literature may be as much as 15 dB.

22.7.3 Balancing dampers

The pressure in a supply air duct system must be sufficient to move air through the entire length of the system. Pressures near the fan will be considerably higher than those near the end of the system. Therefore, volume control dampers are used in ducts to obtain an even distribution of air throughout the system. If a damper is located near the terminal outlet with no sound absorptive lining between the damper and the outlet, the noise generated by a partially closed damper can be 15 to 20 dB higher than the noise being generated by the terminal outlet.

22.7.4 Mixing boxes

One of the functions of mixing boxes is to regulate airflow by reducing air pressure; noise is then generated. The noise has two transmission paths:

1. Through the orifices of the box containing the valve and into the duct system

2. Through the casing (usually sheet metal) of the box containing the valve and into space outside of the duct system

Most manufacturers of these mixing boxes or terminal units publish noise ratings indicating the sound power levels that are discharged from the low-pressure end of the device. Some also indicate the requirements, if any, for sound attenuation that is recommended in the low-pressure ductwork between the terminal unit and outlet devices.

Some manufacturers also test the noise radiated from the box containing the valves; however, these data are not usually published. Obviously, if the terminal unit is located away from occupied spaces, its noise radiation may be of no concern. If, however, the unit is located over a suspended acoustical ceiling (which has little or no sound transmission loss properties at low frequencies), the noise from the unit may exceed specified noise criteria in the space below.

22.7.5 Terminal device noise

In general, there are three classifications of terminal devices:

1. Air distribution supply outlets—ceiling diffuser, side wall, sill, or floor outlet

2. Return air grilles or registers

3. Terminal boxes—dual duct mixing boxes, variable air volume boxes

The manufacturers of air distribution terminal devices publish comprehensive data of the sound power ratings of their products when used within the recommended ranges of airflow and static pressures. A ceiling, side wall, sill, or floor diffuser should be selected to meet the acoustical requirements specified. However, a diffuser can be properly selected and still exceed specified levels if the duct leading to the diffuser is not well designed and installed, or if the system is not balanced properly.

23

Troubleshooting

23.1 HVAC System Problems

23.1.1 Introduction

Theoretically, HVAC system problems should not be the job of the testing, adjusting, and balancing (TAB) technician to solve. However, to the experienced, well-educated TAB technician, helping solve problems generally may allow the TAB work to proceed without too much interruption.

The first step in solving a problem is to understand the symptoms. If the symptom points to a system design error, the condition and solution may become obvious rather quickly; although the solution could be both costly and time consuming.

If the problem is a result of confirmed TAB measurement readings, that do not match the specified equipment, then the solution often belongs to the installing contractor. If the problem is traced to inoperative equipment, the fault generally can be isolated with the help of all parties involved.

If a simple adjustment solves the problem, this often can be made by the TAB technician or the responsible contractor. If the problem is more complicated, a procedure must be established to determine responsibility and to direct a solution.

23.1.2 Problem solving

23.1.2.1 Know the systems. A thorough knowledge of the HVAC systems being balanced often allows TAB technicians to go to the problem source. Often some of these simple problems are anticipated, so the TAB technicians are prepared when they occur.

23.1.2.2 Complaint or problem? The TAB technician should first determine whether a complaint is *real* or *imaginary*. Compare "too hot" complaints to actual drybulb and wetbulb temperatures in the space, and compare these to design conditions, considering outside ambient conditions also. Maybe the complainant's inexpensive desk thermometer is inaccurate, or maybe the space is at 76 °F (24 °C) when design is 78 °F (26 °C), but the occupant wants 72 °F (22 °C).

23.1.2.3 Symptom analysis. Often the problem or symptom may disappear when the investigation is being made during a later time period. Conditions at the time may need to be reconstructed if possible. Sometimes there may be overreactions that solve themselves.

23.1.2.4 Isolate probable causes. Determine whether the problem is from the system, from individual equipment, or from some other cause. For example, damage from a suspected system water leak may be from a well-traveled roof leak. Begin with investigating the most obvious possible cause, progressing toward the more complex. Representatives from the responsible firms should become involved.

23.1.2.5 Solution actions. Some problems may be self-solving. Many are solved by simple adjustments. Remedial action may allow more complex problems to be minimized so that the TAB work can proceed and be completed.

23.2 Air Systems

23.2.1 System or fan noise

The most common duct system complaints, other than "too hot" or "too cold," are about "noise." Duct system noise can be generated by the fan, regenerated within the duct system, caused by duct expansion or "oil canning," and caused by air velocity through the grilles and diffusers. Obviously, there are many other noise-producing sources to investigate.

23.2.2 Fans

A fan should be selected to operate at the *peak* of the static efficiency (SE) curve where it is the quietest (Figure 23.1). However, there are many fan problems other than noise. The experienced TAB technician can usually trace the source of mechanical or vibrational noise quickly. A stethoscope should be the only instrument needed. If the problem

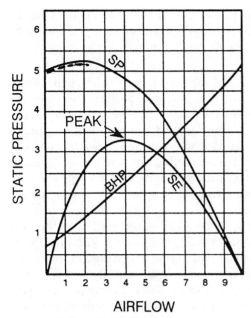

Figure 23.1 Centrifugal fan curves.

cannot be isolated, the possibility of surge, system effect, resonance, or unbalance must then be checked in that sequence.

23.2.2.1 Surge. *Surge* is a common condition that occurs when a fan is operating somewhere in the unstable area to the left of or at the fan curve "hump" or peak of the static pressure curve (see the dashed line part of the SP curve in Figure 23.1). The surge also creates a characteristic noise that is easily recognized by the experienced TAB technician. A static pressure problem may be relieved by reducing the system static pressure. Duct blockage, jammed fire dampers, plugged filters, or improper fan selection can be the cause. An analysis of the fan curve, using the actual measured air volume and static pressure, should determine the solution, assuming the correct fan has been installed.

23.2.2.2 System effect. A visual inspection of the duct connections to the fan should indicate whether *system effect* may be reducing fan capacities (Chapter 6).

In addition to fan capacity losses, poor fan–duct connection design and/or installation may create other problems. Fan performance may also be seriously impeded by a drive guard restriction. Belt and drive guards should be kept as far as possible from the inlet openings and, wherever possible, fabricated from expanded metal mesh rather than solid sheet steel. Be alert for fan *inlet restrictions*.

23.2.2.3 Inlet spin. *Inlet spin*, resulting from a poor inlet connection, is probably the most frequent cause of poor fan performance and/or system effect. When air is introduced into a fan plenum it should be directed at the center line of the fan inlet. If the return air velocity spin is imparted in the direction of wheel rotation, the fan volume, static pressure, and horsepower are lowered. If the air spin is opposite to the wheel rotation, the volume, static pressure, and horsepower will be greater than expected. Either situation of spin will cause a reduction in fan efficiency or capacity.

By inserting a Pitot tube through the flexible connection at the fan inlet it is easy to demonstrate the condition of spin. With the Pitot tube connected with two tubes to a draft gauge or inclined manometer to read *velocity pressure*, probe the connection carefully, holding the Pitot tube parallel to the fan shaft (Figure 23.2). Eccentric flow will be indicated by higher readings on the top, bottom, or side of the connection. The angle of airflow will be indicated by slowly oscillating the Pitot tube back and forth. By carefully observing the pressure readings as the Pitot tube is slowly twisted back and forth, it is possible to determine the angle of spin. To figure the loss in capacity equivalent to the angle of spin, the TAB technician may consult a reliable inlet damper characteristic curve.

23.2.2.4 Resonance. *Resonance* usually is caused by transmitted vibration to the HVAC unit casing or attached apparatus panels. By changing the fan speed in either direction by more than 10%, the resonance should vanish. Correction or alteration can then be made by

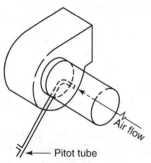

Pitot tube

Figure 23.2 Probing for return air inlet spin.

damping, absorbing, or stiffening the noisy panels if the original fan speed must be used.

23.2.2.5 Fan unbalance. Another serious and somewhat common problem is *fan unbalance.* The only solution is to have the fan properly balanced after the vibration isolation has been checked and confirmed to be satisfactory.

23.2.3 Duct system air leakage

In addition to duct leakage discussed in Chapter 6, there may be air leakage in the HVAC units, in the volume controlling air terminal units, and in the connections to the supply air diffusers or outlets. The loses in the ductwork may be as low as 2 to 3%, but all other air leakage may increase the total to 8 to 12% of the duct system airflow. The difference in airflow from Pitot tube traverses at the fan and the total outlet airflow should account for this leakage.

23.2.4 High–pressure drop fittings

Duct systems rarely are installed exactly as shown on the project mechanical drawings and as specified. Extra fittings may be added to avoid obstacles, turning vanes may be left out of square turn elbows, and unusually shaped fittings may be required, especially under rooftop HVAC units. Some of these duct fittings may be hidden from view and have very high pressure drops. The solution to this problem is for the TAB technician to review the duct system installation before the ceilings are installed. If changes cannot be made, at least calculations can be made to see whether the specified fan, motor, and drives are adequate.

23.3 Fan Troubleshooting Chart

Table 23.1 has been extracted with permission from the Air Movement and Control Association (AMCA) *Fan Application Manual*, Publication 202-88.

23.4 Hydronic Systems

23.4.1 System and pump noise

Other than the most usual lack of heating or cooling complaint about hydronic system supplied HVAC units, noise from air bubbles flowing in the system, piping or fin-tube expansion noises, and pump vibration or motor hum noise top the list.

TABLE 23.1   Fan Troubleshooting Chart [Courtesy of the Air Movement and Control Association (AMCA)]

A. Fan noise	
Source	Probable cause
1. Impeller hitting inlet or housing	a. Impeller not centered in inlet or housing
	b. Inlet or housing damage
	c. Crooked or damaged impeller
	d. Shaft loose in bearing
	e. Impeller loose on shaft
	f. Bearing loose in bearing support
	g. Bent shaft
	h. Misaligned shaft and bearings
2. Impeller hitting cutoff	a. Cutoff not secure in housing
	b. Cutoff damaged
	c. Cutoff improperly positioned
3. Drive	a. Sheave not tight on shaft (motor or fan)
	b. Belts hitting belt tube
	c. Belts too loose. Adjust for belt stretching after 48 hours operating
	d. Belts too tight
	e. Belts wrong cross-section
	f. Belts not "matched" in length on multibelt drive
	g. Variable-pitch sheaves not adjusted so each groove has same pitch diameter (multibelt drive)
	h. Misaligned sheaves
	i. Belts worn
	j. Motor, motor base, or fan not securely anchored
	k. Belts oily or dirty
	l. Improper drive selection
	m. Loose key
4. Coupling	a. Coupling unbalanced, misaligned, or loose, or may need lubricant
	b. Loose key
5. Bearing	a. Defective bearing
	b. Needs lubrication
	c. Loose on bearing support
	d. Loose on shaft
	e. Seals misaligned
	f. Foreign material inside bearing
	g. Worn bearing
	h. Fretting corrosion between inner race and shaft
	i. Bearing not sitting on flat surface
6. Shaft seal squeal	a. Needs lubrication
	b. Misaligned
	c. Bent shaft
	d. Bearing loose on support
7. Impeller	a. Loose on shaft
	b. Defective impeller. **Do not run fan. Contact the manufacturer**
	c. Unbalance
	d. Coating loose

TABLE 23.1   Fan Troubleshooting Chart (Courtesy of AMCA) (continued)

A. Fan noise	
Source	Probable cause
	e. Worn as result of abrasive or corrosive material moving through flow pasages
	f. Blades rotating close to structural member
	g. Blades coinciding with an equal number of structural members
8. Housing	a. Foreign material in housing
	b. Cutoff or other part loose (rattling during operation)
9. Motor	a. Lead-in cable not secure
	b. AC hum in motor or relay
	c. Starting relay chatter
	d. Noisy motor bearings
	e. Single phasing on a three-phase motor
	f. Low voltage
	g. Cooling fan striking shroud
10. Shaft	a. Bent
	b. Undersized. May cause noise at impeller, bearings, or sheave
11. High air velocity	a. Ductwork too small for application
	b. Fan selection too small for application
	c. Registers or grilles too small for application
	d. Heating or cooling coil with insufficient face area for application
12. Obstruction in high-velocity airstream may cause rattle or pure tone whistle	a. Dampers
	b. Registers
	c. Grilles
	d. Sharp elbows
	e. Sudden expansion in ductwork
	f. Sudden contraction in ductwork
	g. Turning vanes
13. Pulsation or surge	a. Restricted system causes fan to operate at poor point of rating
	b. Fan too large for application
	c. Ducts vibrate at same frequency as fan pulsations
	d. Rotating stall
	e. Inlet vortex surge
	f. Distorted inlet flow
14. Air velocity through cracks, holes or past obstructions	a. Leaks in ductwork
	b. Fins on coils
	c. Registers or grilles
15. Rattles and/or rumbles	a. Vibrating ductwork
	b. Vibrating cabinet parts
	c. Vibrating parts not isolated from building

B. Insufficient fan airflow	
1. Fan	a. Impeller installed backwards
	b. Impeller running backwards
	c. Improper blade angle setting
	d. Cutoff missing or improperly installed
	e. Impeller not centered with inlet collar(s)

TABLE 23.1   Fan Troubleshooting Chart (Courtesy of AMCA) (continued)

	f. Fan speed too slow
	g. Impeller or inlet dirty or clogged
	h. Improper running clearance
	i. Improper inlet cone to wheel fit
	j. Improperly set inlet vane or damper
2. Duct system	a. Actual system is more restrictive (more resistance to flow) than expected
	b. Dampers closed
	c. Registers closed
	d. Leaks in supply ducts
	e. Insulating duct liner loose
3. Filters	a. Dirty or clogged
	b. Replacement filter with greater than specified pressure drop
4. Coils	a. Dirty or clogged
	b. Incorrect fin spacing
5. Recirculation	a. Internal cabinet leaks in bulkhead separating fan outlet (pressure zone) from fan inlets (suction zone)
	b. Leaks around fan outlet at connection through cabinet bulkhead
6. Obstructed fan inlets	a. Elbows, cabinet walls, or other obstructions restrict airflow. Inlet obstructions cause more restrictive systems but do not cause increased negative pressure readings near the fan inlet(s) (see *System Effects* in Section 6.5). Fan speed may be increased to counteract the effect of restricted fan inlet(s). CAUTION: **Do not increase fan speed beyond the manufacturer's recommendations.**
7. No straight duct at fan outlet	a. Fans that are normally used in duct system are tested with a length of straight duct at the fan outlet. If there is no straight duct at the fan outlet, decreased performance may result. If it is not practical to install a straight section of duct at the fan outlet, the fan speed may be increased to overcome this pressure loss (see *System Effects* in Section 6.5). CAUTION: **Do not increase fan speed beyond the fan manufacturer's recommendations.**
8. Obstructions in high-velocity airstream	a. Obstruction near fan outlet or inlet
	b. Sharp elbows near fan outlet or inlet
	c. Improperly designed turning vanes
	d. Projections, dampers, or other obstruction in a part of the system where air velocity is high

C. Fan airflow too high	
Source	Probable cause
1. System	a. Oversized ductwork
	b. Access door open

TABLE 23.1   Fan Troubleshooting Chart (Courtesy of AMCA) (continued)

	c. Registers or grills not installed
	d. Dampers set to bypass coils
	e. Filter(s) not in place
	f. System resistance low
2. Fan	a. Fan speed too fast
	b. Improper blade angle setting

D. Static pressure low, airflow correct

Air density	Pressures will be less with high-temperature air or at high altitudes

E. Static pressure low, airflow high

Source	Probable cause
1. System	a. Fan inlet and/or outlet conditions not same as tested. (See discussion and *System Effect Factors* in Section 6.5).
2. Fan	a. Impeller installed backwards
	b. Impeller running backwards
	c. Improper blade angle setting
	d. Cutoff missing or improperly installed
	e. Impeller not centered with inlet collar(s)
	f. Fan speed too slow
	g. Impeller or inlet dirty or clogged
	h. Improper running clearance
	i. Improper inlet cone to wheel fit
	j. Improperly set inlet vane or damper

F. Static pressure high, airflow low

1. Duct system	a. Actual system is more restrictive (more resistance to flow) than designed
	b. Dampers closed
	c. Registers closed
	d. Insulating duct liner loose
2. Filters	a. Dirty or clogged
	b. Replacement filter with greater than specified pressure drop
3. Coils	a. Dirty or clogged
	b. Fin spacing too close

G. Fan power high

Source	Probable cause
1. Fan	a. Backward inclined impeller installed backwards
	b. Fan speed too high
	c. Forward curve or radial blade fan operating below design pressures
	d. Blade angle not set properly
2. System	a. Oversized ductwork
	b. Face and bypass dampers oriented so coil dampers are open at same time bypass dampers are open
	c. Filter(s) left out

TABLE 23.1   Fan Troubleshooting Chart (Courtesy of AMCA) (continued)

	d. Access door open NOTE: The causes listed in this table section pertain primarily to radial blade, radial tip, and forward curve centrifugal fans, i.e., fans that exhibit rising horsepower curves. Normally, backward inclined, backward curve, or backward inclined airfoil centrifugal fans and axial flow fans do not fall into this category.
3. Air density	a. Calculated horsepower requirements based on less dense (e.g., high-temperature) air, but actual air is denser (e.g., cold startup)
4. Fan selection	a. Fan not selected at efficient point of rating

<table>
<tr><td colspan="2" align="center">H. Fan does not operate</td></tr>
<tr><td align="center">Source</td><td align="center">Probable cause</td></tr>
<tr><td>Electrical or mechanical</td><td>Mechanical and electrical problems are usually straightforward and are normally analyzed in a routine manner by service personnel. In this category are such items as:
a. Blown fuses or circuit breakers
b. Broken belts
c. Loose pulleys
d. Electricity turned off
e. Impeller touching housing
f. Wrong or low voltage
g. Motor too small or load inertia too large
h. Seized bearing</td></tr>
</table>

23.4.2 Pumps

Much like fans, pumps operate the quietest at the point on the pump curve of highest efficiency. Unlike fans, pumps often work in parallel with other similar sized pumps. Depending on the system designer's choice of efficiencies based on the number of pumps operating at the same time, pumps may operate at several different capacities and efficiencies. Variable-flow systems also can impose excessive wear on pumps, especially when operating near the shutoff condition and when fluid flow is beyond the maximum efficiency point. See Chapter 15 for additional information on pumps.

23.4.2.1 Cavitation. When a pump sounds like it is pumping "marbles," the chances are that *cavitation* is occurring. Cavitation is the formation of vapor bubbles in a pump and occurs when the vapor pressure of the water is equal to or greater than the absolute pressure of the water stream. Cavitation causes hammering in a pump and can quickly destroy a pump impeller. It must be avoided! Cavitation is avoided by ensuring that the absolute pressure is always greater than the vapor pressure. This is true for closed systems, such as hot water heating systems, and open systems that include cooling towers.

In a closed system, it is easy to control cavitation simply by ensuring that the system pressure is adequate throughout the water system. Cavitation is more difficult to handle with open water systems, since the system is exposed to atmospheric pressure.

Centrifugal pumps, used in most HVAC applications, generally cannot draw water into themselves; the water must be "pushed" into them. The amount of push needed is called the *net positive suction head required* (NPSHR).

Net positive suction head available (NPSHA) is the pressure available at a pump suction to push the water into the pump. Cavitation cannot occur if the NPSHA is always greater than the NPSHR.

Cavitation can be a problem that TAB technicians need to look for, especially when the system is operating at maximum capacity.

23.4.2.2 Misalignments. Although pump *piping stresses* and *motor misalignment* are not the responsibility of TAB technicians, they should be reported and corrected immediately.

Piping stresses should never exist on pump connections. If they do, the pump volute may be knocked out of alignment and wear will occur on bearings, sleeves, and case rings. Flexible joints on the pump connections should not be counted on to eliminate this problem. They can be extended to their limits, thus imposing stress on the pump connections. The only way to avoid this problem is to ensure that the piping is supported in a manner that prevents the exertion of force on the pump connections.

There is a very simple procedure that will verify whether there is pressure on pump connections. With the pump shut down, the pipe flange bolts should be removed. If the mating flanges come apart or shift, there is pressure on the connections, and the piping supports should be adjusted until the flanges mate without any force.

The pump and its motor must be in alignment; otherwise the flexible coupling insert and the bearings in the pump and motor will wear rapidly. The instrumentation and procedures for aligning pumps and motors are well known in the industry, and there is no excuse for such misalignment.

23.4.2.3 Entrained air in pump. Air exists in water in two forms, *dissolved* and *entrained.* Entrained air exists as bubbles in the water stream. Dissolved air does not bother pumps mechanically; entrained air does and can create the same problems as cavitation. It must be remembered that air converts from dissolved to the entrained state when the water temperature is increased or the system pressure is decreased.

23.4.2.4 Pump inlet piping. Review the suction pipe sizes, strainer screen sizes, and fitting and valve configurations to assure that the pump will operate at full capacity without cavitation.

23.4.3 Hydronic System Air Control

Entrained air or air bubbles affect water flow throughout the hydronic system. Even though air is vented from an empty system when it is filled, residual air often remains. Entrained air always exists in makeup water because the pressure of the water normally is reduced and the temperature increased (in hot water systems) as it enters the system. When the water velocity is not high enough to keep air entrained, the air rises to the high points in the piping or coils.

23.4.3.1 Air venting. Entrained air can be removed from HVAC hydronic systems by the use of mechanical separators and air vents. All air vents should be equipped with an air chamber ahead of them, and automatic or manual air vents may be used.

Once the air has been separated from the system fluid, it must be eliminated from the system. This assumes that in a system utilizing a compression tank, the tank has the proper air charge to compensate for fluid expansion caused by the temperature rise. Any additional air may be detrimental to system operation. During the initial system fill, most of the air that occupied the system volume must be vented off through manual vent valves or by loosening fittings.

The problem arises when *all* of the air is not initially removed from the system. The air continues to travel and it has a tendency to stop piping circuit flow by collecting in unvented high points, or the air may partially or totally stop heat transfer in coils. TAB technicians often waste considerable amounts of time tracking down unvented air. **However, air venting is the responsibility of the installing contractor.**

23.4.3.2 Compression tank. A *compression tank(s)* (or expansion tank), if too small or waterlogged (no air charge), will cause the pressure relief valve to discharge extra pressure and water as a hot water system heats up. When the system cools, the makeup water pressure reducing valve will add new water to the system along with additional entrained air. Often air will enter backwards through automatic air vents as the system water cools and the pressure is lowered. Sometimes the system water level is lowered, causing air to enter.

A similar problem may exist if the compression tank is connected to piping on the discharge side of the pump instead of the suction side as shown in Figure 23.3.

23.4.3.3 Fill valve. The pressure gauge on the boiler should read, when cold, about 9 ft. w.g. or 4 psi (27.6 kPa) above the highest point in the system, normally a heating or cooling coil. This pressure setting is used for the automatic fill valve or makeup water pressure reducing valve. For example, if the highest radiation is 42 feet above the fill valve or boiler:

$$42 \text{ ft.} \times 0.433 = 18.2 \text{ psi}$$

$$\text{Cold} = \underline{4.0 \text{ psi}}$$

$$\text{Fill pressure} = 22.2 \text{ psi (cold boiler)}$$

23.5 HYDRONIC TROUBLE-SHOOTING CHART

Table 23.2 contains some frequent pump and hydronic system problems. Many field experience problems can be added to this list by experienced TAB technicians.

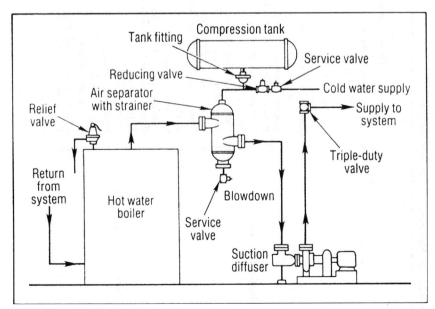

Figure 23.3 Typical hydronic boiler and pump installation (Courtesy of Heating/Piping/Air Conditioning).

TABLE 23.2 Hydronic Troubleshooting Chart

A. Pump and system

Problem	Possible cause	Action
1. No circulation	a. Set screw not tight, coupler loose on shaft	Tighten set screw in recess in the shaft
	b. Impeller slipping on shaft	Check to see if impeller is placed on the keyway of the shaft
		Tighten impeller nut
	c. Air-bound system	Vent system
	d. Air-bound pump	Vent pump casing
	e. Broken pump coupler	Replace; check alignment
	f. Clogged impeller or piping	Locate and remove obstruction
	g. System valve closed	Open
	h. Pump electrical circuit broken	Check all related low and line voltage circuits
2. Inadequate circulation	a. Air-bound system	Vent system
	b. Air-bound pump	Vent pump casing
	c. Clogged impeller or piping	Locate and remove obstruction
	d. Clogged strainer	Remove and clean screen
	e. Pump impeller damaged	Replace
	f. Insufficient NPSH (net positive suction head)	Lower pump or raise pressure or relocate
	g. Pump too small	Replace pump or impeller
	h. Partially air-bound pump	Vent pump casing
	i. Pump running backwards (three-phase)	Reverse any two motor leads
	j. Improper motor speed	Check wiring and voltage
3. Pump or system noise	a. Entrained air	Vent system
	b. Pump cavitation	Lower pump or raise pressure or relocate
	c. Pump misalignment	Realign pump
	d. Worn pump coupler	Replace; check alignment
	e. Excessive water velocity	Install balancing cocks or parallel piping
	f. Poor foundation (base-mounted)	Provide rigid foundation with adequate grouting
	g. Pipe vibration	Provide adequate pipe support

TABLE 23.2 Hydronic Troubleshooting Chart (Continued)

A. Pump and system

Problem	Possible cause	Action
4. Premature failure of pump components	a. Improper pump (size) (type)	Replace
	b. Improper pump location	Relocate
	c. Pump misalignment	Realign
	d. Excessive water treatment	Check manufacturer's instructions
	e. Over oiling of pump	Check manufacturer's instructions
	f. Under oiling of pump	Check manufacturer's instructions
	g. Pump operating close to or beyond end point of curve	Balance system
	h. Excessive piping load	Provide proper pipe support
5. Seal failures within 1 year period or less in a closed system	a. Excessive dirt, sand, and oxides	Clean system
	b. Excessive or improper water treatment	Check for proper water treatment recommendations from pump manufacturer
	c. Pump cavitation:	
	1. Improper selection	Check pump operation on its curve—overloading
	2. Compression tank location	High-head pump (compression tank on suction side of pump)
	d. Air seal without lubricant (water)	Vent air from pump volute
	e. Excessive temperatures	Check type of seal and maximum operating temperature from manufacturer
	f. Pumps run without fluid	Pumps must be primed before operation
6. Seal pitting	a. Caused by wear and excessive amounts of free oxygen	Check system, may have a constant leak
	b. Oxygen corrosion	Fresh water feeding carries oxygen into the system
	c. Magnetic iron oxide	

B. Air control

Problem	Possible cause	Action
1. Waterlogged compression tank	a. Gravity circulation between boiler and tank	Install air control system
	b. Leak in tank	Check with soap solution—replace
	c. Leak in gauge glass top gasket dries out allowing air to escape	Check tapping—would most likely be in upper tapping

363

TABLE 23.2 Hydronic Troubleshooting Chart (Continued)

B. Air control

Problem	Possible cause	Action
2. Insufficient air control in air control devices, boiler top outlet fittings	a. Dip tube not 2½ in. below water line in boiler	This is almost impossible to find without taking the supply piping on the boiler apart
c. In-line fitting	a. Too high a velocity through fitting for air separation	Check size of fitting—it should be the same as the pipe size
	b. Initial system startup not performed properly	Check manufaturer's instructions
	c. Proper pitch in piping to the tank	Horizontal pipe to be pitched toward the compression tank
	d. Leaks in system piping	Check for leaks
4. No heat radiation No circulation	a. Air bound	Vent coil or piping

C. Valves

Problem	Possible cause	Action
1. Relief valve opens	a. Defective relief valve	Replace
	b. Compression tank undersized	Check for proper size; replace
	c. Waterlogged compression tank	Install air control system or drain tank
	d. Runaway burner	Check controls
	e. Fuel valve stuck in open position	Check valve
	f. Pump not operating	Check pump
	g. High-limit control fails	Check control
	h. Defective reducing valve or setting	Clean or replace
	i. System operating pressure too high	Check static pressure and temperature operation
2. Valve drips	a. Dirt on seat	Open rapidly
3. Reducing valve failures	a. Valve does not feed	Check if valve is scaled
	b. Strainer plugged	Clean or replace
	c. Valve seat scaled shut	Turn adjustment screw all the way down to free—if it doesn't, replace valve
4. Reducing valve does not reduce pressure	a. City pressure too high for valve	Check valve limitation and replace with higher pressure rated valve

Problem		Possible cause	Action
5. Relief valve pops—hot or cold	a.	Reducing valve sticking in open position	Check valve, replace if necessary
6. Flow-control valve problems Gravity circulation	a. b. c.	Dirt on seat Stem not turned down all the way Valve body not installed in a horizontal position	Take cover off and wipe seat Turn handle on stem all the way and seat If in a vertical position, change to horizontal, use straight or angle valve
7. Zone valves will not operate (see note)	a. b. c.	Thermostat Power unit burned out Sticking seal assembly, valve will not seat	Check thermostat Check and replace—check electrical connections Check and clean stem and seat for buildup of mineral scale—clean

D. Miscellaneous

Problem		Possible cause	Action
1. Insufficient heat in one or more zones	a. b. c. d. e. f. g. h.	Air binding Clogged zone piping Defective zone valve Unbalanced circuits Undersized radiation Broken coupler on pump Motor burnout Power off	Install air control Locate and remove obstruction Repair or replace Balance Add radiation or more insulation—increase water temperature Check pump Check motor of pump Check electrical connections—power source
2. Overheating—cold or mild weather	a. b. c. d. e. f. g.	Gravity circulation Defective flow control Zone valve stuck in open position Thermostat not operating Flow-control stem in open position Fuel valve stuck in open position Controls not operating properly	Install flow—control valve Clean or repair or replace Check; repair or replace Check; replace Close valve Check; replace if needed Check
3. Pounding or waterhammer	a. b. c. d. e.	Lack of system pressure Oversized compression tank Excessive boiler temperature Pumping into the boiler using high head pumps Solenoid valves	Check static pressure of system—see if it is correct for system Check to see if water is circulating through boiler—stuck fuel valve 1. Increase static pressure above pump head if possible 2. Move pump to pump out of the boiler and discharge into the system Do not use on hydronic systems
4. Cracking sound	a.	Boiler full of lime (mineral compounds)	Clean and flush boiler

Tab Cost Estimating

24.1 Task Man-Hours

24.1.1 Introduction

As stated in Chapter 1, TAB work is essential to the proper performance of building HVAC systems and the resultant indoor air quality. It is the responsibility of the TAB firm to review the project plans and specifications carefully when preparing a proposal for the TAB work.

If accurate records of man-hour requirements for each task are kept for each project, this chapter could be used only to verify the data. Unfortunately, some TAB firms cut corners when doing balancing work by using unqualified technicians and supervisors, and labor costs vary considerably.

24.1.2 Man-hour variables

Task man-hours also may vary considerably because of difficulty of access to the building area, systems, and equipment; height from the floor; whether space is occupied or unoccupied; type of system and/or equipment installed; repetitive sizes and functions; type of controls; and cooperation of other trades, building owner, HVAC system operator, or tenant occupying the space. The experienced TAB firm will have *multipliers* for each of the above items to use on the average calculated man-hours for each function. Figure 24.1 may be used for a rough labor estimate for TAB work for *ordinary* HVAC systems.

24.2 Estimating Man-Hours

24.2.1 Report forms

Sample test report forms are published by many organizations in the TAB field, such as NEBB test report forms used only by NEBB Cer-

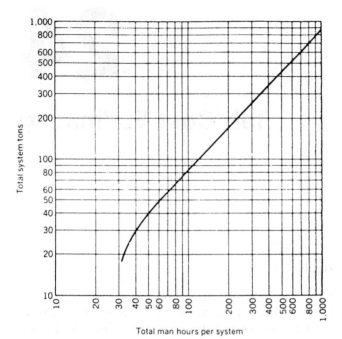

Figure 24.1   Rough labor estimate for ordinary HVAC systems.

tified TAB Firms. ANSI/ASHRAE Standard 111-1988, *Practices for Measurement, Testing, Adjusting and Balancing of Building Heating, Ventilation, Air-Conditioning and Refrigeration Systems*, does not contain sample forms but does contain a list of items for each type of equipment or system being tested. Good TAB procedures used with accurately and honestly reported data on the TAB test report forms will assure the HVAC system owner that the systems perform as specified. Also they will provide a permanent record that may be used when building system alterations or additions are made.

24.2.2   Using man-hour tables

The man-hour tables in this chapter are average, basic values to be modified by factors discussed in Section 24.1.2 and found in Table 24.13 and *experience factors* based on the experience and qualifications of the TAB technicians used on the project. Where a specific task or operation is not found, the data for a similar task or operation should be used. All normal activities for each TAB task, such as moving ladders, drilling holes, calibrating instruments, taking readings, and recording data, are included in the man-hour estimates.

For example, using data from Table 24.1 for a basic HVAC unit (up to 25 HP or 19 kW), essential TAB items should total 7.0 man-hours. A return air fan, if used, would add 2.0 man-hours for a total of 9.0 man-hours. A basic multizone unit with five zones (no return air fan) would total 7.0 plus 5.0 (zones) or 12.0 man-hours.

24.3 Tab Man-Hour Tables—Air Systems

TABLE 24.1 HVAC Units (Air Side)

Operation	Man-hours
Base fan–coil–filter unit, TAB (up to 25 HP or 19 kW)	3.0[a]
Base fan–coil–filter unit, TAB (30–75 HP or 22–56 kW)	4.0[a]
Base fan–coil–filter unit, TAB (100 HP or 75 kW up)	5.0[a]
Supply air duct traverse, TAB	1.5
Return air duct traverse, TAB	1.0
Mixing box, OA dampers, TAB	0.8–1.5
Vortex dampers, elec./electronic	0.3–0.8
Vortex dampers, pneumatic	0.1–0.2
Variable frequency drives	0.3–1.3
Final adjustments	0.7
Return air fan, TAB	2.0
Face and bypass section	1.0
Multizone unit (per zone)	1.0
Roof top unit (add)	2.0

[a]Does not include changing drive components.

TABLE 24.2 Fans

Operation	Man-hours
Centrifugal fan, TAB:	
Direct drive	1.2
To 5000 cfm (2500 L /s)—belt drive	2.5
5100 to 25,000 cfm (2550–12,500 L/s)	3.2
25,100 to 50,000 cfm (12,550–25,000 L/s	5.0
Roof exhauster, TAB:	
Direct drive	1.5
Belt drive	3.0
Propeller fan, airflow and electric	1.0
Ceiling exhaust fan	0.8

TABLE 24.3 Ventilation

Operation	Man-hours
Duct traverse:	
Small	1.2
Medium	1.9
Large	2.5
Hood general or dishwasher:	
Small	1.2
Large	2.0
Hood, range or oven:	
Small	2.0
Large	3.0
Cabinet hood (lab):	
Without fan	1.5–3.0
With fan	2.5–4.5

TABLE 24.4 Terminal Units

Operation	Man-hours
Constant volume boxes (TAB):	
Single duct	1.0–2.0
Dual duct	1.5–3.0
VAB boxes (TAB):	
DDC controls	1.0–2.0
Electric controls	1.0–1.5
Pneumatic controls	0.8–1.5
Fan powered	1.3–2.5
Fan powered—DDC	1.5–3.0
Reheat coil (add)	0.5–0.8

TABLE 24.5 Terminal Devices

Operation	Man-hours
Ceiling diffusers (TAB):	
Flow measuring hood	0.4
Small—anemometer	0.6
Large—anemometer	0.8
Linear slot diffusers (TAB):	
Flow measuring hood (unit length)	0.5
Anemometer to 10 ft. (3 m)	0.8
Grilles and registers (TAB):	
Flow measuring hood	0.3
Anemometer	0.5

TABLE 24.6   Air-to-Air Exchangers

Operation	Man-hours
Plate exhanger (TAB):	
Small	3.0
Large	4.5
Heat pipe exchanger (TAB):	
Small	3.5
Large	5.5
Heat Wheel (TAB):	
Small—sensible heat	3.0
Large—sensible heat	5.0
Small—total heat	4.0
Large—total heat	6.0

24.4   TAB Man-Hour Tables— Hydronic Systems

TABLE 24.7   HVAC Units (Water Side)

Operation	Man-hours
Hot or chilled water coil (TAB):	
With flow valve or station	1.2
Without flow valve or station	2.0
DX coil	1.5
Steam coil	1.0

TABLE 24.8   Pumps

Operation	Man-hours
Circulator, in line (TAB)	1.0
Water pump (TAB):	
5 to 75 HP (56 kW) with flow valve or station	2.0–2.5
100 to 200 HP (150 kW) with flow valve or station	3.0–3.5
5 to 75 HP (56 kW) without flow valve or station	2.8–3.0
100 to 200 HP (150 kW) without flow valve or station	4.0–5.0
Basin pump (evaporative condenser or cooler)	1.5

TABLE 24.9   Cooling and Heating Equipment

Operation	Man-hours
Boiler (TAB)	1.8
Boiler and burner package (TAB)	3.6
Chiller package (TAB)	5.5
Chiller only	2.0
Condenser only	2.0
Cooling tower (TAB):	
Airflow or electric	4.0
Water flow (to 750 tons)	5.0
Water flow (over 750 tons)	8.0
Air cooled condenser—airflow or electric	3.0
Heat exchangers (steam or water to water)	2.5

TABLE 24.10   Terminal Units (Hydronic)

Operation	Man-hours
Convectors and fin radiation (per unit)	0.5
Fan coil units and unit ventilators—2 pipe	0.7
Fan coil units and unit ventilators—4 pipe	1.0
Induction units (includes airflow)	1.2
Ceiling units—add each	0.3
Duct-mounted coils	0.8
Deduct 30% with flow valves	

24.5   Room Conditions and TAB Reports

TABLE 24.11   Indoor and Outdoor Air

Operation	Man-hours
Small room or area—DB/WB conditions	0.5–0.8
Large room or area—DB/WB conditions	1.0–1.2
Outdoor—DB/WB conditions	0.5–0.8
Room pressurization to ambient	1.5–2.0
Room to room pressurization	1.0–1.6

TABLE 24.12   TAB Reports

Operation	Man-hours
Review, calculations, report assembly:	
Small project	12.0–20.0
Medium project	20.0–28.0
Large project	24.0–36.0

24.6 Job Conditions

TABLE 24.13 Multiplier Factors

Condition	Factor range
Poor access to building	1.05–1.20
Poor access to floor or area	1.05–1.25
Poor access to equipment or system	1.05–1.25
Work height over 12 ft. (3.36 m)	1.10–1.30
Work height over 18 ft. (5.4 m)	1.20–1.50
Occupied by people	1.00–1.15
Furniture and equipment in place	1.00–1.10
Partitions in place	1.00–1.10
Open area	0.80–1.00
Repetition of size or type	0.90–1.00
Season (rooftop units), ice or extreme heat	1.10–1.25
Off-season TAB work	1.00–1.20
Overtime conditions	1.20–1.50
Poor coordination with others	1.05–1.25
Construction incomplete	1.15–1.35
Familiar or favorite equipment used	0.85–1.00
5% tolerance in lieu of 10%	1.10–1.30

Notes:
1. Factors can be accumulative
2. Factors may apply to part or all of the TAB work.

Chapter

25

TAB Mathematics

25.1 Areas

25.1.1 Squares and rectangles

25.1.1.1 U.S. units. The cross-sectional area of a duct must be calculated to determine the amount of air flowing in a duct. Duct dimensions are always given in inches, but duct area must be expressed in square feet (Figures 25.1 and 25.2). This is because airflow is not simply a measurement of a plane but is a measurement of volume, and volume within a duct is always stated in cubic feet per minute (cfm).

To obtain the area in square feet, the duct dimensions in inches must be changed to feet. If the duct dimensions are left in inches, the result in square inches must then be converted to square feet by dividing the square inches by 144.

Equation 25.1 (U.S.)

$$A = W \times H$$

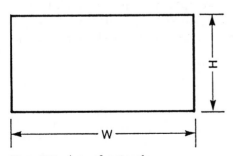

Figure 25.1 Area of rectangle.

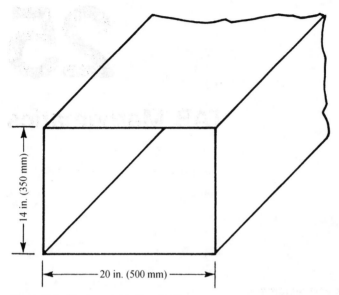

Figure 25.2 Duct cross-sectional area.

Where: A = area (square inches or square feet)
   W = width (inches or feet)
   H = height (inches or feet)

Example 25.1 (U.S.)   Find the cross-sectional area of a 20 × 14 in. rectangular duct.

Solution

$$A = W \times H = 20 \times 14 = 280 \text{ in.}^2$$

$$280/144 = 1.944 \text{ ft.}^2$$

or $A = 20/12 \times 14/12 = 1.167 \times 1.667 = 1.945 \text{ ft.}^2$

25.1.1.2   Metric units.   Duct dimensions generally are given in millimeters. Although some countries use cubic meters per second (m^3/s) for airflow, NEBB and ASHRAE use liters per second (L/s). As 1 m^3/s = 1000 L/s, the following equation is used as a base for both answers:

Equation 25.1 (Metric)

$A = W \times H/1,000,000$

Where: A = area (m^2)
   W = width (mm)
   H = height (mm)

Example 25.1 (Metric) Find the cross-sectional area of a 500 × 350 mm rectangular duct (in square meters).

Solution

$$A = W \times H/1,000,000 = 500 \times 350/1,000,000$$

$$A = 0.175 \text{ m}^2$$

25.1.2 Circles

25.1.2.1 Full circle area. The mathematical equation for the area of a circle is:

Equation 25.2

$$A = \pi R^2$$

Where: A = area (in.^{2} or mm^2)
  R = radius (in. or mm)

Example 25.2 (U.S.) Find the area of a 6 in. radius circle in square inches and square feet.

Solution

$$A = \pi R^2$$

$$A = 3.1416 \times (6)^2 = 113.1 \text{ in.}^2$$

$$A = 113.1/144 = 0.785 \text{ ft}^2$$

Example 25.2 (Metric) Find the area of a 250 mm radius circle in mm^2 and m^2.

Solution

$$A = \pi R^2$$

$$A = 3.1416 \times (150)^2 = 70,685.8 \text{ mm}^2$$

$$A = 70,685.8/1,000,000 = 0.071 \text{ m}^2$$

Often only the circumference of a circle can be measured. Use the following equation to calculate the radius.

Equation 25.3

$$C = \pi D = \pi 2R;$$

$$R = \frac{C}{2\pi}$$

Where: R = radius (in. or mm)
  D = diameter (in. or mm)
  C = circumference (in. or mm)

25.1.2.2 Circle sector area. The following equations may be used for the area of a sector of a circle with angle Θ (Figure 25.3):

<table>
<tr><td align="center">**Equation 25.4**</td><td align="center">**Equation 25.5**</td></tr>
<tr><td align="center">$A_s = \tfrac{1}{2}RS$</td><td align="center">$A_s = \dfrac{\Theta°A}{360°} = \dfrac{\Theta°\pi R^2}{360°}$</td></tr>
</table>

Where: A_s = area of circle sector (in.^{2} or mm^2)
  A = area of full circle (radius = R) (in^2 or mm^2)
  R = radius (in. or mm)
  S = length of arc (in. or mm)
  $\Theta°$ = angle in degrees

Example 25.3 (U.S.) Find the area of a circle segment with an angle of 45° and a radius of 10 in. Calculate the arc length(s) and use both Equation 25.4 and Equation 25.5.

Solution 45°/360° = 0.125; circumference = $\pi D = \pi 20 = 62.83$ in.

$$S = 0.125 \times 62.83 = 7.854 \text{ in.}$$

$$A_s = \tfrac{1}{2}RS = \tfrac{1}{2} \times 10 \times 7.854 = 39.27 \text{ in.}^2$$

$$A_s = \frac{\Theta°\pi R^2}{360°} = \frac{45° \times \pi \times (10)^2}{360°} = 39.27 \text{ in.}^2$$

Example 25.3 (Metric) Find the area of a circle segment with an angle of 45° and a radius of 250 mm. Use both Equation 25.4 and Equation 25.5.

Solution 45°/360° = 0.125; circumference = $\pi D = \pi 500 = 1570.8$ mm

$$S = 0.125 \times 1570.8 = 196.3 \text{ mm}$$

$$A_s = \tfrac{1}{2}RS = \tfrac{1}{2} \times 250 \times 196.3 = 24,543.7 \text{ mm}^2$$

$$A_s = \frac{\Theta°A}{360°} = \frac{\Theta°\pi R^2}{360°} = \frac{45° \times \pi \times (250)^2}{360°} = 24,543.7 \text{ mm}^2$$

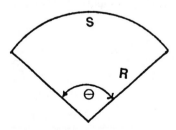

Figure 25.3 Circle sector.

25.1.2.3 Circle segment. The following equations may be used for the area of a segment of a circle with angle Θ (Figure 25.4):

Equation 25.6

$$A_s = \frac{\Theta° \pi R^2}{360°} - \frac{\sin \Theta R^2}{2}$$

Example 25.4 (U.S.) Find the area of a circle segment with an angle of 45° and a radius of 10 in.

Solution sin 45° = 0.7071 (from Table 25.1)

$$A_s = \frac{\Theta° \pi R^2}{360°} - \frac{\sin \Theta R^2}{2} = \frac{45° \pi (10)^2}{360°} - \frac{0.7071(10)^2}{2}$$

$$A_s = 39.27 - 35.35 = 3.02 \text{ in.}^2$$

Example 25.4 (Metric) Find the area of a circle segment with an angle of 45° and a radius of 250 mm.

Solution sin 45° = 0.7071 (from Table 25.1)

$$A_s = \frac{\Theta° \pi R^2}{360°} - \frac{\sin \Theta° R^2}{2} = \frac{45 \pi (250)^2}{360°} - \frac{0.7071(250)^2}{2}$$

$$A_s = 24{,}543.7 - 22{,}093.8 = 2449.9 \text{ mm}^2$$

25.1.3 Triangles

The area of a triangle is equal to one-half of the base times the altitude (Figures 25.5 to 25.7). The altitude is the height of the triangle.

Equation 25.7

$$A = \tfrac{1}{2}ab$$

Where: A = area (in.² or mm²)
$\quad\quad a$ = altitude or height (in. or mm)
$\quad\quad b$ = base (in. or mm)

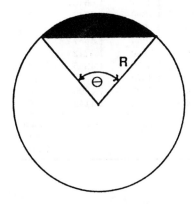

Figure 25.4 Circle segment.

TABLE 25.1   Natural Trigonometric Functions

Angle in degrees	sin	cos	tan	cot	sec	csc	Angle in degrees
1	0.0175	0.9999	0.0175	57.2900	1.0001	67.299	89
2	0.0349	0.9994	0.0349	28.6363	1.0006	28.654	88
3	0.0523	0.9986	0.0524	19.0811	1.0014	19.107	87
4	0.0698	0.9976	0.0699	14.3007	1.0024	14.335	86
5	0.0872	0.9962	0.0875	11.4301	1.0038	11.474	85
6	0.1045	0.9945	0.1051	9.5144	1.0055	9.5668	84
7	0.1219	0.9926	0.1228	8.1443	1.0075	8.2055	83
8	0.1392	0.9903	0.1405	7.1154	1.0098	7.1853	82
9	0.1564	0.9877	0.1584	6.3137	1.0125	6.3924	81
10	0.1737	0.9848	0.1763	5.6713	1.0154	5.7588	80
11	0.1908	0.9816	0.1944	5.1445	1.0187	5.2408	79
12	0.2079	0.9782	0.2126	4.7046	1.0223	4.8097	78
13	0.2250	0.9744	0.2309	4.3315	1.0263	4.4454	77
14	0.2419	0.9703	0.2493	4.0108	1.0306	4.1336	76
15	0.2588	0.9659	0.2680	3.7320	1.0353	3.8637	75
16	0.2756	0.9613	0.2867	3.4874	1.0403	3.6279	74
17	0.2924	0.9563	0.3057	3.2708	1.0457	3.4203	73
18	0.3090	0.9511	0.3249	3.0777	1.0515	3.2361	72
19	0.3256	0.9445	0.3443	2.9042	1.0576	3.0715	71
20	0.3420	0.9397	0.3640	2.7475	1.0642	2.9238	70
21	0.3584	0.9336	0.3839	2.6051	1.0711	2.7904	69
22	0.3746	0.9272	0.4040	2.4751	1.0785	2.6695	68
23	0.3907	0.9205	0.4245	2.3558	1.0864	2.5593	67
24	0.4067	0.9135	0.4452	2.2460	1.0946	2.4586	66
25	0.4226	0.9063	0.4663	2.1445	1.1034	2.3662	65
26	0.4384	0.8988	0.4877	2.0503	1.1126	2.2812	64
27	0.4540	0.8910	0.5095	1.9626	1.1223	2.2027	63
28	0.4695	0.8830	0.5317	1.8807	1.1326	2.1300	62
29	0.4848	0.8746	0.5543	1.8040	1.1433	2.0627	61
30	0.5000	0.8660	0.5774	1.7320	1.1547	2.0000	60
31	0.5150	0.8572	0.6009	1.6643	1.1666	1.9416	59
32	0.5299	0.8481	0.6249	1.6003	1.1792	1.8871	58
33	0.5446	0.8387	0.6494	1.5399	1.1924	1.8361	57
34	0.5592	0.8290	0.6745	1.4826	1.2062	1.7883	56
35	0.5736	0.8192	0.7002	1.4281	1.2208	1.7434	55
36	0.5878	0.8090	0.7265	1.3764	1.2361	1.7013	54
37	0.6018	0.7986	0.7536	1.3270	1.2521	1.6616	53
38	0.6157	0.7880	0.7813	1.2799	1.2690	1.6243	52
39	0.6293	0.7772	0.8098	1.2349	1.2867	1.5890	51
40	0.6428	0.7660	0.8391	1.1917	1.3054	1.5557	50
41	0.6560	0.7547	0.8693	1.1504	1.3250	1.5242	49
42	0.6691	0.7431	0.9004	1.1106	1.3456	1.4945	48
43	0.6820	0.7314	0.9325	1.0724	1.3673	1.4662	47
44	0.6947	0.7193	0.9657	1.0355	1.3902	1.4395	46
45	0.7071	0.7071	1.0000	1.0000	1.4142	1.4142	45
Angle in degrees	cos	sin	cot	tan	csc	sec	Angle in degrees

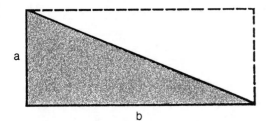

Figure 25.5 Area of a triangle.

The logic of this equation can be seen in Figure 25.5. The triangle is one half of the rectangle shown by the dashed lines. Since the area of the rectangle is $A = ab$, then the area of one triangle is $\frac{1}{2}ab$.

Example 25.5 (U.S.)   Find the area of the right triangle in Figure 25.6.

Solution

$$A = \frac{1}{2}ab = \frac{1}{2} \times 6 \times 8$$

$$A = 24 \text{ in.}^2$$

Example 25.5 (Metric)   Find the area of the triangle in Figure 25.7 if a equals 250 mm and b equals 150 mm.

Solution

$$A = \frac{1}{2}ab = \frac{1}{2} \times 250 \times 150 \text{ mm}$$

$$A = 18,750 \text{ mm}^2$$

25.1.4 Trapezoids

The area of a trapezoid (Figure 25.8) (sides b_1 and b_2 parallel) may be found from the following equation.

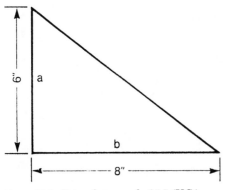

Figure 25.6 Triangle example 25.5 (U.S.).

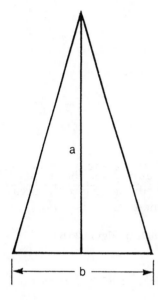

Figure 25.7 Triangle for Example 25.5 (Metric).

Equation 25.8

$A = \frac{1}{2}(b_1 + b_2)h$

Where: A = area (in.^{2} or mm^2)
  b_1 = top side (in. or mm)
  b_2 = base (in. or mm)
  h = height (in. or mm)

Example 25.6 (U.S.) Find the area of a trapezoid where $b_1 = 10$ in., $b_2 = 16$ in., $h = 12$ in.

Solution

$$A = \frac{1}{2}(b_1 + b_2)h = \frac{1}{2}(10 + 16)12$$

$$A = 156 \text{ in.}^2$$

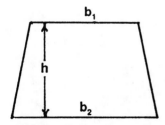

Figure 25.8 Trapezoid.

Example 25.6 (Metric)   Find the area of a trapezoid where b_1 = 250 mm, b_2 = 400 mm, h = 300 mm.

Solution

$$A = \tfrac{1}{2}(b_1 + b_2)h = \tfrac{1}{2}(250 + 400)300$$

$$A = 97{,}500 \text{ mm}^2$$

25.1.5 Flat oval

To find the cross-sectional area of flat oval ducts (Figure 25.9), use Equation 25.9. To find the surface area of a flat oval duct, use Equation 25.10.

Equation 25.9

$$A_c = wh + \tfrac{1}{4}\pi h^2$$

Equation 25.10 (U.S.)

$$A_p = \frac{(2w + \pi h)L}{12}$$

Equation 25.10 (Metric)

$$A_p = \frac{(2w + \pi h)L}{1000}$$

Where: A_c = area of cross-section (in.^{2} or mm^2)
A_p = area of perimeter times length (ft^2 or m^2)
w = width of flat side (in. or mm)
a = overall width ($w + h$) (in. or mm)
L = length (ft. or m)

Example 25.7 (U.S.)   Find the cross-sectional area of a flat oval duct where a = 40 in. and h = 12 in.

Solution

$$w = a - h = 40 - 12 = 28 \text{ in.}$$

$$A_c = wh + \tfrac{1}{4}\pi h^2 = 28 \times 12 + \tfrac{1}{4}\pi(12)^2$$

$$A_c = 336 + 113.1 = 449.1 \text{ in.}^2$$

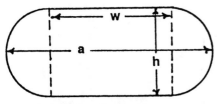

Figure 25.9 Flat oval area.

Example 25.7 (Metric)   Find the cross-sectional area of a flat oval duct in square millimeters and square meters where $a = 1000$ mm and $h = 300$ mm. $(1 \text{ m}^2 = 1{,}000{,}000 \text{ mm}^2)$

Solution

$$w = a - h = 1000 - 300 = 700 \text{ mm}$$

$$A_c = wh + \tfrac{1}{4}\pi h^2 = 700 \times 300 + \tfrac{1}{4}\pi(300)^2$$

$$A_c = 210{,}000 + 70685.8 = 280{,}685.8 \text{ mm}^2$$

$$A_c = 0.281 \text{ m}^2$$

25.2   Volumes

Volumes used in HVAC and TAB work normally would be room or space volumes (for air changes, etc.), rectangular sump volumes, and tanks, which generally are spherical.

25.2.1   Cubic volumes

For any volume measurements (Figure 25.10) with parallel sides and right angles, Equation 25.11 may be used.

Equation 25.11

$$V = HWL$$

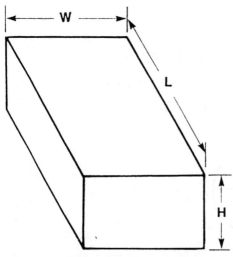

Figure 25.10   Cubic volume measurements.

Where: V = volume (ft³ or m³)
$\quad\quad H$ = height (ft or m)
$\quad\quad W$ = width (ft or m)
$\quad\quad L$ = length (ft or m)

Example 25.8 (U.S.) Find the volume of a room 25 ft. by 12 ft. with a 9 ft. ceiling. Calculate the ventilation airflow (cfm) to the room if two air changes per hour are required.

Solution

$$V = HWL = 9 \times 12 \times 25 = 2700 \text{ ft}^3$$

Airflow = $2 \times 2700/60$ min/h = 90 cfm ventilation air

Example 25.8 (Metric) Find the volume of a room 7.5 m × 3.6 m with a 2.7 m ceiling. Calculate the ventilation airflow (L/s) to the room if two air changes per hour are required.

Solution

$$V = HWL = 2.7 \times 3.6 \times 7.5 = 72.9 \text{ m}^3$$

Airflow = $2 \times 72.9/3600$ s/h = 0.0405 m³/s
Airflow = 40.5 L/s (1 m³/s = 1000 L/s)

25.2.2 Cylindrical volumes

A vertical round tank with flat heads shown in Figure 25.11 is a cylinder that uses Equation 25.12:

Equation 25.12

$$V = h\pi R^2$$

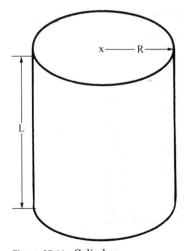

Figure 25.11 Cylinder.

Where: V = volume (ft^2 or m^3)
l = length (ft or m)
R = radius (ft or m)

If the heads of the tanks are true hemispheres, then the volume for one full sphere must be added to the above volume. The volume of standard tank dished heads may be found in Table 25.2.

Example 25.9 (U.S.) Find the volume of a horizontal 5 ft. diameter fuel oil tank that is 15 ft. long (Figure 25.11). a) How many gallons does the tank hold? b) Find the capacity if the tank has dished heads.

Solution

$$V = l\pi R^2 = 15\pi(2.5)^2 = 294.5 \text{ ft}^3$$

a) 294.5 × 7.49 gal/ft^3 = 2206 gal (1 ft^3 = 7.49 gal; see Table C-4 in Appendix C)
b) 50.37 (Table 25.2) + 2206 = 2256 gal

Example 25.9 (Metric) Find the volume of a horizontal 1.5 m diameter fuel oil tank that is 4.5 m long. a) How many liters does the tank hold? b) Find the capacity if the tank has dished heads.

Solution

$$V = l\pi R^2 = 4.5\pi(0.75)^2 = 7.95 \text{ m}^3$$

a) 7.95 m^3 × 1000 L/m^3 = 7950 L
b) 190 (approx. − Table 25.2) + 7950 = 8140 L

TABLE 25.2   Liquid Volumes of Horizontal Standard Tank Dished Heads

Head diameter		Tank full		Tank ¼ full	
Inches	Millimeters	Gallons	Liters	Gallons	Liters
12	305	0.40	1.51	0.05	0.19
18	457	1.36	5.15	0.17	0.64
24	610	3.22	12.19	0.41	1.55
30	762	6.30	23.85	0.79	2.99
36	914	10.88	41.18	1.39	5.26
42	1067	17.28	65.40	2.16	8.18
48	1219	25.79	97.62	3.31	12.53
60	1524	50.37	190.65	6.49	24.56
72	1829	87.04	329.45	11.15	42.20
84	2134	138.22	523.16	17.78	67.30
96	2438	206.32	780.92	26.60	100.68

NOTE: Tank half full, use half of full tank values.

25.2.3 Duct surface

The surface area of a round duct, which is a cylinder, is:

Equation 25.13

$$A_p = L\pi D$$

Where: A_p = area of perimeter (ft^2 or m^2)
L = length of duct (ft or m)
D = diameter (ft or m)

25.2.4 Sphere

The volume for a sphere may be obtained by using Equation 25.14.

Equation 25.14

$$V = \frac{4\pi R^3}{3}$$

Where: V = volume (ft^3 or m^3)
R = radius (ft or m)

25.3 Equivalent Areas

25.3.1 Piping

TAB technicians must understand the principles of equivalent areas in order to determine whether a problem exists in environmental systems. For example, what arrangement of pipes would be required to carry the fluid flow where the main pipe is 10 in. in diameter? Would two 5 in. pipes provide the equivalent area of the 10 in. diameter pipe? The answer is "No."

Figures 25.12 and 25.13 illustrate the equivalent area for a 10 in. (250 mm) diameter pipe. The area of a 10 in. (250 mm) diameter circle is 78.5 in.^{2} (49,087 mm^2) according to the equation $A = \pi R^2$. The combined area of two 5 in. (125 mm) diameter circles is 39.2 in.^{2} (24,544 mm^2). An 8 in. (200 mm) diameter pipe and a 6 in. (150 mm) diameter pipe connected to the 10 in. (250 mm) diameter pipe provide equal areas. If technicians see branches that are not the equivalent area of the main pipe or duct, they should be noted and reported. It is the design engineer's or installing contractor's responsibility to determine whether they were intended to be the sizes shown for some special reason.

In SMACNA and ASHRAE manuals, tables can be found for "circular equivalents of rectangular ducts for equal friction and capacity." These tables are to be used *only* for determining duct sizes and friction

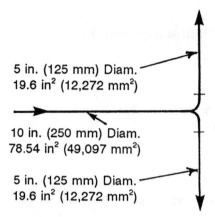

Figure 25.12   Incorrect equivalent areas.

losses. They are *not* to be used to determine the area used for duct average velocities and for calculating duct fitting losses.

Duct average velocities (fpm or m/s) are obtained by dividing the airflow (cfm or L/s) by the cross-sectional area in square feet or square meters and using appropriate constants (see Chapter 2).

Example 25.10 (U.S.)   If the 10 in. diameter duct in Figure 25.12 is split into two equal airflows, calculate the theoretical size of the two branches. What would the closest standard size duct be with total area?

Solution   ½ area of 10 in. duct = 78.54/2 = 39.27 in.^{2}

$$A = \pi R^2;$$

$$R = \sqrt{A/\pi} = \sqrt{39.27/\pi}$$

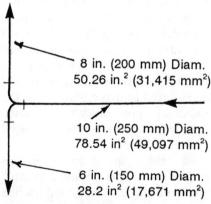

Figure 25.13   Equivalent areas.

$$R = 3.54 \text{ in.}; D = 7.07 \text{ in.}$$

Two each 7 inch Std. size $= 2\pi R^2 = 2\pi(3.5)^2 = 76.97 \text{ in.}^2$

Example 25.10 (Metric) If the 250 mm diameter duct in Figure 25.12 is split into two equal airflows, calculate the theoretical size of the two branches. What would the closest standard size duct be with total area?

Solution ½ area of 250 mm duct $= 49,087/2 = 24,544 \text{ mm}^2$

$$A = \pi R^2;$$

$$R = \sqrt{A/\pi} = \sqrt{24,544/\pi}$$

$$R = 88.39 \text{ mm}; D = 176.78 \text{ mm}$$

Two each 175 mm Std. size $= 2\pi R^2 = 2\pi(87.5)^2 = 48,106 \text{ mm}^2$
or
Two each 180 mm Std. size $= 2\pi R^2 = 2\pi(90)^2 = 50,894 \text{ mm}^2$

25.4 Fractions

Dimensions in U.S. units often include fractions of inches that must be converted to decimals of inches or feet to use in equations or when converting to metric units. Table 25.3 is a convenient table to use for this purpose.

TABLE 25.3 Inch–Foot–Decimal Conversion

Inches, fractions	Inches, decimals	Feet, decimals	Inches, fractions	Inches, decimals	Feet, decimals	Inches, fractions	Inches, decimals	Feet, decimals
1/64	0.0156	0.0013	11/32	0.3438	0.0287	43/64	0.6719	0.0560
1/32	0.0313	0.0026	23/64	0.3594	0.0299	11/16	0.6875	0.0573
3/64	0.0469	0.0039				45/64	0.7031	0.0586
1/16	0.0625	0.0052	3/8	0.3750	0.0313	23/32	0.7188	0.0599
5/64	0.0781	0.0065	25/64	0.3906	0.0326	47/64	0.7344	0.0612
3/32	0.0938	0.0078	13/32	0.4063	0.0339			
7/64	0.1094	0.0091	27/64	0.4219	0.0352	3/4	0.7500	0.0625
			7/16	0.4375	0.0365	49/64	0.7656	0.0638
1/8	0.1250	0.0104	29/64	0.4531	0.0378	25/32	0.7813	0.0651
9/64	0.1406	0.0117	15/32	0.4688	0.0391	51/64	0.7969	0.0664
5/32	0.1563	0.0130	31/64	0.4844	0.0404	13/16	0.8125	0.0677
11/64	0.1719	0.0143				53/64	0.8281	0.0690
3/16	0.1875	0.0156	1/2	0.5000	0.0417	27/32	0.8437	0.0703
13/64	0.2031	0.0169	33/64	0.5156	0.0430	55/64	0.8594	0.0716
7/32	0.2188	0.0182	17/32	0.5313	0.0443			
15/64	0.2343	0.0195	35/64	0.5469	0.0456	7/8	0.8750	0.0729
			9/16	0.5625	0.0469	57/64	0.8906	0.0742
1/4	0.2500	0.0208	37/64	0.5781	0.0482	29/32	0.9063	0.0755
17/64	0.2656	0.0221	19/32	0.5938	0.0495	59/64	0.9219	0.0768
9/32	0.2813	0.0234	39/64	0.6094	0.0508	15/16	0.9375	0.0781
19/64	0.2969	0.0247	5/8	0.6250	0.0521	61/64	0.9531	0.0794
5/16	0.3125	0.0260	41/64	0.6406	0.0534	31/32	0.9688	0.0807
21/64	0.3281	0.0273	21/32	0.6563	0.0547	63/64	0.9844	0.0820

Duct Design Data

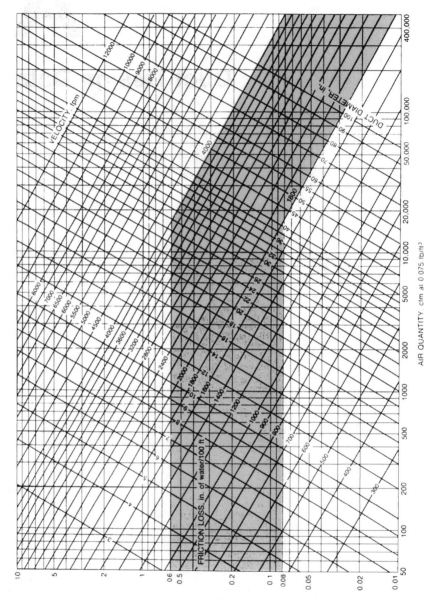

Figure A.1 Duct friction loss chart (U.S. units).

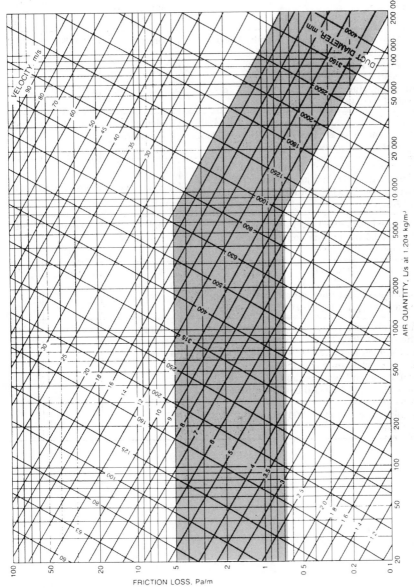

Figure A.2 Duct friction loss chart (metric units).

TABLE A.1   Duct Material Roughness Factors

Duct material	Roughness category	Absolute roughness, ϵ_1	
		ft	mm
Uncoated carbon steel, clean (Moody, 1944) (0.00015 ft) (0.05 mm) PVC plastic pipe (Swim, 1982) (0.0003 to 0.00015 ft) (0.01 to 0.05 mm) Aluminum (Hutchinson, 1953) (0.00015 to 0.0002 ft) (0.04 to 0.06 mm)	Smooth	0.0001	0.03
Galvanized steel, longitudinal seams, 4 ft (1200 mm) joints (Griggs, 1987) (0.00016 to 0.00032 ft) (0.05 to 0.1 mm)	Medium smooth	0.0003	0.09
Galvanized steel, spiral seam with 1, 2, and 3 ribs, 12 ft (3600 mm) joints (Jones, 1979; Griggs, 1987) (0.00018 to 0.00038 ft) (0.05 to 0.12 mm)	(New Duct Friction Loss Chart)		
Hot-dipped galvanized steel, longitudinal seams, 2.5 ft (760 mm) joints (Wright, 1945) (0.0005 ft) (0.15 mm)	Old average	0.0005	0.15
Fibrous glass duct, rigid Fibrous glass duct liner, air side with facing material (Swim, 1978) (0.005 ft) (1.5 mm)	Medium rough	0.003	0.9
Fibrous glass duct liner, air side spray coated (Swim, 1978) (0.015 ft) (4.5 mm) Flexible duct, metallic (0.004 to 0.007 ft) (1.2 to 2.1 mm) when fully extended Flexible duct, all types of fabric and wire (0.0035 to 0.015 ft) (1.0 to 4.6 mm) when fully extended Concrete (Moody, 1944) (0.001 to 0.01 ft) (0.3 to 3.0 mm)	Rough	0.01	3.0

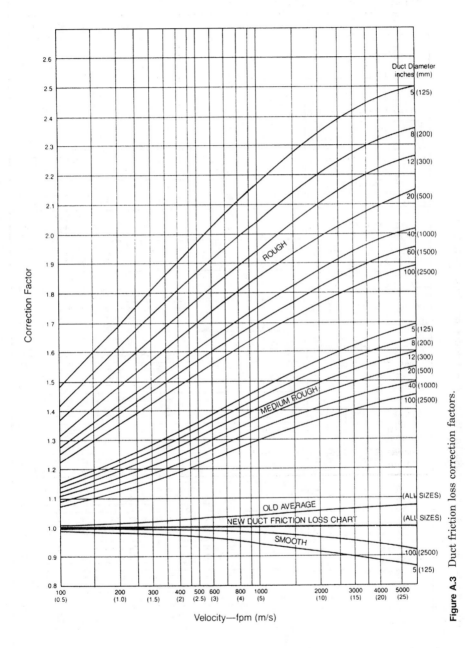

Figure A.3 Duct friction loss correction factors.

Velocity—fpm (m/s)

Correction Factor

Duct Diameter
inches (mm)

5 (125)
8 (200)
12 (300)
20 (500)
40 (1000)
60 (1500)
100 (2500)

ROUGH

5 (125)
8 (200)
12 (300)
20 (500)
40 (1000)
100 (2500)

MEDIUM ROUGH

(ALL SIZES)
OLD AVERAGE
NEW DUCT FRICTION LOSS CHART
(ALL SIZES)
SMOOTH
100 (2500)
5 (125)

TABLE A.2 Circular Equivalents of Rectangular Ducts for Equal Friction and Capacity (U.S. Units)

Dimensions in Inches

Side rectangular duct	4.0	4.5	5.0	5.5	6.0	6.5	7.0	7.5	8.0	9.0	10.0	11.0	12.0	13.0	14.0	15.0	16.0
3.0	3.8	4.0	4.2	4.4	4.6	4.7	4.9	5.1	5.2	5.5	5.7	6.0	6.2	6.4	6.6	6.8	7.0
3.5	4.1	4.3	4.6	4.8	5.0	5.2	5.3	5.5	5.7	6.0	6.3	6.5	6.8	7.0	7.2	7.5	7.7
4.0	4.4	4.6	4.9	5.1	5.3	5.5	5.7	5.9	6.1	6.4	6.7	7.0	7.3	7.6	7.8	8.1	8.3
4.5	4.6	4.9	5.2	5.4	5.7	5.9	6.1	6.3	6.5	6.9	7.2	7.5	7.8	8.1	8.4	8.6	8.8
5.0	4.9	5.2	5.5	5.7	6.0	6.2	6.4	6.7	6.9	7.3	7.6	8.0	8.3	8.6	8.9	9.1	9.4
5.5	5.1	5.4	5.7	6.0	6.3	6.5	6.8	7.0	7.2	7.6	8.0	8.4	8.7	9.0	9.3	9.6	9.9

Side rectangular duct	6	7	8	9	10	11	12	13	14	15	16	17	18	19	20	22	24	26	28	30
6	6.6																			
7	7.1	7.7																		
8	7.6	8.2	8.7																	
9	8.0	8.7	9.3	9.8																
10	8.4	9.1	9.8	10.4	10.9															
11	8.8	9.5	10.2	10.9	11.5	12.0														
12	9.1	9.9	10.7	11.3	12.0	12.6	13.1													
13	9.5	10.3	11.1	11.8	12.4	13.1	13.7	14.2												
14	9.8	10.7	11.5	12.2	12.9	13.5	14.2	14.7	15.3											
15	10.1	11.0	11.8	12.6	13.3	14.0	14.6	15.3	15.8	16.4										
16	10.4	11.3	12.2	13.0	13.7	14.4	15.1	15.7	16.4	16.9	17.5									
17	10.7	11.6	12.5	13.4	14.1	14.9	15.6	16.2	16.8	17.4	18.0	18.6								
18	11.0	11.9	12.9	13.7	14.5	15.3	16.0	16.7	17.3	17.9	18.5	19.1	19.7							
19	11.2	12.2	13.2	14.1	14.9	15.7	16.4	17.1	17.8	18.4	19.0	19.6	20.2	20.8						
20	11.5	12.5	13.5	14.4	15.2	16.0	16.8	17.5	18.2	18.9	19.5	20.1	20.7	21.3	21.9					
22	12.0	13.0	14.1	15.0	15.9	16.8	17.6	18.3	19.1	19.8	20.4	21.1	21.7	22.3	22.9	24.0				
24	12.4	13.5	14.6	15.6	16.5	17.4	18.3	19.1	19.9	20.6	21.3	22.0	22.7	23.3	23.9	25.1	26.2			
26	12.8	14.0	15.1	16.2	17.1	18.1	19.0	19.8	20.6	21.4	22.1	22.9	23.5	24.2	24.9	26.1	27.3	28.4		
28	13.2	14.5	15.6	16.7	17.7	18.7	19.6	20.5	21.3	22.1	22.9	23.7	24.4	25.1	25.8	27.1	28.3	29.5	30.6	
30	13.6	14.9	16.1	17.2	18.3	19.3	20.2	21.1	22.0	22.9	23.7	24.4	25.2	25.9	26.6	28.0	29.3	30.5	31.7	32.8

Dimensions in Inches

Side rectangular duct	6	7	8	9	10	11	12	13	14	15	16	17	18	19	20	22	24	26	28	30	Side rectangular duct
32	14.0	15.3	16.5	17.7	18.8	19.8	20.8	21.8	22.7	23.5	24.4	25.2	26.0	26.7	27.5	28.9	30.2	31.5	32.7	33.9	32
34	14.4	15.7	17.0	18.2	19.3	20.4	21.4	22.4	23.3	24.2	25.1	25.9	26.7	27.5	28.3	29.7	31.0	32.4	33.7	34.9	34
36	14.7	16.1	17.4	18.6	19.8	20.9	21.9	22.9	23.9	24.8	25.7	26.6	27.4	28.2	29.0	30.5	32.0	33.3	34.6	35.9	36
38	15.0	16.5	17.8	19.0	20.2	21.4	22.4	23.5	24.5	25.4	26.4	27.2	28.1	28.9	29.8	31.3	32.8	34.2	35.6	36.8	38
40	15.3	16.8	18.2	19.5	20.7	21.8	22.9	24.0	25.0	26.0	27.0	27.9	28.8	29.6	30.5	32.1	33.6	35.1	36.4	37.8	40
42	15.6	17.1	18.5	19.9	21.1	22.3	23.4	24.5	25.6	26.6	27.6	28.5	29.4	30.3	31.2	32.8	34.4	35.9	37.3	38.7	42
44	15.9	17.5	18.9	20.3	21.5	22.7	23.9	25.0	26.1	27.1	28.1	29.1	30.0	30.9	31.8	33.5	35.1	36.7	38.1	39.5	44
46	16.2	17.8	19.3	20.6	21.9	23.2	24.4	25.5	26.6	27.7	28.7	29.7	30.6	31.6	32.5	34.2	35.9	37.4	38.9	40.4	46
48	16.5	18.1	19.6	21.0	22.3	23.6	24.8	26.0	27.1	28.2	29.2	30.2	31.2	32.2	33.1	34.9	36.6	38.2	39.7	41.2	48
50	16.8	18.4	19.9	21.4	22.7	24.0	25.2	26.4	27.6	28.7	29.8	30.8	31.8	32.8	33.7	35.5	37.2	38.9	40.5	42.0	50
52	17.1	18.7	20.2	21.7	23.1	24.4	25.7	26.9	28.0	29.2	30.3	31.3	32.3	33.3	34.3	36.2	37.9	39.6	41.2	42.8	52
54	17.3	19.0	20.6	22.0	23.5	24.8	26.1	27.3	28.5	29.7	30.8	31.8	32.9	33.9	34.9	36.8	38.6	40.3	41.9	43.5	54
56	17.6	19.3	20.9	22.4	23.8	25.2	26.5	27.7	28.9	30.1	31.2	32.3	33.4	34.4	35.4	37.4	39.2	41.0	42.7	44.3	56
58	17.8	19.5	21.2	22.7	24.2	25.5	26.9	28.2	29.4	30.6	31.7	32.8	33.9	35.0	36.0	38.0	39.8	41.6	43.3	45.0	58
60	18.1	19.8	21.5	23.0	24.5	25.9	27.3	28.6	29.8	31.0	32.2	33.3	34.4	35.5	36.5	38.5	40.4	42.3	44.0	45.7	60
62		20.1	21.7	23.3	24.8	26.3	27.6	28.9	30.2	31.5	32.6	33.8	34.9	36.0	37.1	39.1	41.0	42.9	44.7	46.4	62
64		20.3	22.0	23.6	25.1	26.6	28.0	29.3	30.6	31.9	33.1	34.3	35.4	36.5	37.6	39.6	41.6	43.5	45.3	47.1	64
66		20.6	22.3	23.9	25.5	26.9	28.4	29.7	31.0	32.3	33.5	34.7	35.9	37.0	38.1	40.2	42.2	44.1	46.0	47.7	66
68		20.8	22.6	24.2	25.8	27.3	28.7	30.1	31.4	32.7	33.9	35.2	36.3	37.5	38.6	40.7	42.8	44.7	46.6	48.4	68
70		21.1	22.9	24.5	26.1	27.6	29.1	30.4	31.8	33.1	34.4	35.6	36.8	37.9	39.1	41.2	43.3	45.3	47.2	49.0	70
72			23.1	24.8	26.4	27.9	29.4	30.8	32.2	33.5	34.8	36.0	37.2	38.4	39.5	41.7	43.8	45.8	47.8	49.6	72
74			23.3	25.1	26.7	28.2	29.7	31.2	32.5	33.9	35.2	36.4	37.7	38.8	40.0	42.2	44.4	46.4	48.4	50.3	74
76			23.6	25.3	27.0	28.5	30.0	31.5	32.9	34.3	35.6	36.8	38.1	39.3	40.5	42.7	44.9	47.0	48.9	50.9	76
78			23.8	25.6	27.3	28.8	30.4	31.8	33.3	34.6	36.0	37.2	38.5	39.7	40.9	43.2	45.4	47.5	49.5	51.4	78
80			24.1	25.8	27.5	29.1	30.7	32.2	33.6	35.0	36.3	37.6	38.9	40.2	41.4	43.7	45.9	48.0	50.1	52.0	80
82				26.1	27.8	29.4	31.0	32.5	34.0	35.4	36.7	38.0	39.3	40.6	41.8	44.1	46.4	48.5	50.6	52.6	82
84				26.4	28.1	29.7	31.3	32.8	34.3	35.7	37.1	38.4	39.7	41.0	42.2	44.6	46.9	49.0	51.1	53.2	84
86				26.6	28.3	30.0	31.6	33.1	34.6	36.1	37.4	38.8	40.1	41.4	42.6	45.0	47.3	49.6	51.7	53.7	86
88				26.9	28.6	30.3	31.9	33.4	34.9	36.4	37.8	39.2	40.5	41.8	43.1	45.5	47.8	50.0	52.2	54.3	88
90				27.1	28.9	30.6	32.2	33.8	35.3	36.7	38.2	39.5	40.9	42.2	43.5	45.9	48.3	50.5	52.7	54.8	90
92					29.1	30.8	32.5	34.1	35.6	37.1	38.5	39.9	41.3	42.6	43.9	46.4	48.7	51.0	53.2	55.3	92
96					29.6	31.4	33.0	34.7	36.2	37.7	39.2	40.6	42.0	43.3	44.7	47.2	49.6	52.0	54.2	56.4	96

TABLE A.2 continued Circular Equivalents of Rectangular Ducts for Equal Friction and Capacity (U.S. Units)

Dimensions in Inches

Side rectangular duct	32	34	36	38	40	42	44	46	48	50	52	56	60	64	68	72	76	80	84	88	Side rectangular duct
32	35.0																				32
34	36.1	37.2																			34
36	37.1	38.2	39.4																		36
38	38.1	39.3	40.4	41.5																	38
40	39.0	40.3	41.5	42.6	43.7																40
42	40.0	41.3	42.5	43.7	44.8	45.9															42
44	40.9	42.2	43.5	44.7	45.8	47.0	48.1														44
46	41.8	43.1	44.4	45.7	46.9	48.0	49.2	50.3													46
48	42.6	44.0	45.3	46.6	47.9	49.1	50.2	51.4	52.5												48
50	43.6	44.9	46.2	47.5	48.8	50.0	51.2	52.4	53.6	54.7											50
52	44.3	45.7	47.1	48.4	49.7	51.0	52.2	53.4	54.6	55.7	56.8										52
54	45.1	46.5	48.0	49.3	50.7	52.0	53.2	54.4	55.6	56.6	57.9										54
56	45.8	47.3	48.8	50.2	51.6	52.9	54.2	55.4	56.6	57.8	59.0	61.2									56
58	46.6	48.1	49.6	51.0	52.4	53.8	55.1	56.4	57.6	58.8	60.0	62.3									58
60	47.3	48.9	50.4	51.9	53.3	54.7	56.0	57.3	58.6	59.8	61.0	63.4	65.6								60
62	48.0	49.6	51.2	52.7	54.1	55.5	56.9	58.2	59.5	60.8	62.0	64.4	66.7								62
64	48.7	50.4	51.9	53.5	54.9	56.4	57.8	59.1	60.4	61.7	63.0	65.4	67.7	70.0							64
66	49.4	51.1	52.7	54.2	55.7	57.2	58.6	60.0	61.3	62.6	63.9	66.4	68.8	71.0							66
68	50.1	51.8	53.4	55.0	56.5	58.0	59.4	60.8	62.2	63.6	64.9	67.4	69.8	72.1	74.3						68
70	50.8	52.5	54.1	55.7	57.3	58.8	60.3	61.7	63.1	64.4	65.8	68.3	70.8	73.2	75.4						70
72	51.4	53.2	54.8	56.5	58.0	59.6	61.1	62.5	63.9	65.3	66.7	69.3	71.8	74.2	76.5	78.7					72
74	52.1	53.8	55.5	57.2	58.8	60.3	61.9	63.3	64.8	66.2	67.5	70.2	72.7	75.2	77.5	79.8					74
76	52.7	54.5	56.2	57.9	59.5	61.1	62.6	64.1	65.6	67.0	68.4	71.1	73.7	76.2	78.6	80.9	83.1				76
78	53.3	55.1	56.9	58.6	60.2	61.8	63.4	64.9	66.4	67.9	69.3	72.0	74.6	77.1	79.6	81.9	84.2				78
80	53.9	55.8	57.5	59.3	60.9	62.6	64.1	65.7	67.2	68.7	70.1	72.9	75.4	78.1	80.6	82.9	85.2	87.5			80
82	54.5	56.4	58.2	59.9	61.6	63.3	64.9	66.5	68.0	69.5	70.9	73.7	76.4	79.0	81.5	84.0	86.3	88.5			82
84	55.1	57.0	58.8	60.6	62.3	64.0	65.6	67.2	68.7	70.3	71.7	74.6	77.3	80.0	82.5	85.0	87.3	89.6	91.8		84
86	55.7	57.6	59.4	61.2	63.0	64.7	66.3	67.9	69.5	71.0	72.5	75.4	78.2	80.9	83.5	85.9	88.3	90.7	92.9		86
88	56.3	58.2	60.1	61.9	63.6	65.4	67.0	68.7	70.2	71.8	73.3	76.3	79.1	81.8	84.5	86.9	89.3	91.7	94.0	96.2	88
90	56.8	58.8	60.7	62.5	64.3	66.0	67.7	69.4	71.0	72.6	74.1	77.1	79.9	82.7	85.3	87.9	90.3	92.7	95.0	97.3	90
92	57.4	59.3	61.3	63.1	64.9	66.7	68.4	70.1	71.7	73.3	74.9	77.9	80.8	83.5	86.2	88.8	91.3	93.7	96.1	98.4	92
94	57.9	59.9	61.9	63.7	65.6	67.3	69.1	70.8	72.4	74.0	75.6	78.7	81.6	84.4	87.1	89.7	92.3	94.7	97.1	99.4	94
96	58.4	60.5	62.4	64.3	66.2	68.0	69.7	71.5	73.1	74.8	76.3	79.4	82.4	85.3	88.0	90.7	93.2	95.7	98.1	100.5	96

Equation for circular equivalent of a rectangular duct:

$$D_e = 1.30[(ab)^{0.625}/(a + b)^{0.250}]$$

where a = length of one side of rectangular duct, inches

TABLE A.3 Circular Equivalents of Rectangular Ducts for Equal Friction and Capacity (Metric Units)

Dimensions in Millimeters

Side rectangular duct	100	125	150	175	200	225	250	275	300	350	400	450	500	550	600	650	700	750	800	900	Side rectangular duct
100	109																				100
125	122	137																			125
150	133	150	164																		150
175	143	161	177	191																	175
200	152	172	189	204	219																200
225	161	181	200	216	232	246															225
250	169	190	210	228	244	259	273														250
275	176	199	220	238	256	272	287	301													275
300	183	207	229	248	266	283	299	314	328												300
350	195	222	245	267	286	305	322	339	354	383											350
400	207	235	260	283	305	325	343	361	378	409	437										400
450	217	247	274	299	321	343	363	382	400	433	464	492									450
500	227	258	287	313	337	360	381	401	420	455	488	518	547								500
550	236	269	299	326	352	375	398	419	439	477	511	543	573	601							550
600	245	279	310	339	365	390	414	436	457	496	533	567	598	628	656						600
650	253	289	321	351	378	404	429	452	474	515	553	589	622	653	683	711					650
700	261	298	331	362	391	418	443	467	490	533	573	610	644	677	708	737	765				700
750	268	306	341	373	402	430	457	482	506	550	592	630	666	700	732	763	792	820			750
800	275	314	350	383	414	442	470	496	520	567	609	649	687	722	755	787	818	847	875		800
900	289	330	367	402	435	465	494	522	548	597	643	686	726	763	799	833	866	897	927	984	900

TABLE A.3 *continued* **Circular Equivalents of Rectangular Ducts for Equal Friction and Capacity (Metric Units)**

Dimensions in Millimeters

Side rectangular duct	100	125	150	175	200	225	250	275	300	350	400	450	500	550	600	650	700	750	800	900	Side rectangular duct
1000	301	344	384	420	454	486	517	546	574	626	674	719	762	802	840	876	911	944	976	1037	1000
1100	313	358	399	437	473	506	538	569	598	652	703	751	795	838	878	916	953	988	1022	1086	1100
1200	324	370	413	453	490	525	558	590	620	677	731	780	827	872	914	954	993	1030	1066	1133	1200
1300	334	382	426	468	506	543	577	610	642	701	757	808	857	904	948	990	1031	1069	1107	1177	1300
1400	344	394	439	482	522	559	595	629	662	724	781	835	886	934	980	1024	1066	1107	1146	1220	1400
1500	353	404	452	495	536	575	612	648	681	745	805	860	913	963	1011	1057	1100	1143	1183	1260	1500
1600	362	415	463	508	551	591	629	665	700	766	827	885	939	991	1041	1088	1133	1177	1219	1298	1600
1700	371	425	475	521	564	605	644	682	718	785	849	908	964	1018	1069	1118	1164	1209	1253	1335	1700
1800	379	434	485	533	577	619	660	698	735	804	869	930	988	1043	1096	1146	1195	1241	1286	1371	1800
1900	387	444	496	544	590	633	674	713	751	823	889	952	1012	1068	1122	1174	1224	1271	1318	1405	1900
2000	395	453	506	555	602	646	688	728	767	840	908	973	1034	1092	1147	1200	1252	1301	1348	1438	2000
2100	402	461	516	566	614	659	702	743	782	857	927	993	1055	1115	1172	1226	1279	1329	1378	1470	2100
2200	410	470	525	577	625	671	715	757	797	874	945	1013	1076	1137	1195	1251	1305	1356	1406	1501	2200
2300	417	478	534	587	636	683	728	771	812	890	963	1031	1097	1159	1218	1275	1330	1383	1434	1532	2300
2400	424	486	543	597	647	695	740	784	826	905	980	1050	1116	1180	1241	1299	1355	1409	1461	1561	2400
2500	430	494	552	606	658	706	753	797	840	920	996	1068	1136	1200	1262	1322	1379	1434	1488	1589	2500
2600	437	501	560	616	668	717	764	810	853	935	1012	1085	1154	1220	1283	1344	1402	1459	1513	1617	2600
2700	443	509	569	625	678	728	776	822	866	950	1028	1102	1173	1240	1304	1366	1425	1483	1538	1644	2700
2800	450	516	577	634	688	738	787	834	879	964	1043	1119	1190	1259	1324	1387	1447	1506	1562	1670	2800
2900	456	523	585	643	697	749	798	845	891	977	1058	1135	1208	1277	1344	1408	1469	1529	1596	1696	2900
Side rectangular duct	100	125	150	175	200	225	250	275	300	350	400	450	500	550	600	650	700	750	800	900	Side rectangular duct

TABLE A.3 continued Circular Equivalents of Rectangular Ducts for Equal Friction and Capacity (Metric Units)

Dimensions in Millimeters

Side rectangular duct	1000	1100	1200	1300	1400	1500	1600	1700	1800	1900	2000	2100	2200	2300	2400	2500	2600	2700	2800	2900	Side rectangular duct
1000	1093																				1000
1100	1146	1202																			1100
1200	1196	1256	1312																		1200
1300	1244	1306	1365	1421																	1300
1400	1289	1354	1416	1475	1530																1400
1500	1332	1400	1464	1526	1584	1640															1500
1600	1373	1444	1511	1574	1635	1693	1749														1600
1700	1413	1486	1555	1621	1684	1745	1803	1858													1700
1800	1451	1527	1598	1667	1732	1794	1854	1912	1968												1800
1900	1488	1566	1640	1710	1778	1842	1904	1964	2021	2077											1900
2000	1523	1604	1680	1753	1822	1889	1952	2014	2073	2131	2186										2000
2100	1558	1640	1719	1793	1865	1933	1999	2063	2124	2183	2240	2296									2100
2200	1591	1676	1756	1833	1906	1977	2044	2110	2173	2233	2292	2350	2405								2200
2300	1623	1710	1793	1871	1947	2019	2088	2155	2220	2283	2343	2402	2459	2514							2300
2400	1655	1744	1828	1909	1986	2060	2131	2200	2266	2330	2393	2453	2511	2568	2624						2400
2500	1685	1776	1862	1945	2024	2100	2173	2243	2311	2377	2441	2502	2562	2621	2678	2733					2500
2600	1715	1808	1896	1980	2061	2139	2213	2285	2355	2422	2487	2551	2612	2672	2730	2787	2842				2600
2700	1744	1839	1929	2015	2097	2177	2253	2327	2398	2466	2533	2598	2661	2722	2782	2840	2896	2952			2700
2800	1772	1869	1961	2048	2133	2214	2292	2367	2439	2510	2578	2644	2708	2771	2832	2891	2949	3006	3061		2800
2900	1800	1898	1992	2081	2167	2250	2329	2406	2480	2552	2621	2689	2755	2819	2881	2941	3001	3058	3115	3170	2900
Side rectangular duct	1000	1100	1200	1300	1400	1500	1600	1700	1800	1900	2000	2100	2200	2300	2400	2500	2600	2700	2800	2900	Side rectangular duct

Equation for circular equivalent of a rectangular duct:

$$D_e = 1.30[(ab)^{0.625}/(a + b)^{0.250}]$$

where a = length of one side of rectangular duct, millimeters.
b = length of adjacent side of rectangular duct, millimeters.
D_e = circular equivalent of rectangular duct for equal friction and capacity, millimeters.

TABLE A.4   Velocities and Velocity Pressures (U.S. Units)

Velocity (fpm)	Velocity pressure (in.w.g.)	Velocity (fpm)	Velocity pressure (in.w.g.)	Velocity (fpm)	Velocity pressure (in.w.g.)	Velocity (fpm)	Velocity pressure (in.w.g.)	Velocity (fpm)	Velocity pressure (in.w.g.)
300	0.01	2050	0.26	3800	0.90	5550	1.92	7300	3.32
350	0.01	2100	0.27	3850	0.92	5600	1.95	7350	3.37
400	0.01	2150	0.29	3900	0.95	5650	1.99	7400	3.41
450	0.01	2200	0.30	3950	0.97	5700	2.02	7450	3.46
500	0.02	2250	0.32	4000	1.00	5750	2.06	7500	3.51
550	0.02	2300	0.33	4050	1.02	5800	2.10	7550	3.55
600	0.02	2350	0.34	4100	1.05	5850	2.13	7600	3.60
650	0.03	2400	0.36	4150	1.07	5900	2.17	7650	3.65
700	0.03	2450	0.37	4200	1.10	5950	2.21	7700	3.70
750	0.04	2500	0.39	4250	1.13	6000	2.24	7750	3.74
800	0.04	2550	0.41	4300	1.15	6050	2.28	7800	3.79
850	0.05	2600	0.42	4350	1.18	6100	2.32	7850	3.84
900	0.05	2650	0.44	4400	1.21	6150	2.36	7900	3.89
950	0.06	2700	0.45	4450	1.23	6200	2.40	7950	3.94
1000	0.06	2750	0.47	4500	1.26	6250	2.43	8000	3.99
1050	0.07	2800	0.49	4550	1.29	6300	2.47	8050	4.04
1100	0.08	2850	0.51	4600	1.32	6350	2.51	8100	4.09
1150	0.08	2900	0.52	4650	1.35	6400	2.55	8150	4.14
1200	0.09	2950	0.54	4700	1.38	6450	2.59	8200	4.19
1250	0.10	3000	0.56	4750	1.41	6500	2.63	8250	4.24
1300	0.11	3050	0.58	4800	1.44	6550	2.67	8300	4.29
1350	0.11	3100	0.60	4850	1.47	6600	2.71	8350	4.35
1400	0.12	3150	0.62	4900	1.50	6650	2.76	8400	4.40
1450	0.13	3200	0.64	4950	1.53	6700	2.80	8450	4.45
1500	0.14	3250	0.66	5000	1.56	6750	2.84	8500	4.50
1550	0.15	3300	0.68	5050	1.59	6800	2.88	8550	4.56
1600	0.16	3350	0.70	5100	1.62	6850	2.92	8600	4.61
1650	0.17	3400	0.72	5150	1.65	6900	2.97	8650	4.66
1700	0.18	3450	0.74	5200	1.69	6950	3.01	8700	4.72
1750	0.19	3500	0.76	5250	1.72	7000	3.05	8750	4.77
1800	0.20	3550	0.79	5300	1.75	7050	3.10	8800	4.83
1850	0.21	3600	0.81	5350	1.78	7100	3.14	8850	4.88
1900	0.22	3650	0.83	5400	1.82	7150	3.19	8900	4.94
1950	0.24	3700	0.85	5450	1.85	7200	3.23	8950	4.99
2000	0.25	3750	0.88	5500	1.89	7250	3.28	9000	5.05

$$\text{Velocity} = 4005 \ \sqrt{V_p} \ \text{or} \ V_p = \left(\frac{\text{velocity}}{4005}\right)^2$$

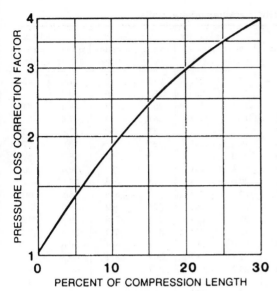

Figure A.4 Correction factor for unextended flexible duct.

TABLE A.5   Velocities and Velocity Pressures (Metric Units)

Velocity (m/s)	Velocity pressure (Pa)	Velocity (m/s)	Velocity pressure (Pa)	Velocity (m/s)	Velocity pressure (Pa)	Velocity (m/s)	Velocity pressure (Pa)	Velocity (m/s)	Velocity pressure (Pa)
1.0	0.6	10.0	60	19.0	217	28.0	472	37.0	824
1.2	0.9	10.2	63	19.2	222	28.2	479	37.2	833
1.4	1.2	10.4	65	19.4	227	28.4	486	37.4	842
1.6	1.5	10.6	68	19.6	231	28.6	493	37.6	851
1.8	2.0	10.8	70	19.8	236	28.8	499	37.8	860
2.0	2.4	11.0	73	20.0	241	29.0	506	38.0	870
2.2	2.9	11.2	76	20.2	246	29.2	513	38.2	879
2.4	3.5	11.4	78	20.4	251	29.4	421	38.4	888
2.6	4.1	11.6	81	20.6	256	29.6	528	38.6	897
2.8	4.7	11.8	84	20.8	261	29.8	535	38.8	907
3.0	5.4	12.0	87	21.0	266	30.0	542	39.0	916
3.2	6.2	12.2	90	21.2	271	30.2	549	39.2	925
3.4	7.0	12.4	93	21.4	276	30.4	557	39.4	935
3.6	7.8	12.6	96	21.6	281	30.6	564	39.6	944
3.8	8.7	12.8	99	21.8	286	30.8	571	39.8	954
4.0	9.6	13.0	102	22.0	291	31.0	579	40.0	963
4.2	10.6	13.2	105	22.2	297	31.2	586	40.2	973
4.4	11.7	13.4	108	22.4	302	31.4	594	40.4	983
4.6	12.7	13.6	111	22.6	308	31.6	601	40.6	993
4.8	13.9	13.8	115	22.8	313	31.8	609	40.8	1002
5.0	15.1	14.0	118	23.0	319	32.0	617	41.0	1012
5.2	16.3	14.2	121	23.2	324	32.2	624	41.2	1022
5.4	17.6	14.4	125	23.4	330	32.4	632	41.4	1032
5.6	18.9	14.6	128	23.6	335	32.6	640	41.6	1042
5.8	20.3	14.8	132	23.8	341	32.8	648	41.8	1052
6.0	21.7	15.0	135	24.0	347	33.0	656	42.0	1062
6.2	23.1	15.2	139	24.2	353	33.2	664	42.2	1072
6.4	24.7	15.4	143	24.4	359	33.4	672	42.4	1083
6.6	26.2	15.6	147	24.6	364	33.6	680	42.6	1093
6.8	27.8	15.8	150	24.8	370	33.8	688	42.8	1103
7.0	29.5	16.0	154	25.0	376	34.0	696	43.0	1113
7.2	31.2	16.2	158	25.2	382	34.2	704	43.2	1124
7.4	33.0	16.4	162	25.4	389	34.4	713	43.4	1134
7.6	34.8	16.6	166	25.6	395	34.6	721	43.6	1145
7.8	36.6	16.8	170	25.8	401	34.8	729	43.8	1155
8.0	38.5	17.0	174	26.0	407	35.0	738	44.0	1166
8.2	40.5	17.2	178	26.2	413	35.2	746	44.2	1176
8.4	42.5	17.4	182	26.4	420	35.4	755	44.4	1187
8.6	44.5	17.6	187	26.6	426	35.6	763	44.6	1198
8.8	46.6	17.8	191	26.8	433	35.8	772	44.8	1209
9.0	48.8	18.0	195	27.0	439	36.0	780	45.0	1219
9.2	51.0	18.2	199	27.2	446	36.2	789	45.2	1230
9.4	53.2	18.4	204	27.4	452	36.4	798	45.4	1241
9.6	55.5	18.6	208	27.6	459	36.6	807	45.6	1252
9.8	57.8	18.8	213	27.8	465	36.8	815	45.8	1263

TABLE A.6 Angular Conversion

Degrees	Radians	Degrees	Radians
10°	0.175	70°	1.22
20°	0.349	80°	1.40
30°	0.524	90°	1.57 ($\pi/2$)
40°	0.698	135°	2.36
50°	0.873	180°	3.14 (π)
60°	1.05	360°	6.28 (2 π)

TABLE A.7 Loss Coefficients for Straight-Through Flow

V_s/V_c	C	
	75% regain	90% regain
0.95	0.03	0.01
.91	0.04	0.02
.87	0.06	0.02
.83	0.08	0.03
.80	0.09	0.04
0.77	0.10	0.04
.74	0.11	0.04
.71	0.12	0.05
.69	0.13	0.05
.67	0.14	0.06
0.65	0.15	0.06
.63	0.15	0.06
.61	0.16	0.06
.59	0.16	0.07
.57	0.17	0.07
0.56	0.18	0.07
.54	0.18	0.07
.53	0.18	0.07
.51	0.18	0.07
.50	0.19	0.07

V_s = downstream velocity; V_c = upstream velocity.

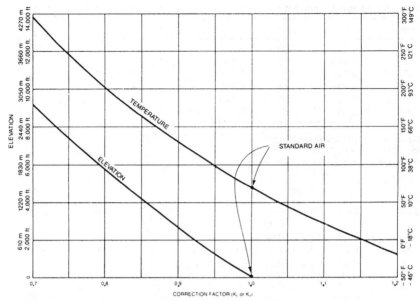

Figure A.5 Air density friction chart correction factors.

When an air distribution system is designed to operate above 2000 ft (610 m) altitude, below 32 °F (0 °C), or above 120 °F (49 °C) temperature, the duct friction loss obtained must be corrected for the air density. The actual air flow (cfm or L/s) is used to find the duct friction loss, which is multiplied by the correction factor or factors from Figure A.5 to obtain the actual friction loss.

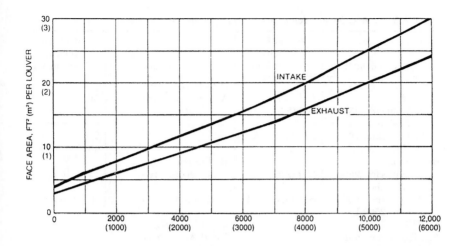

AIRFLOW, cfm (L/s) PER LOUVER		
Parameters Used Above	Intake Louver	Exhaust Louver
Minimum Free Area (48-Inch Square Test Section)	45%	45%
Water Penetration, Oz/(ft²-15 min.) Maximum Static Pressure Drop, in.w.g. (Pa)	Negligible (less than 0.2) 0.15 (37)	Not Applicable 0.25 (62)

Figure A.6 Recommended Criteria for Louver Sizing

TABLE A.8 Typical Design Velocities for Duct Components

Duct element	Actual face velocity [fpm (m/s)]	Duct element	Actual face velocity [fpm (m/s)]
LOUVERS		HEATING COILS	
A. Intake:		A. Steam and hot	500–600
1. 7000 cfm (3500	400 (2.0)	water	(2.5–3.0)
L/s) and greater			(most common)
2. Less than 7000	See Figure A.6		200 (1.0) min,
cfm (3500 L/s)			1500 (7.5) max.
B. Exhaust:		B. Electric:	
1. 5000 cfm (2500	500 (2.5)	1. Open wire	Refer to mfg.
L/s) and up			data
2. Less than 5000	See Figure A.6	2. Finned tubular	Refer to mfg.
cfm (2500 L/s)			data
FILTERS		COOLING OR	
A. Fibrous media unit		DEHUMIDIFYING	
filters:		COILS	
1. Viscous	250–700	A. Without eliminators	500–600
impingement	(1.2–3.5)		(2.5–3.0)
2. Dry type	Up to 750 (3.5)	B. With eliminators	600–800
3. HEPA	250 (1.2)		(3.0–4.0)
B. Renewable media		AIR WASHERS	
filters:		A. Spray-type	300–700
1. Moving curtain	500 (2.5)		(1.5–3.5)
viscous		B. Cell-type	Refer to mfg.
impingement			data
2. Moving curtain	200 (1.0)	C. High-velocity spray-	Refer to mfg.
dry media		type	data
C. Electronic air			
cleaners:			
1. Ionizing plate-	300–500		
type	(1.5–2.5)		

TABLE A.9   Elbow Loss Coefficients

A. Elbow, Smooth Radius (Die Stamped), Round

Fitting loss $(TP) = C \times V_p$ Use the velocity pressure V_p of the upstream section.

Coefficients for 90° Elbows: (see Note 1)

R/D	0.5	0.75	1.0	1.5	2.0	2.5
C	0.71	0.33	0.22	0.15	0.13	0.12

B. Elbow, Round, 3 to 5 pc—90°

Fitting loss $(TP) = C \times V_p$ Use the velocity pressure V_p of the upstream section.

Coefficient C

No. of pieces	R/D				
	0.5	0.75	1.0	1.5	2.0
5	—	0.46	0.33	0.24	0.19
4	—	0.50	0.37	0.27	0.24
3	0.98	0.54	0.42	0.34	0.33

C. Elbow, Round, Mitered

Fitting loss $(TP) = C \times V_p$ Use the velocity pressure V_p of the upstream section.

Coefficient C

θ	20°	30°	45°	60°	75°	90°
C	0.08	0.16	0.34	0.55	0.81	1.2

D. Elbow, Rectangular, Mitered

Fitting loss $(TP) = C \times V_p$ Use the velocity pressure V_p of the upstream section.

Coefficient C

θ	H/W								
	0.25	0.5	0.75	1.0	1.5	2.0	4.0	6.0	8.0
20°	0.08	0.08	0.08	0.07	0.07	0.07	0.06	0.05	0.05
30°	0.18	0.17	0.17	0.16	0.15	0.15	0.13	0.12	0.11
45°	0.38	0.37	0.36	0.34	0.33	0.31	0.27	0.25	0.24
60°	0.60	0.59	0.57	0.55	0.52	0.49	0.43	0.39	0.38
75°	0.89	0.87	0.84	0.81	0.77	0.73	0.63	0.58	0.57
90°	1.3	1.3	1.2	1.2	1.1	1.1	0.92	0.85	0.83

E. Elbow, Rectangular, Smooth Radius Without Vanes

Fitting loss $(TP) = C \times V_p$ Use the velocity pressure V_p of the upstream section.

Coefficients for 90° Elbows: (see Note 1)

	Coefficient C								
				H/W					
R/W	0.25	0.5	0.75	1.0	1.5	2.0	4.0	6.0	8.0
0.5	1.5	1.4	1.3	1.2	1.1	1.0	1.1	1.2	1.2
0.75	0.57	0.52	0.48	0.44	0.40	0.39	0.40	0.43	0.44
1.0	0.27	0.25	0.23	0.21	0.19	0.18	0.19	0.27	0.21
1.5	0.22	0.20	0.19	0.17	0.15	0.14	0.15	0.17	0.17
2.0	0.20	0.18	0.16	0.15	0.14	0.13	0.14	0.15	0.15

F. Elbow, Rectangular, Mitered with Turning Vanes

Fitting loss $(TP) = C \times V_p$ Use the velocity pressure V_p of the upstream section.

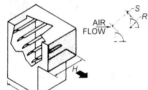

Loss Coefficients (C) for Single-thickness Vanes

Dimensions, inches (mm)		Velocity, fpm (m/s)]			
R	S	1000 (5)	1500 (7.5)	2000 (10)	2500 (12.5)
2.0 (50)	1.5 (38)	0.24	0.23	0.22	0.20
4.5 (114)	3.25 (83)	0.26	0.24	0.23	0.22

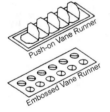

Loss Coefficients (C) for Double-thickness Vanes

Dimensions, inches (mm)		Velocity [fpm (m/s)]			
R	S	1000 (5)	1500 (7.5)	2000 (10)	2500 (12.5)
2.0 (50)	1.5 (38)	0.43	0.42	0.41	0.40
2.0 (50)	2.25 (56)	0.53	0.52	0.50	0.49
4.5 (114)	3.25 (83)	0.27	0.25	0.24	0.23

Note 1: For angles other than 90° multiply by the following factors:

θ	0°	20°	30°	45°	60°	75°	90°	110°	130°	150°	180°
K	0	0.31	0.45	0.60	0.78	0.90	1.00	1.13	1.20	1.28	1.40

TABLE A.10   Transition Loss Coefficients

A. Transition, Round, Conical

Use V_p of the upstream section.

$R_e = 8.56DV$

where:

D = Upstream Diameter (inches)
V = Upstream Velocity (fpm)

When: $\theta = 180°$

Coefficient C (see Note 1)

R_e	A_1/A	θ 16°	20°	30°	45°	60°	90°	120°	180°
0.5×10^5	2	0.14	0.19	0.32	0.33	0.33	0.32	0.31	0.30
	4	0.23	0.30	0.46	0.61	0.68	0.64	0.63	0.62
	6	0.27	0.33	0.48	0.66	0.77	0.74	0.73	0.72
	10	0.29	0.38	0.59	0.76	0.80	0.83	0.84	0.83
	≥16	0.31	0.38	0.60	0.84	0.88	0.88	0.88	0.88
2×10^5	2	0.07	0.12	0.23	0.28	0.27	0.27	0.27	0.26
	4	0.15	0.18	0.36	0.55	0.59	0.59	0.58	0.57
	6	0.19	0.28	0.44	0.90	0.70	0.71	0.71	0.69
	10	0.20	0.24	0.43	0.76	0.80	0.81	0.81	0.81
	≥16	0.21	0.28	0.52	0.76	0.87	0.87	0.87	0.87
≥6×10^5	2	0.05	0.07	0.12	0.27	0.27	0.27	0.27	0.27
	4	0.17	0.24	0.38	0.51	0.56	0.58	0.58	0.57
	6	0.16	0.29	0.46	0.60	0.69	0.71	0.70	0.70
	10	0.21	0.33	0.52	0.60	0.76	0.83	0.84	0.83
	≥16	0.21	0.34	0.56	0.72	0.79	0.85	0.87	0.89

B. Transition, Rectangular, Pyramidal

Use V_p of the upstream section.

When $\theta = 180°$

Coefficient C (see Note 1)

A_1/A	θ 16°	20°	30°	45°	60°	90°	120°	180°
2	0.18	0.22	0.25	0.29	0.31	0.32	0.33	0.30
4	0.36	0.43	0.50	0.56	0.61	0.63	0.63	0.63
6	0.42	0.47	0.58	0.68	0.72	0.76	0.76	0.75
≥10	0.42	0.49	0.59	0.70	0.80	0.87	0.85	0.86

Note 1: A = area of entering airstream, A_1 = area of leaving airstream.

C. Contraction, Round and Rectangular, Gradual to Abrupt

Use the velocity pressure (V_p) of the downstream section.

Coefficient C (see Note 2)

A_1/A	θ						
	10°	15°–40°	50°–60°	90°	120°	150°	180°
2	0.05	0.05	0.06	0.12	0.18	0.24	0.26
4	0.05	0.04	0.07	0.17	0.27	0.35	0.41
6	0.05	0.04	0.07	0.18	0.28	0.36	0.42
10	0.05	0.05	0.08	0.19	0.29	0.37	0.43

When $\theta = 180°$

Note 2: A_1 = area of entering airstream, A = area of leaving airstream.

TABLE A.11   Rectangular Branch Connection Loss Coefficients

A. Tee, 45° Entry, Rectangular Main and Branch

Use V_p of the upstream section.

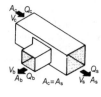

Branch, Coefficient C (see Note 1)

V_b/V_c	Q_b/Q_c						
	0.1	0.2	0.3	0.4	0.5	0.6	0.7
0.2	0.91						
0.4	0.81	0.79					
0.6	0.77	0.72	0.70				
0.8	0.78	0.73	0.69	0.66			
1.0	0.78	0.98	0.85	0.79	0.74		
1.2	0.90	1.11	1.16	1.23	1.03	0.86	
1.4	1.19	1.22	1.26	1.29	1.54	1.25	0.92
1.6	1.35	1.42	1.55	1.59	1.63	1.50	1.31
1.8	1.44	1.50	1.75	1.74	1.72	2.24	1.63

For main loss coefficient (C) see Table K below.

B. Tee, Rectangular Main to Round Branch

Use V_p of the upstream section.

Branch, Coefficient C (see Note 1)

V_b/V_c	Q_b/Q_c						
	0.1	0.2	0.3	0.4	0.5	0.6	0.7
0.2	1.00						
0.4	1.01	1.07					
0.6	1.14	1.10	1.08				
0.8	1.18	1.31	1.12	1.13			
1.0	1.30	1.38	1.20	1.23	1.26		
1.2	1.46	1.58	1.45	1.31	1.39	1.48	
1.4	1.70	1.82	1.65	1.51	1.56	1.64	1.71
1.6	1.93	2.06	2.00	1.85	1.70	1.76	1.80
1.8	2.06	2.17	2.20	2.13	2.06	1.98	1.99

For main loss coefficient (C) see Table K below.

C. Tee, Rectangular Main and Branch

Use V_p of the upstream section.

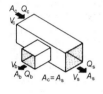

Branch, Coefficient C (see Note 1)

V_b/V_c	Q_b/Q_c						
	0.1	0.2	0.3	0.4	0.5	0.6	0.7
0.2	1.03						
0.4	1.04	1.01					
0.6	1.11	1.03	1.05				
0.8	1.16	1.21	1.17	1.12			
1.0	1.38	1.40	1.30	1.36	1.27		
1.2	1.52	1.61	1.68	1.91	1.47	1.66	
1.4	1.79	2.01	1.90	2.31	2.28	2.20	1.95
1.6	2.07	2.28	2.13	2.71	2.99	2.81	2.09
1.8	2.32	2.54	2.64	3.09	3.72	3.48	2.21

For main loss coefficient (C) see Table K below.

D. Tee, Rectangular Main and Branch with Extractor

Use V_p of the upstream section.

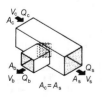

Branch, Coefficient C (see Note 1)

V_b/V_c	Q_b/Q_c						
	0.1	0.2	0.3	0.4	0.5	0.6	0.7
0.2	0.60						
0.4	0.62	0.69					
0.6	0.74	0.80	0.82				
0.8	0.99	1.10	0.95	0.90			
1.0	1.48	1.12	1.41	1.24	1.21		
1.2	1.91	1.33	1.43	1.52	1.55	1.64	
1.4	2.47	1.67	1.70	2.04	1.86	1.98	2.47
1.6	3.17	2.40	2.33	2.53	2.31	2.51	3.13
1.8	3.85	3.37	2.89	3.23	3.09	3.03	3.30

For main loss coefficient (C) see Table K below.

E. Wye, Rectangular
Use V_p of the upstream section.

$\dfrac{R}{W} = 1.0$
90° Branch

Branch, Coefficient C (see Note 1)

A_b/A_s	A_b/A_c	Q_b/Q_c						
		0.1	0.2	0.3	0.4	0.5	0.6	0.7
0.25	0.25	0.55	0.50	0.60	0.85	1.2	1.8	3.1
0.33	0.25	0.35	0.35	0.50	0.80	1.3	2.0	2.8
0.5	0.5	0.62	0.48	0.40	0.40	0.48	0.60	0.78
0.67	0.5	0.52	0.40	0.32	0.30	0.34	0.44	0.62
1.0	0.5	0.44	0.38	0.38	0.41	0.52	0.68	0.92
1.0	1.0	0.67	0.55	0.46	0.37	0.32	0.29	0.29
1.33	1.0	0.70	0.60	0.51	0.42	0.34	0.28	0.26
2.0	1.0	0.60	0.52	0.43	0.33	0.24	0.17	0.15

Main, Coefficient C (see Note 1)

A_b/A_s	A_b/A_c	Q_b/Q_c						
		0.1	0.2	0.3	0.4	0.5	0.6	0.7
0.25	0.25	−.01	−.03	−.01	0.05	0.13	0.21	0.29
0.33	0.25	0.08	0	−.02	−.01	0.02	0.08	0.16
0.5	0.5	−.03	−.06	−.05	0	0.06	0.12	0.19
0.67	0.5	0.04	−.02	−.04	−.03	−.01	0.04	0.12
1.0	0.5	0.72	0.48	0.28	0.13	0.05	0.04	0.09
1.0	1.0	−.02	−.04	−.04	−.01	0.06	0.13	0.22
1.33	1.0	0.10	0	0.01	−.03	−.01	0.03	0.10
2.0	1.0	0.62	0.38	0.23	0.13	0.08	0.05	0.06

F. Converging Tee, 45° Entry Branch to Rectangular Main
Use V_p of the downstream section.

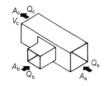

When:	A_b/A_s	A_s/A_c	A_b/A_c
	0.5	1.0	0.5

Branch, coefficient C (see Note 1)

V_c	Q_b/Q_c						
	0.1	0.2	0.3	0.4	0.5	0.6	0.7
< 1200 fpm (6 m/s)	−.83	−.68	−.30	0.28	0.55	1.03	1.50
> 1200 fpm (6 m/s)	−.72	−.52	−.23	0.34	0.76	1.14	1.83

For main loss coefficient (C) see Table K below.

G. Converging Tee, Round Branch to Rectangular Main
Use V_p of the downstream section.

Branch, Coefficient C (see Note 1)

V_c	Q_b/Q_c						
	0.1	0.2	0.3	0.4	0.5	0.6	0.7
< 1200 fpm (6 m/s)	−.63	−.55	0.13	0.23	0.78	1.30	1.93
> 1200 fpm (6 m/s)	−.49	−.21	0.23	0.60	1.27	2.06	2.75

For main loss coefficient (C) see Table K below.

When:

A_b/A_s	A_s/A_c	A_b/A_c
0.5	1.0	0.5

H. Converging Tee, Rectangular Main and Branch
Use V_p of the downstream section.

Branch, Coefficient C (see Note 1)

V_c	Q_b/Q_c						
	0.1	0.2	0.3	0.4	0.5	0.6	0.7
< 1200 fpm (6 m/s)	−.75	−.53	−.03	0.33	1.03	1.10	2.15
> 1200 fpm (6 m/s)	−.69	−.21	0.23	0.67	1.17	1.66	2.67

For main loss coefficient (C) see Table K below.

When:

A_b/A_s	A_s/A_c	A_b/A_c
0.5	1.0	0.5

I. Converging Wye, Rectangular

Use V_p of the downstream section.

Branch, Coefficient C (see Note 1)

		Q_b/Q_c						
A_b/A_s	A_b/A_c	0.1	0.2	0.3	0.4	0.5	0.6	0.7
0.25	0.25	−.50	0	0.50	1.2	2.2	3.7	5.8
0.33	0.25	−1.2	−.40	0.40	1.6	3.0	4.8	6.8
0.5	0.5	−.50	−.20	0	0.25	0.45	0.70	1.0
0.67	0.5	−1.0	−.60	−.20	0.10	0.30	0.60	1.0
1.0	0.5	−2.2	−1.5	−.95	−.50	0	0.40	0.80
1.0	1.0	−.60	−.30	−.10	−.04	0.13	0.21	0.29
1.33	1.0	−1.2	−.80	−.40	−.20	0	0.16	0.24
2.0	1.0	−2.1	−1.4	−.90	−.50	−.20	0	0.20

Main, Coefficient C (see Note 1)

		Q_b/Q_c						
A_s/A_c	A_b/A_c	0.1	0.2	0.3	0.4	0.5	0.6	0.7
0.75	0.25	0.30	0.30	0.20	−.10	−.45	−.92	−1.5
1.0	0.5	0.17	0.16	0.10	0	−0.08	−.18	−.27
0.75	0.5	0.27	0.35	0.32	0.25	0.12	−.03	−.23
0.5	0.5	1.2	1.1	0.90	0.65	0.35	0	−.40
1.0	1.0	0.18	0.24	0.27	0.26	0.23	0.18	0.10
0.75	1.0	0.75	0.36	0.38	0.35	0.27	0.18	0.05
0.5	1.0	0.80	0.87	0.80	0.68	0.55	0.40	0.25

J. Tee, Rectangular Main to Conical Branch

Use V_p of the upstream section.

Branch, Coefficient C (see Note 1)

V_b/V_c	0.40	0.50	0.75	1.0	1.3	1.5
C	0.80	0.83	0.90	1.0	1.1	1.4

K. Main Duct Loss Coefficient for Table A.11 Fittings

Main, coefficient C (see Note 1)

V_b/V_c	0.2	0.4	0.6	0.8	1.0	1.2	1.4	1.6	1.8
C	0.03	0.04	0.07	0.12	0.13	0.14	0.27	0;30	0.25

Note 1: A = area (in.^{2} or mm^2), Q = airflow (cfm or L/s), V = Velocity (fpm or m/s).

TABLE A.12   Round Branch Connection Loss Coefficients

A. Tee or Wye, 30° to 90°, Round
Use the V_p of the upstream section.

Wye $\theta = 30°$:

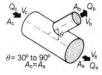

Branch, Coefficient C (see Note 1)

A_b/A_c	Q_b/Q_c						
	0.1	0.2	0.3	0.4	0.5	0.6	0.7
0.8	0.75	0.55	0.40	0.28	0.21	0.16	0.15
0.7	0.72	0.51	0.36	0.25	0.18	0.15	0.16
0.6	0.69	0.46	0.31	0.21	0.17	0.16	0.20
0.5	0.65	0.41	0.26	0.19	0.18	0.22	0.32
0.4	0.59	0.33	0.21	0.20	0.27	0.40	0.62
0.3	0.55	0.28	0.24	0.38	0.76	1.3	2.0
0.2	0.40	0.26	0.58	1.3	2.5	—	—
0.1	0.28	1.5	—	—	—	—	—

Wye $\theta = 45°$:

Branch, Coefficient C (see Note 1)

A_b/A_c	Q_b/Q_c						
	0.1	0.2	0.3	0.4	0.5	0.6	0.7
0.8	0.78	0.62	0.49	0.40	0.34	0.31	0.32
0.7	0.77	0.59	0.47	0.38	0.34	0.32	0.35
0.6	0.74	0.56	0.44	0.37	0.35	0.36	0.43
0.5	0.71	0.52	0.41	0.38	0.40	0.45	0.59
0.4	0.66	0.47	0.40	0.43	0.54	0.69	0.95
0.3	0.66	0.48	0.52	0.73	1.2	1.8	2.7
0.2	0.56	0.56	1.0	1.8	—	—	—
0.1	0.60	2.1	—	—	—	—	—

Wye $\theta = 60°$:

Branch, Coefficient C (see Note 1)

A_b/A_c	Q_b/Q_c						
	0.1	0.2	0.3	0.4	0.5	0.6	0.7
0.8	0.83	0.71	0.62	0.56	0.52	0.50	0.53
0.7	0.82	0.69	0.61	0.56	0.54	0.54	0.60
0.6	0.81	0.68	0.60	0.58	0.58	0.61	0.72
0.5	0.79	0.66	0.61	0.62	0.68	0.76	0.94
0.4	0.76	0.65	0.65	0.74	0.89	1.1	1.4
0.3	0.80	0.75	0.89	1.2	1.8	2.6	3.5
0.2	0.77	0.96	1.6	2.5	—	—	—
0.1	1.0	2.9	—	—	—	—	—

Tee $\theta = 90°$:

Branch, Coefficient C (see Note 1)

A_b/A_c	Q_b/Q_c						
	0.1	0.2	0.3	0.4	0.5	0.6	0.7
0.8	0.95	0.92	0.92	0.93	0.94	0.95	1.1
0.7	0.95	0.94	0.95	0.98	1.0	1.1	1.2
0.6	0.96	0.97	1.0	1.1	1.1	1.2	1.4
0.5	0.97	1.0	1.1	1.2	1.4	1.5	1.8
0.4	0.99	1.1	1.3	1.5	1.7	2.0	2.4
0.3	1.1	1.4	1.8	2.3	—	—	—
0.2	1.3	1.9	2.9	—	—	—	—
0.1	2.1	—	—	—	—	—	—

B. 90° Conical Tee, Round
Use the V_p of the upstream section.

Branch, Coefficient C (see Note 1)

V_b/V_c	0	0.2	0.4	0.6	0.8	1.0	1.2	1.4	1.6
C	1.0	0.85	0.74	0.62	0.52	0.42	0.36	0.32	0.32

For main loss coefficient (C) see Table D below.

C. 45° Conical Wye, Round
Use the V_p of the upstream section.

Branch, Coefficient C (see Note 1)

V_b/V_c	0	0.2	0.4	0.6	0.8	1.0	1.2	1.4	1.6
C	1.0	0.84	0.61	0.41	0.27	0.17	0.12	0.12	0.14

For main loss coefficient (C) see Table D below.

D. Diverging Fitting Main Duct Loss Coefficients

Main, coefficient C (see Note 1)

V_s/V_c	0	0.1	0.2	0.3	0.4	0.5	0.6	0.8
C	0.35	0.28	0.22	0.17	0.13	0.09	0.06	0.02

E. Converging Wye, Round
Use the V_p of the downstream section.

Branch, Coefficient C (see Note 1)

$\dfrac{V_b}{V_c}$	A_b/A_c						
	0.1	0.2	0.3	0.4	0.6	0.8	1.0
0.4	−.56	−.44	−.35	−.28	−.15	−.04	0.05
0.5	−.48	−.37	−.28	−.21	−.09	0.02	0.11
0.6	−.38	−.27	−.19	−.12	0	0.10	0.18
0.7	−.26	−.16	−.08	−.01	0.10	0.20	0.28
0.8	−.21	−.02	0.05	0.12	0.23	0.32	0.40
0.9	0.04	0.13	0.21	0.27	0.37	0.46	0.53
1.0	0.22	0.31	0.38	0.44	0.53	0.62	0.69
1.5	1.4	1.5	1.5	1.6	1.7	1.7	1.8
2.0	3.1	3.2	3.2	3.2	3.3	3.3	3.3
2.5	5.3	5.3	5.3	5.4	5.4	5.4	5.4
3.0	8.0	8.0	8.0	8.0	8.0	8.0	8.0

Main, Coefficient C

$\dfrac{V_s}{V_c}$	A_b/A_c						
	0.1	0.2	0.3	0.4	0.6	0.8	1.0
0.1	−8.6	−4.1	−2.5	−1.7	−.97	−.58	−.34
0.2	−6.7	−3.1	−1.9	−1.3	−.67	−.36	−.18
0.3	−5.0	−2.2	−1.3	−.88	−.42	−.19	−.05
0.4	−3.5	−1.5	−.88	−.55	−.21	−.05	0.05
0.5	−2.3	−.95	−.51	−.28	−.06	0.06	0.13
0.6	−1.3	−.50	−.22	−.09	0.05	0.12	0.17
0.7	−.63	−.18	−.03	0.04	0.12	0.16	0.18
0.8	−.18	0.01	0.07	0.10	0.13	0.15	0.17
0.9	0.03	0.07	0.08	0.09	0.10	0.11	0.13
1.0	−0.01	0	0	0.10	0.02	0.04	0.05

F. Converging Tee, 90°, Round
Use the V_p of the downstream section.

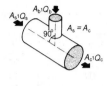

	Branch, Coefficient C (see Note 1)					
	A_b/A_c					
Q_b/Q_c	0.1	0.2	0.3	0.4	0.6	0.8
0.1	0.40	−.37	−.51	−.46	−.50	−.51
0.2	3.8	0.72	0.17	−.02	−.14	−.18
0.3	9.2	2.3	1.0	0.44	0.21	0.11
0.4	16	4.3	2.1	0.94	0.54	0.40
0.5	26	6.8	3.2	1.1	0.66	0.49
0.6	37	9.7	4.7	1.6	0.92	0.69
0.7	43	13	6.3	2.1	1.2	0.88
0.8	65	17	7.9	2.7	1.5	1.1
0.9	82	21	9.7	3.4	1.8	1.2
1.0	101	26	12	4.0	2.1	1.4

For main loss coefficient (C) see Table H below.

G. Converging Wye, Conical, Round

Use the V_p of the downstream section.

Branch, Coefficient C (see Note 1)

$\dfrac{A_s}{A_c}$	$\dfrac{A_b}{A_c}$	Q_b/Q_c								
		0.2	0.4	0.6	0.8	1.0	1.2	1.4	1.6	1.8
0.3	0.2	−2.4	−.01	2.0	3.8	5.3	6.6	7.8	8.9	9.8
	0.3	−2.8	−1.2	0.12	1.1	1.9	2.6	3.2	3.7	4.2
0.4	0.2	−1.2	0.93	2.8	4.5	5.9	7.2	8.4	9.5	10
	0.3	−1.6	−.27	0.81	1.7	2.4	3.0	3.6	4.1	4.5
	0.4	−1.8	−.72	0.07	0.66	1.1	1.5	1.8	2.1	2.3
0.5	0.2	−.46	1.5	3.3	4.9	6.4	7.7	8.8	9.9	11
	0.3	−.94	0.25	1.2	2.0	2.7	3.3	3.8	4.2	4.7
	0.4	−1.1	−.24	0.42	0.92	1.3	1.6	1.9	2.1	2.3
	0.5	−1.2	−.38	0.18	0.58	0.88	1.1	1.3	1.5	1.6
0.6	0.2	−.55	1.3	3.1	4.7	6.1	7.4	8.6	9.6	11
	0.3	−1.1	0	0.88	1.6	2.3	2.8	3.3	3.7	4.1
	0.4	−1.2	−.48	0.10	0.54	0.89	1.2	1.4	1.6	1.8
	0.5	−1.3	−.62	−.14	0.21	0.47	0.68	0.85	0.99	1.1
	0.6	−1.3	−.69	−.26	0.04	0.26	0.42	0.57	0.66	0.75
0.8	0.2	0.06	1.8	3.5	5.1	6.5	7.8	8.9	10	11
	0.3	−.52	0.35	1.1	1.7	2.3	2.8	3.2	3.6	3.9
	0.4	−.67	−.05	0.43	0.80	1.1	1.4	1.6	1.8	1.9
	0.5	−.73	−.19	0.18	0.46	0.68	0.85	0.99	1.1	1.2
	0.6	−.75	−.27	0.05	0.28	0.45	0.58	0.68	0.76	0.83
	0.7	−.77	−.31	−.02	0.18	0.32	0.43	0.50	0.56	0.61
	0.8	−.78	−.34	−.07	0.12	0.24	0.33	0.39	0.44	0.47
1.0	0.2	—	2.1	3.7	5.2	6.6	7.8	9.0	11	11
	0.3	—	.54	1.2	1.8	2.3	2.7	3.1	3.7	3.7
	0.4	—	.21	0.62	0.96	1.2	1.5	1.7	2.0	2.0
	0.5	—	.05	0.37	0.60	0.79	0.93	1.1	1.2	1.2
	0.6	—	−.02	0.23	0.42	0.55	0.66	0.73	0.80	0.85
	0.8	—	−.10	0.11	0.24	0.33	0.39	0.43	0.46	0.47
	1.0	—	−.14	0.05	0.16	0.23	0.27	0.29	0.30	0.30

H. Converging Fitting Main Duct Loss Coefficients

Main, coefficient C (see Note 1)

Q_b/Q_c	0.1	0.2	0.3	0.4	0.5	0.6	0.7	0.8	0.9
C	0.16	0.27	0.38	0.46	0.53	0.57	0.59	0.60	0.59

Note 1: A = area (in.^{2} or mm^2), Q = airflow (cfm or L/s), V = velocity (fpm or m/s).

TABLE A.13   Miscellaneous Fitting Coefficients

A. Damper, Butterfly, Thin Plate, Round
Use the V_p of the upstream section.

			Coefficient C				
θ	0°	10°	20°	30°	40°	50°	60°
C	0.20	0.52	1.5	4.5	11	29	108

0° is full open

B. Damper, Butterfly, Thin Plate, Rectangular
Use the V_p of the upstream section.

			Coefficient C				
θ	0°	10°	20°	30°	40°	50°	60°
C	0.04	0.33	1.2	3.3	9.0	26	70

0° is full open

C. Damper, Rectangular, Parallel Blades
Use the V_p of the upstream section.

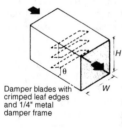

Damper blades with crimped leaf edges and 1/4" metal damper frame

	Coefficient C								
					θ				
L/R	80°	70°	60°	50°	40°	30°	20°	10°	0° Fully open
0.3	116	32	14	9.0	5.0	2.3	1.4	0.79	0.52
0.4	152	38	16	9.0	6.0	2.4	1.5	0.85	0.52
0.5	188	45	18	9.0	6.0	2.4	1.5	0.92	0.52
0.6	245	45	21	9.0	5.4	2.4	1.5	0.92	0.52
0.8	284	55	22	9.0	5.4	2.5	1.5	0.92	0.52
1.0	361	65	24	10	5.4	2.6	1.6	1.0	0.52
1.5	576	102	28	10	5.4	2.7	1.6	1.0	0.52

$$\frac{L}{R} = \frac{NW}{2(H + W)}$$

where:
N is number of damper blades
W is duct dimension parallel to blade axis
L is sum of damper blade lengths
R is perimeter of duct
H is duct dimension perpendicular to blade axis

D. Damper, Rectangular, Opposed Blades
Use the V_p of the upstream section.

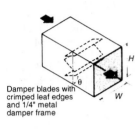

Damper blades with
crimped leaf edges
and 1/4" metal
damper frame

Coefficient C

L/R	θ									
	80°	70°	60°	50°	40°	30°	20°	10°	0°	Fully open
0.3	807	284	73	21	9.0	4.1	2.1	0.85		0.52
0.4	915	332	100	28	11	5.0	2.2	0.92		0.52
0.5	1045	377	122	33	13	5.4	2.3	1.0		0.52
0.6	1121	411	148	38	14	6.0	2.3	1.0		0.52
0.8	1299	495	188	54	18	6.6	2.4	1.1		0.52
1.0	1521	547	245	65	21	7.3	2.7	1.2		0.52
1.5	1654	677	361	107	28	9.0	3.2	1.4		0.52

$$\frac{L}{R} = \frac{NW}{2(H + W)}$$ where:

N is number of damper blades
W is duct dimension parallel to blade axis
L is sum of damper blade lengths
R is perimeter of duct
H is duct dimension perpendicular to blade axis

E. Perforated Plate in Duct, Thick, Round and Rectangular
Use the V_p of the upstream section.

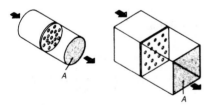

Coefficient C

t/d	n								
	0.20	0.25	0.30	0.40	0.50	0.60	0.70	0.80	0.90
0.015	52	30	18	8.2	4.0	2.0	0.97	0.42	0.13
0.2	48	28	17	7.7	3.8	1.9	0.91	0.40	0.13
0.4	46	27	17	7.4	3.6	1.8	0.88	0.39	0.13
0.6	42	24	15	6.6	3.2	1.6	0.80	0.36	0.13

$t/d > 0.015$ Where: t = plate thickness
$n = \dfrac{A_p}{A}$ d = diameter of perforated holes
n = free area ratio of plate
A_p = total flow area of perforated plate
A = area of duct

F. Rectangular Duct with Four 90° Mitered Ells To Avoid an Obstruction
Use the V_p of the upstream section.

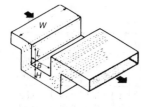

L/H Ratio	0.5	1.0	1.5	2
Single blade turning vanes	—	0.85	0.83	0.77
Double blade turning vanes	—	1.85	2.84	2.91
"S" type splitter vanes	0.61	0.65	—	—
No vanes—Up to 1200 fpm (6 m/s)	0.88	5.26	6.92	7.56
No vanes—Over 1200 fpm (6 m/s)	1.26	6.22	8.82	9.24

Coefficient C

Where: W/H = 1.0 to 3.0
        B = 12 in. to 24 in. (300 to 600 mm)

G. Rectangular Duct, Depressed To Avoid an Obstruction
Use of V_p of the upstream section.

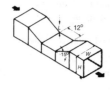

Coefficient C

	L/H			
W/H	0.125	0.15	0.25	0.30
1.0	0.26	0.30	0.33	0.35
4.0	0.10	0.14	0.22	0.30

H. Round Duct, Depressed To Avoid an Obstruction
Use the V_p of the upstream section.

Where: L/D = 0.33
        C = 0.24

I. Exit, Abrupt, Round and Rectangular, with or without a Wall
Use the V_p of the upstream section.

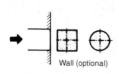

C = 1.0
With Screen: C_s = 1 + (C from Table K)

Wall (optional)

J. Duct Mounted in Wall, Round and Rectangular

Use the V_p of the upstream section.

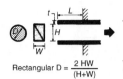

Rectangular $D = \dfrac{2\,HW}{(H+W)}$

	Coefficient C						
	L/D						
t/D	0	0.002	0.01	0.05	0.2	0.5	≥ 1.0
≈ 0	0.50	0.57	0.68	0.80	0.92	1.0	1.0
0.02	0.50	0.51	0.52	0.55	0.66	0.72	0.72
≥ 0.05	0.50	0.50	0.50	0.50	0.50	0.50	0.50

Rectangular: $D = \dfrac{2HW}{(H + W)}$

With screen or perforated plate:
a. Sharp edge $(t/D_e \leq 0.05)$; $C_s = 1 + C_1$
b. Thick edge $(t/D_e > 0.05)$; $C_s = C + C_1$
where:
C_s is new coefficient of fitting with a screen or perforated plate at the entrance.
C is from above table
C_1 is from Table K (screen) or Table E (perforated plate)

K. Screen in Duct, Round and Rectangular

Use the V_p of the upstream section.

$n = A_s/A$

Where:
n = Free area ratio of screen
A_s = Total flow area of screen
A = Area of Duct

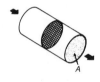

Coefficient C

n	0.30	0.40	0.50	0.55	0.60	0.65	0.70	0.75	0.80	0.90
C	6.2	3.0	1.7	1.3	0.97	0.75	0.58	0.44	0.32	0.14

Hydronic Design Data

1B   Water Piping—U.S. Units

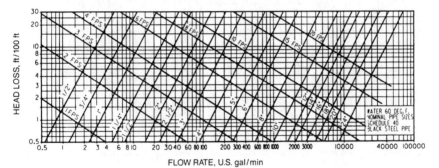

Figure B.1 Friction loss for water in commercial steel pipe (Schedule 40).

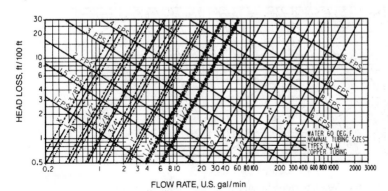

Figure B.2 Friction loss for water in copper tubing (types K, L, M).

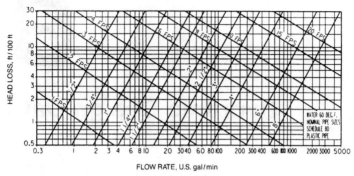

FLOW RATE, U.S. gal/min

Figure B.3   Friction loss for water in plastic pipe (Schedule 80).

2B   Water Piping—Metric

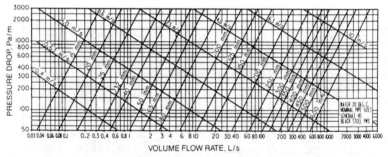

VOLUME FLOW RATE, L/s

Figure B.4   Friction loss for water in commercial steel pipe (Schedule 40).

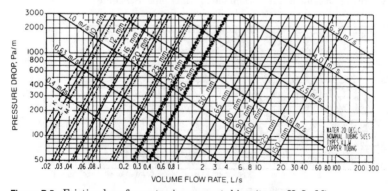

VOLUME FLOW RATE, L/s

Figure B.5   Friction loss for water in copper tubing (types K, L, M).

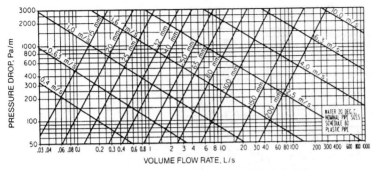

Figure B.6 Friction loss for water in plastic pipe (Schedule 80).

3B Fitting Equivalents (Water)

TABLE B.1 Equivalent Length in Feet of Pipe for 90° Elbows

Velocity (ft/s)	Pipe size (inches)														
	$\frac{1}{2}$	$\frac{3}{4}$	1	$1\frac{1}{4}$	$1\frac{1}{2}$	2	$2\frac{1}{2}$	3	$3\frac{1}{2}$	4	5	6	8	10	12
1	1.2	1.7	2.2	3.0	3.5	4.5	5.4	6.7	7.7	8.6	10.5	12.2	15.4	18.7	22.2
2	1.4	1.9	2.5	3.3	3.9	5.1	6.0	7.5	8.6	9.5	11.7	13.7	17.3	20.8	24.8
3	1.5	2.0	2.7	3.6	4.2	5.4	6.4	8.0	9.2	10.2	12.5	14.6	18.4	22.3	26.5
4	1.5	2.1	2.8	3.7	4.4	5.6	6.7	8.3	9.6	10.6	13.1	15.2	19.2	23.2	27.6
5	1.6	2.2	2.9	3.9	4.5	5.9	7.0	8.7	10.0	11.1	13.6	15.8	19.8	24.2	28.8
6	1.7	2.3	3.0	4.0	4.7	6.0	7.2	8.9	10.3	11.4	14.0	16.3	20.5	24.9	29.6
7	1.7	2.3	3.0	4.1	4.8	6.2	7.4	9.1	10.5	11.7	14.3	16.7	21.0	25.5	30.3
8	1.7	2.4	3.1	4.2	4.9	6.3	7.5	9.3	10.8	11.9	14.6	17.1	21.5	26.1	31.0
9	1.8	2.4	3.2	4.3	5.0	6.4	7.7	9.5	11.0	12.2	14.9	17.4	21.9	26.6	31.6
10	1.8	2.5	3.2	4.3	5.1	6.5	7.8	9.7	11.2	12.4	15.2	17.7	22.2	27.0	32.0

TABLE B.2 Equivalent Length in Meters of Pipe for 90° Elbows

Velocity (m/s)	Pipe Size (mm)													
	15	20	25	32	40	50	65	75	100	125	150	200	250	300
0.33	0.4	0.5	0.7	0.9	1.1	1.4	1.6	2.0	2.6	3.2	3.7	4.7	5.7	6.8
0.67	0.4	0.6	0.8	1.0	1.2	1.5	1.8	2.3	2.9	3.6	4.2	5.3	6.3	7.6
1.00	0.5	0.6	0.8	1.1	1.3	1.6	1.9	2.5	3.1	3.8	4.5	5.6	6.8	8.0
1.33	0.5	0.6	0.8	1.1	1.3	1.7	2.0	2.5	3.2	4.0	4.6	5.8	7.1	8.4
1.67	0.5	0.7	0.9	1.2	1.4	1.8	2.1	2.6	3.4	4.1	4.8	6.0	7.4	8.8
2.00	0.5	0.7	0.9	1.2	1.4	1.8	2.2	2.7	3.5	4.3	5.0	6.2	7.6	9.0
2.33	0.5	0.7	0.9	1.2	1.5	1.9	2.2	2.8	3.6	4.4	5.1	6.4	7.8	9.2
2.67	0.5	0.7	0.9	1.3	1.5	1.9	2.3	2.8	3.6	4.5	5.2	6.5	8.0	9.4
3.00	0.5	0.7	0.9	1.3	1.5	1.9	2.3	2.9	3.7	4.5	5.3	6.7	8.1	9.6
3.33	0.5	0.8	0.9	1.3	1.5	1.9	2.4	3.0	3.8	4.6	5.4	6.8	8.2	9.8

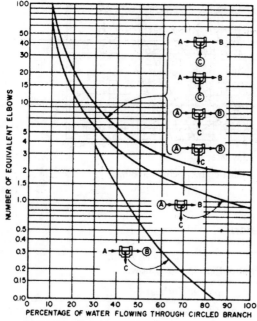

Figure B.7 Elbow equivalents of tees at various flow conditions.

TABLE B.3 Iron and Copper Elbow Equivalents

Fitting	Iron pipe	Copper tubing
Elbow, 90°	1.0	1.0
Elbow, 45°	0.7	0.7
Elbow, 90° long turn	0.5	0.5
Elbow, welded, 90°	0.5	0.5
Reduced coupling	0.5	0.4
Open return bend	1.0	1.0
Angle radiator valve	2.0	3.0
Radiator or convector	3.0	4.0
Boiler or heater	3.0	4.0
Open gate valve	0.5	0.7
Open globe valve	12.0	17.0

4B   Properties of Steam

TABLE B.4   Properties of Saturated Steam (U.S. Units)

Pressure		Saturation temperature (°F)	Specific volume (ft³/lb)		Enthalpy (Btu/lb)		
Gauge (psi)	Absolute (psia)		Liquid, V_f	Steam, V_g	Liquid, h_f	Evap., h_{fg}	Steam, h_g
Vacuum							
25 in.Hg	2.4	134	0.0163	146.4	101	1018	1119
9.56 in.Hg	10	193	0.0166	38.4	161	982	1143
0	14.7	212	0.0167	26.8	180	970	1150
2	16.7	218	0.0168	23.8	187	966	1153
5	19.7	227	0.0168	20.4	195	961	1156
15	29.7	250	0.0170	13.9	218	946	1164
50	64.7	298	0.0174	6.7	267	912	1179
100	114.7	338	0.0179	3.9	309	881	1190
150	164.7	366	0.0182	2.8	339	857	1196
200	214.7	388	0.0185	2.1	362	837	1179

TABLE B.5   Properties of Saturated Steam (Metric Units)

Pressure (kPa)	Saturation temperature (°C)	Specific volume (L/kg)		Enthalpy (kJ/kg)		
		Liquid, V_f	Steam, V_g	Liquid, h_f	Evap., h_{fg}	Steam, h_g
19.9	60	1.02	7669	251	2358	2609
47.4	80	1.03	3405	335	2308	2643
101	100	1.04	1672	419	2256	2775
199	120	1.06	891	504	2202	2706
362	140	1.08	508	589	2144	2733
618	160	1.10	307	676	2082	2758
1003	180	1.13	194	763	2015	2778
1555	200	1.16	127	852	1941	2793

5B   Steam Piping (U. S. Units)

TABLE B.6   Flow Rate of Steam in Schedule 40 Pipe at Initial Saturation Pressure of 3.5 and 12 psig

(Flow rate expressed in pounds per hour)

Nom. pipe size (in.)	$\frac{1}{16}$ psi (1 oz) Sat. press. (psig)		$\frac{1}{8}$ psi (2 oz) Sat. press. (psig)		$\frac{1}{4}$ psi (4 oz) Sat. press. (psig)		$\frac{1}{2}$ psi (8 oz) Sat. press. (psig)		$\frac{3}{4}$ psi (12 oz) Sat. press. (psig)		1 psi Sat. press. (psig)		2 psi Sat. press. (psig)	
	3.5	12	3.5	12	3.5	12	3.5	12	3.5	12	3.5	12	3.5	12
$\frac{3}{4}$	9	11	14	16	20	24	29	35	36	43	42	50	60	73
1	17	21	26	31	37	46	54	66	68	82	81	95	114	137
$1\frac{1}{4}$	36	45	53	66	78	96	111	138	140	170	162	200	232	280
$1\frac{1}{2}$	56	70	84	100	120	147	174	210	218	260	246	304	360	430
2	108	134	162	194	234	285	336	410	420	510	480	590	710	850
$2\frac{1}{2}$	174	215	258	310	378	460	540	660	680	820	780	950	1,150	1,370
3	318	380	465	550	660	810	960	1,160	1,190	1,430	1,380	1,670	1,950	2,400
$3\frac{1}{2}$	462	550	670	800	990	1,218	1,410	1,700	1,740	2,100	2,000	2,420	2,950	3,450
4	640	800	950	1,160	1,410	1,690	1,980	2,400	2,450	3,000	2,880	3,460	4,200	4,900
5	1,200	1,430	1,680	2,100	2,440	3,000	3,570	4,250	4,380	5,250	5,100	6,100	7,500	8,600
6	1,920	2,300	2,820	3,350	3,960	4,850	5,700	5,700	7,000	8,600	8,400	10,000	11,900	14,200
8	3,900	4,800	5,570	7,000	8,100	10,000	11,400	14,300	14,500	17,700	16,500	20,500	24,000	29,500
10	7,200	8,800	10,200	12,600	15,000	18,200	21,000	26,000	26,200	32,000	30,000	37,000	42,700	52,000
12	11,400	13,700	16,500	19,500	23,400	28,400	33,000	40,000	41,000	49,500	48,000	57,500	67,800	81,000

Pressure drop (psi per 100 ft in length)

TABLE B.7 Comparative Capacity of Steam Lines at Various Pitches for Steam and Condensate Flowing in Opposite Directions in Pounds per Hour

(Pitch of pipe in inches per 10 ft; velocity in feet per second)

Pitch of pipe	$\frac{1}{4}$ in.		$\frac{1}{2}$ in.		1 in.		$1\frac{1}{2}$ in.		2 in.		3 in.		4 in.		5 in.	
Pipe size inches	lb/h	Max. vel.	lb/h	Max. vel.	lb/h	Max. vel.	lb/h	Max. vel.	lb/h	Max. vel.	lb/h	Max. vel.	lb/h	Max. vel.	lb/h	Max. vel.
$\frac{3}{4}$	3.2	8	4.1	11	5.7	13	6.4	14	7.1	16	8.3	17	9.9	22	10.5	22
1	6.8	9	9.0	12	11.7	15	12.8	17	14.8	19	17.3	22	19.2	24	20.5	25
$1\frac{1}{4}$	11.8	11	15.9	13	19.9	17	24.6	20	27.0	22	31.3	25	33.4	26	38.1	31
$1\frac{1}{2}$	19.8	12	25.9	16	33.0	19	37.4	22	42.0	24	46.8	26	50.8	28	59.2	33
2	42.9	15	54.0	18	68.8	24	83.3	27	92.9	30	99.6	32	102.4	32	115.0	33

TABLE B.8 Pressure Drops in Common Use for Sizing Steam Pipe

(For corresponding initial steam pressure)

Initial steam pressure (psig)	Pressure drop per 100 ft	Total pressure drop in steam supply piping
Subatmos. or vacuum return	2–4 oz	1–2 psi
0	$\frac{1}{2}$ oz	1 oz
1	1 oz	1–4 oz
2	2 oz	8 oz
5	4 oz	$1\frac{1}{2}$ psi
10	8 oz	3 psi
15	1 psi	4 psi
30	2 psi	5–10 psi
50	2–5 psi	10–15 psi
100	2–5 psi	15–25 psi
150	2–10 psi	25–30 psi

TABLE B.9 Length in Feet of Pipe To Be Added to Actual Length of Run for Fittings To Obtain Equivalent Length

Size of pipe (in.)	Length in feet to be added to run				
	Standard elbow	Side outlet tee[b]	Gate valve[a]	Globe valve[a]	Angle valve[a]
$\frac{1}{2}$	1.3	3	0.3	14	7
$\frac{3}{4}$	1.8	4	0.4	18	10
1	2.2	5	0.5	23	12
$1\frac{1}{4}$	3.0	6	0.6	29	15
$1\frac{1}{2}$	3.5	7	0.8	34	18
2	4.3	8	1.0	46	22
$2\frac{1}{2}$	5.0	11	1.1	54	27
3	6.5	13	1.4	66	34
$3\frac{1}{2}$	8	15	1.6	80	40
4	9	18	1.9	92	45
5	11	22	2.2	112	56
6	13	27	2.8	136	67
8	17	35	3.7	180	92
10	21	45	4.6	230	112
12	27	53	5.5	270	132
14	30	63	6.4	310	152

[a]Valve in full position.
[b]Values given apply only to a tee used to divert the flow in the main to the last riser.

TABLE B.10   Steam Pipe Capacities for Low-Pressure Systems

(For Use on one-pipe systems or two-pipe systems in which condensate flows against the steam flow)

	Capacity in pounds per hour				
	Two-pipe systems			One-pipe systems	
Nominal pipe size (in.)	Condensate flowing against steam		Supply risers up-feed	Radiator valves and vertical connections	Radiator and riser runouts
	Vertical	Horizontal			
A	B[a]	C[c]	D[b]	E	F[c]
$\frac{3}{4}$	8	7	6	—	7
1	14	14	11	7	7
$1\frac{1}{4}$	31	27	20	16	16
$1\frac{1}{2}$	48	42	38	23	16
2	97	93	72	42	23
$2\frac{1}{2}$	159	132	116	—	42
3	282	200	200	—	65
$3\frac{1}{2}$	387	288	286	—	119
4	511	425	380	—	186
5	1,050	788	—	—	278
6	1,800	1,400	—	—	545
8	3,750	3,000	—	—	—
10	7,000	5,700	—	—	—
12	11,500	9,500	—	—	—
16	22,000	19,000	—	—	—

NOTE: Steam at an average pressure of 1 psig is used as a basis of calculating capacities.

[a]Do not use column B for pressure drops of less than 1/16 psi per 100 ft of equivalent run.

[b]Do not use column D for pressure drops of less than 1/24 psi per 100 ft of equivalent run except on sizes 3 in. and over.

[c]Pitch of horizontal runouts to risers and radiators should be not less than $\frac{1}{2}$ in. per ft. Where this pitch cannot be obtained, runouts over 8 ft in length should be one pipe size larger than called for in this table.

6B Steam Piping (Metric)

TABLE B.11 Return Main and Riser Capacities for Low-Pressure Systems—Pounds per Hour

(Reference to this table will be made by column letters G through V)

Pipe size inches	$\frac{1}{32}$ psi or $\frac{1}{2}$ oz drop per 100 ft			$\frac{1}{24}$ psi or $\frac{2}{3}$ oz drop per 100 ft			$\frac{1}{16}$ psi or 1 oz drop per 100 ft			$\frac{1}{8}$ psi or 2 oz drop per 100 ft			$\frac{1}{4}$ psi or 4 oz drop per 100 ft		
	Wet	Dry	Vac.	Wet	Dry	Vac.	Wet	Dry	Vac.	Wet	Dry	Vac.	Wet	Dry	Vac.
G	H	I	J	K	L	M	N	O	P	Q	R	S	T	U	V
Return main															
$\frac{3}{4}$	—	—	—	—	—	42	—	—	100	—	—	142	—	—	200
1	125	62	—	145	71	143	175	80	175	250	103	249	350	115	350
$1\frac{1}{4}$	213	130	—	248	149	244	300	168	300	425	217	426	600	241	600
$1\frac{1}{2}$	338	206	—	393	236	388	475	265	475	675	340	674	950	378	950
2	700	470	—	810	535	815	1,000	575	1,000	1,400	740	1,420	2,000	825	2,000
$2\frac{1}{2}$	1,180	760	—	1,580	868	1,360	1,680	950	1,680	2,350	1,230	2,380	3,350	1,360	3,350
3	1,880	1,460	—	2,130	1,560	2,180	2,680	1,750	2,680	3,750	2,250	3,800	5,350	2,500	5,350
$3\frac{1}{2}$	2,750	1,970	—	3,300	2,200	3,250	4,000	2,500	4,000	5,500	3,230	5,680	8,000	3,580	8,000
4	3,880	2,930	—	4,580	3,350	4,500	5,500	3,750	5,500	7,750	4,830	7,810	11,000	5,380	11,000
5	—	—	—	—	—	7,880	—	—	9,680	—	—	13,700	—	—	19,400
6	—	—	—	—	—	12,600	—	—	15,500	—	—	22,000	—	—	31,000
Riser															
$\frac{3}{4}$	—	48	—	—	48	143	—	48	175	—	48	249	—	48	350
1	—	113	—	—	113	244	—	113	300	—	113	426	—	113	600
$1\frac{1}{4}$	—	248	—	—	248	388	—	248	475	—	248	674	—	248	960
$1\frac{1}{2}$	—	375	—	—	375	815	—	375	1,000	—	375	1,420	—	375	2,000
2	—	750	—	—	750	1,360	—	750	1,680	—	750	2,380	—	750	3,350
$2\frac{1}{2}$	—	—	—	—	—	2,180	—	—	2,680	—	—	3,800	—	—	5,350
3	—	—	—	—	—	3,250	—	—	4,000	—	—	5,680	—	—	8,000
$3\frac{1}{2}$	—	—	—	—	—	4,480	—	—	5,500	—	—	7,810	—	—	11,000
4	—	—	—	—	—	7,880	—	—	9,680	—	—	13,700	—	—	19,400
5	—	—	—	—	—	12,600	—	—	15,500	—	—	22,000	—	—	31,000

TABLE B.12 Flow Rate in Kilograms per Hour of Steam in Schedule 40 Pipe at Initial Saturation Pressure of 25 and 85 kPa above Atmospheric

Nom. pipe size (mm)	14 Pa/m Sat. press. (kPa)		28 Pa/m Sat. press. (kPa)		58 Pa/m Sat. press. (kPa)		Pressure drop (Pa/m) 113 Pa/m Sat. press. (kPa)		170 Pa/m Sat. press. (kPa)		225 Pa/m Sat. press. (kPa)		450 Pa/m Sat. press. (kPa)	
	25	85	25	85	25	85	25	85	25	85	25	85	25	85
20	4	5	6	7	9	11	13	16	16	20	19	23	27	33
25	8	10	12	14	17	21	24	30	31	37	37	43	52	62
32	16	20	24	30	35	44	50	63	64	77	73	91	105	127
40	25	32	38	45	54	67	79	95	99	118	112	138	163	195
50	49	61	73	88	106	129	152	186	191	231	218	268	322	386
65	79	98	117	141	171	209	245	299	308	372	354	431	522	621
80	144	172	211	249	299	367	435	526	540	649	626	758	885	1,090
90	210	249	304	363	449	552	640	771	789	953	907	1,100	1,340	1,560
100	290	363	431	526	640	767	898	1,090	1,110	1,360	1,310	1,570	1,910	2,220
125	544	649	762	953	1,110	1,360	1,620	1,930	1,990	2,380	2,310	2,770	3,400	3,900
150	871	1,040	1,280	1,520	1,800	2,200	2,590	2,590	3,180	3,900	3,910	4,540	5,400	6,440
200	1,770	2,180	2,530	3,180	3,670	4,540	5,170	6,490	6,580	8,030	7,480	9,300	10,900	13,400
250	3,270	3,990	4,630	5,720	6,800	8,260	9,530	11,800	11,900	14,500	13,600	16,800	19,400	23,600
300	5,170	6,210	7,480	8,850	10,600	12,900	15,000	18,100	18,600	22,500	21,800	26,100	30,800	36,000

TABLE B.13 Comparative Capacity of Steam Lines at Various Pitches for Steam and Condensate Flowing in Opposite Directions

Pipe size (mm)	Pitch of pipe															
	20		40		80		120		170		250		350		420	
	kg/h	Max. vel. (m/s)	kg/h	Max. vel. (m/s)	kg/h	Max. vel. (m/s)	kg/h	Max. vel. (m/s)	kg/h	Max. vel. (m/s)	kg/h	Max. vel. (m/s)	kg/h	Max. vel. (m/s)	kg/h	Max. vel. (m/s)
20	1.5	2.4	1.9	3.4	2.6	4.0	2.9	4.3	3.2	4.9	3.8	5.2	4.5	6.7	4.8	6.7
25	3.1	2.7	4.1	3.7	5.3	4.6	5.8	5.2	6.7	5.8	7.8	6.7	8.7	7.3	9.3	7.6
32	5.4	3.4	6.8	4.3	9.0	5.2	11.2	6.1	12.2	6.7	14.2	7.6	15.2	7.9	17.3	9.4
40	9.0	3.7	11.7	4.9	15.0	5.8	17.0	6.7	19.1	7.3	15.1	7.9	23.0	8.5	26.9	10.1
50	19.5	4.6	24.5	5.5	31.2	7.3	37.8	8.2	42.1	9.1	45.2	9.8	46.4	9.8	52.2	10.1

TABLE B.14 Equivalent Length of Fittings To Be Added to Pipe Run

Size of pipe (mm)	Length in meters to be added to run				
	Standard elbow	Side outlet tee[b]	Gate valve[a]	Globe valve[a]	Angle valve[a]
15	0.4	1	0.1	4	2
20	0.5	1	0.1	5	3
25	0.7	1	0.1	7	4
32	0.9	2	0.2	9	5
40	1.1	2	0.2	10	6
50	1.3	2	0.3	14	8
65	1.5	3	0.3	16	8
80	1.9	4	0.4	20	10
100	2.7	5	0.3	28	14
125	3.3	7	0.7	34	17
150	4.0	8	0.9	41	20
200	5.2	11	1.1	55	28
250	6.4	14	1.4	70	34
300	8.2	16	1.7	82	40
350	9.1	19	1.9	94	46

[a]Valve in full open position.
[b]Values given apply only to a tee used to divert the flow in the main to the last riser.

TABLE B.15 Steam Pipe Capacities for Low-Pressure Systems

(For one-pipe systems or two-pipe systems in which condensate flows against the steam flow)

Nominal pipe size (mm)	Capacity (kg/h)				
	Two-pipe systems		One-pipe system		
	Condensate flowing against steam		Supply risers up-feed	Radiator valves and vertical connections	Radiator and riser runouts
	Vertical	Horizontal			
A	B[a]	C[c]	D[b]	E	F[c]
20	4	3	3	—	3
25	6	6	5	3	3
32	14	12	9	7	7
40	22	19	17	10	7
50	44	42	33	19	10
65	72	60	53	—	19
80	128	91	91	—	29
90	176	131	130	—	54
100	232	193	172	—	84
125	476	357	—	—	126
150	816	635	—	—	247
200	1700	1360	—	—	—
250	3180	2590	—	—	—
300	5220	4310	—	—	—
400	9980	8620	—	—	—

NOTE: Steam at an average pressure of 7 kPa above atmospheric is used as a basis of calculating capacities.

[a]Do not use column B for pressure drops of less than 13 Pa/m of equivalent run.

[b]Do not use column D for pressure drops of less than 9 Pa/m of equivalent run except on sizes 88 mm and over.

[c]Pitch of horizontal runouts to risers and radiators should be not less than 40 mm/m. Where this pitch cannot be obtained, runouts over 2.5 m in length should be one pipe size larger than called for in this table.

TABLE B.16 Return Main and Riser Capacities for Low-Pressure Systems (kg/h)

Pipe size (in.)	7 Pa/m			9 Pa/m			14 Pa/m			28 Pa/m			57 Pa/m		
	Wet	Dry	Vac.	Wet	Dry	Vac.	Wet	Dry	Vac.	Wet	Dry	Vac.	Wet	Dry	Vac.
G	H	I	J	K	L	M	N	O	P	Q	R	S	T	U	V
Return main															
20	—	—	—	—	—	19	—	—	45	—	—	64	—	—	91
25	57	28	—	66	32	65	79	36	79	113	47	113	159	52	159
32	97	59	—	112	68	111	136	76	136	193	98	193	272	109	272
40	153	93	—	178	107	176	215	120	215	306	154	306	431	171	431
50	318	213	—	367	243	370	454	261	454	635	336	644	907	374	907
65	535	345	—	717	394	616	762	431	762	1,070	558	1,080	1,520	617	1,520
80	853	662	—	967	708	989	1,220	794	1,220	1,700	1,020	1,720	2,430	1,130	2,430
90	1,250	894	—	1,500	998	1,400	1,810	1,130	1,810	2,490	1,470	2,580	3,630	1,620	3,630
100	1,760	1,330	—	2,080	1,520	2,040	2,490	1,700	2,490	3,520	2,190	3,540	4,990	2,440	4,990
125	—	—	—	—	—	3,570	—	—	4,390	—	—	6,210	—	—	8,800
150	—	—	—	—	—	5,720	—	—	7,030	—	—	9,980	—	—	14,100
Riser															
20	—	22	—	—	22	65	—	22	79	—	22	113	—	22	159
25	—	51	—	—	51	111	—	51	136	—	51	193	—	51	272
32	—	112	—	—	112	176	—	112	215	—	112	306	—	112	431
40	—	170	—	—	170	370	—	170	454	—	170	644	—	170	907
50	—	340	—	—	340	616	—	340	762	—	340	1,080	—	340	1,520
65	—	—	—	—	—	989	—	—	1,220	—	—	1,720	—	—	2,430
80	—	—	—	—	—	1,470	—	—	1,810	—	—	2,580	—	—	3,630
70	—	—	—	—	—	2,030	—	—	2,490	—	—	3,540	—	—	4,990
100	—	—	—	—	—	3,570	—	—	4,390	—	—	6,210	—	—	8,800
125	—	—	—	—	—	5,720	—	—	7,030	—	—	9,980	—	—	14,100

HVAC Equations—U.S. Units

TABLE C.1   Air Equations (U.S. Units)

a.

$$V_p = 1096 \sqrt{\frac{V_p}{d}}$$

or for standard air ($d = 0.075$ lb/ft^3):

$$V = 4005 \sqrt{V_p}$$

To solve for d:

$$d = 1.325 \frac{P_b}{T}$$

V = Velocity (fpm)
V_p = Velocity pressure (in.w.g.)
d = Density (lb/ft^3)
P_b = Absolute static presure (in.Hg)
(barometric pressure + static pressure)
T = Absolute temperature ($460° +$ °F)

b.

$$Q_{(sens.)} = 60 \times C_p \times d \times \text{cfm} \times \Delta t$$

or for standard air
($C_p = 0.24$ Btu/lb • °F):

$$Q_{(sens.)} = 1.08 \times \text{cfm} \times \Delta t$$

Q = Heat flow (Btu/h)
C_p = Specific heat (Btu/lb • °F)
d = Density (lb/ft^3)
Δt = Temperature difference (°F)

c.

$$Q_{(lat.)} = 4840 \times \text{cfm} \times \Delta W \text{ (lb)}$$
$$Q_{(lat.)} = 0.69 \times \text{cfm} \times \Delta W \text{ (gr.)}$$

ΔW = Humidity ratio (lb or gr. H$_2$O/lb dry air)

d.

$$Q_{(total)} = 4.5 \times \text{cfm} \times \Delta h$$

Δh = Enthalpy diff. (Btu/lb dry air)

e.

$$Q = A \times U \times \Delta t$$

A = Area of surface (ft^2)
U = Heat transfer coefficient (Btu/ft^2 • h • °F)

f.

$$R = \frac{1}{U}$$

R = Sum of thermal resistances (ft^2 • h • °F/Btu)

g.

$$\frac{P_1 V_1}{T_1} = \frac{P_2 V_2}{T_2} = RM$$

P = Absolute pressure (lb/ft^2)
V = Total volume (ft^3)
T = Absolute temp. ($460° +$ °F = °R)
R = Gas constant
M = Mass (lb)

h.
$$TP = V_p + SP$$

TP = Total pressure (in.w.g.)
V_p = Velocity pressure (in.w.g.)
SP = Static pressure (in.w.g.)

i.
$$V_p = \left(\frac{V}{4005}\right)^2$$

V = Velocity (fpm)

j.
$$V = V_m \left[\frac{d\ (\text{other than standard})}{0.075\ (d = \text{Std. air})}\right]$$

V_m = Measured velocity (fpm)
d = Density (lb/ft^3)

k.
$$\text{cfm} = A \times V$$

A = Area of duct cross section (ft^2)

l.
$$TP = C \times V_p$$

C = Duct fitting loss coefficient

TABLE C.2   Fan Equations

a.
$$\frac{\text{cfm}_2}{\text{cfm}_1} = \frac{\text{rpm}_2}{\text{rpm}_1}$$

cfm = Cubic feet per minute
rpm = Revolutions per minute

b.
$$\frac{P_2}{P_1} = \left(\frac{\text{rpm}_2}{\text{rpm}_1}\right)^2$$

P = Static or total pressure (in.w.g.)

c.
$$\frac{\text{BHP}_2}{\text{BHP}_1} = \left(\frac{\text{rpm}_2}{\text{rpm}_1}\right)^3$$

BHP = Brake horsepower

d.
$$\frac{d_2}{d_1} = \left(\frac{\text{rpm}_2}{\text{rpm}_1}\right)^2$$

d = Density (lb/ft^3)

e.
$$\frac{\text{rpm (fan)}}{\text{rpm (motor)}} = \frac{\text{Pitch diam. motor pulley}}{\text{Pitch diam. fan pulley}}$$

TABLE C.3   Pump Equations (U.S. Units)

a.
$$\frac{\text{gpm}_2}{\text{gpm}_1} = \frac{\text{rpm}_2}{\text{rpm}_1}$$

gpm = Gallons per minute
rpm = Revolutions per minute

b.
$$\frac{\text{gpm}_2}{\text{gpm}_1} = \frac{D_2}{D_1}$$

D = Impeller diameter

c.
$$\frac{H_2}{H_1} = \left(\frac{\text{rpm}_2}{\text{rpm}_1}\right)^2$$

H = Head (ft w.g.)

d.
$$\frac{H_2}{H_1} = \left(\frac{D_2}{D_1}\right)^2$$

e.
$$\frac{\text{BHP}_2}{\text{BHP}_1} = \left(\frac{\text{rpm}_2}{\text{rpm}_1}\right)^3$$

BHP = Brake horsepower

f.
$$\frac{\text{BHP}_2}{\text{BHP}_1} = \left(\frac{D_2}{D_1}\right)^3$$

TABLE C.4   Hydronic Equivalents (U.S. Units)

a. One gallon water = 8.33 lb
b. Specific heat (C_p) water = 1.00 Btu/lb • °F (at 68°F)
c. Specific heat (C_p) water vapor = 0.45 Btu/lb • °F (at 68°F)
d. One foot of water = 0.433 psi
e. One psi = 2.3 ft. w.g. = 2.04 in.Hg
f. One cubic foot of water = 62.4 lb = 7.49 gal
g. One inch of mercury (Hg) = 13.6 in.w.g. = 1.13 ft w.g.
h. Atmospheric pressure = 29.92 in.Hg = 14.696 psi

TABLE C.5   Hydronic Equations (U. S. Units)

a. $Q = 500 \times \text{gpm} \times \Delta t$

gpm = Gallons per minute
Q = Heat flow (Btu/h)
Δt = Temperature diff. (°F)

b. $Q = 60 \text{ min/h} \times 8.33 \times C_p \times \text{gpm} \times \Delta t$

C_p = Specific heat (water = 1 Btu/lb • °F)

c. $\dfrac{\Delta P_2}{\Delta P_1} = \left(\dfrac{\text{gpm}_2}{\text{gpm}_1}\right)^2$

ΔP = Pressure diff. (psi)

d. $\Delta P = \left(\dfrac{\text{gpm}}{C_v}\right)^2$

C_v = Valve constant (dimensionless)

e. $\text{WHP} = \dfrac{\text{gpm} \times H \times \text{Sp. Gr.}}{3960}$

WHP = Water horsepower
H = Head (ft w.g.)
Sp.Gr. = Specific gravity (use 1.0 for water)
gpm = Gallons per minute

f. $\text{BHP} = \dfrac{\text{gpm} \times H \times \text{Sp.Gr.}}{3960 \times E_p} = \dfrac{\text{WHP}}{E_p}$

BHP = Brake horsepower

g. $E_p = \dfrac{\text{WHP} \times 100}{\text{BHP}}$ (in percent)

E_p = Efficiency of pump

h. $\text{NPSHA} = P_a \pm P_s + \dfrac{V^2}{2g} - P_{vp}$

NPSHA = Net positive suction head available
P_a = Atm. press. (use 34 ft. w.g.)
P_s = Pressure at pump centerline (ft. w.g.)
P_{vp} = Absolute vapor pressure (ft. w.g.)
g = Gravity acceleration (32.2 ft./s²)

i. $h = f \times \dfrac{L}{D} \times \dfrac{V^2}{2g}$

h = Head loss (ft)
f = Friction factor (dimensionless)
L = Length of pipe (ft)
D = Internal diameter (ft)
V = Velocity (ft/s)

TABLE C.6   Converting Pressure in Inches of Mercury to Feet of Water at Various Water Temperatures

Water temperature (°F)	60°	150°	200°	250°	300°	340°
Feet head differential per in.Hg differential	1.046	1.07	1.09	1.11	1.15	1.165

TABLE C.7   Electric Equations (U.S. Units)

(Single phase)

a. $BHP = \dfrac{I \times E \times PF \times Eff}{746}$ BHP = Brake horsepower

(Three phase)

b. $BHP = \dfrac{I \times E \times PF \times Eff \times 1.73}{746}$ I = Amps (A)
E = Volts (V)
PF = Power factor

c. $E = IR$ Eff = Efficiency

d. $P = EI$ R = Ohms (Ω)

e. $\dfrac{\text{F.L. amps}^* \times \text{voltage}^*}{\text{actual voltage}} = \text{actual}$ P = Watts (W)
full-load amps

*Nameplate ratings.

TABLE C.8   Air Density Correction Factors (U.S. Units)

Altitude (ft)	Sea Level	1000	2000	3000	4000	5000	6000	7000	8000	9000	10,000
Barometer (in.Hg)	29.92	28.86	27.82	26.82	25.84	24.90	23.98	23.09	22.22	21.39	20.58
(in.w.g.)	407.5	392.8	378.6	365.0	351.7	338.9	326.4	314.3	302.1	291.1	280.1

Air Temp (°F)											
40°	1.26	1.22	1.17	1.13	1.09	1.05	1.01	0.97	0.93	0.90	0.87
0°	1.15	1.11	1.07	1.03	0.99	0.95	0.91	0.89	0.85	0.82	0.79
40°	1.06	1.02	0.99	0.95	0.92	0.88	0.85	0.82	0.79	0.76	0.73
70°	1.00	0.96	0.93	0.89	0.86	0.83	0.80	0.77	0.74	0.71	0.69
100°	0.95	0.92	0.88	0.85	0.81	0.78	0.75	0.73	0.70	0.68	0.65
150°	0.87	0.84	0.81	0.78	0.75	0.72	0.69	0.67	0.65	0.62	0.60
200°	0.80	0.77	0.74	0.71	0.69	0.66	0.64	0.62	0.60	0.57	0.55
250°	0.75	0.72	0.70	0.67	0.64	0.62	0.60	0.58	0.56	0.58	0.51
300°	0.70	0.67	0.65	0.62	0.60	0.58	0.56	0.54	0.52	0.50	0.48
350°	0.65	0.62	0.60	0.58	0.56	0.54	0.52	0.51	0.49	0.47	0.45
400°	0.62	0.60	0.57	0.55	0.53	0.51	0.49	0.48	0.46	0.44	0.42
450°	0.58	0.56	0.54	0.52	0.50	0.48	0.46	0.45	0.43	0.42	0.40
500°	0.55	0.53	0.51	0.49	0.47	0.45	0.44	0.43	0.41	0.39	0.38
550°	0.53	0.51	0.49	0.47	0.45	0.44	0.42	0.41	0.39	0.38	0.36
600°	0.50	0.48	0.46	0.45	0.43	0.41	0.40	0.39	0.37	0.35	0.34
700°	0.46	0.44	0.43	0.41	0.39	0.38	0.37	0.35	0.34	0.33	0.32
800°	0.42	0.40	0.39	0.37	0.36	0.35	0.33	0.32	0.31	0.30	0.29
900°	0.39	0.37	0.36	0.35	0.33	0.32	0.31	0.30	0.29	0.28	0.27
1000°	0.36	0.35	0.33	0.32	0.31	0.30	0.29	0.28	0.27	0.26	0.25

Standard air density, sea level, 70 °F = 0.075 lb/ft³ at 29.92 in.Hg.

HVAC Equations—Metric Units

TABLE D.1   Air Equations (Metric)

a.

$$V = 1.414 \sqrt{\frac{V_p}{d}}$$

or for standard air:
$(d = 1.204 \text{ kg/m}^3)$:
$$V = \sqrt{1.66 \, V_p}$$
To solve for d:
$$d = 3.48 \frac{P_b}{T}$$

V = Velocity (m/s)
V_p = Velocity pressure [pascals (Pa)]
d = Density (kg/m³)
P_b = Absolute static pressure (kPa)
  (barometric pressure + static
  pressure)
T = Absolute temp. (273° + °C = K)

b.

$$Q = C_p \times d \times \text{L/s} \times \Delta t$$

or for standard air:
$(C_p = 1.005 \text{ kJ/kg} \cdot °C)$ (dry)

or

$(C_p = 1.025 \text{ kJ/kg} \cdot °C)$ (moist)
$$Q_{(sens.)}(W) = 1.23 \times \text{L/s} \times \Delta t$$
$$Q_{(sens.)}(kW) = 1.23 \times \text{m}^3/\text{s} \times \Delta t$$

Q = Heat flow (watts or kilowatts)
C_p = Specific heat (kJ/kg · °C)
d = Density (kg/m³)
Δt = Temperature difference (°C)
L/s = Airflow (liters per second)
m³/s = Airflow (cubic meters per second)

c.

$$Q_{(lat.)}(W) = 3.01 \times \text{L/s} \times \Delta W$$

Q = Heat flow (watts or kilowatts)
ΔW = Humidity ratio (g H_2O/kg dry air)

d.

$$Q_{(total\ heat)}(W) = 1.20 \times \text{L/s} \times \Delta h$$

Δh = Enthalpy diff. (kJ/kg dry air)

e.

$$Q = A \times U \times \Delta t$$

A = Area of surface (m²)
U = Heat transfer coefficient (W/m² · K)

f.

$$R = \frac{1}{U}$$

R = Sum of thermal resistances
  (m² · K/W)

g.

$$\frac{P_1 V_1}{T_1} = \frac{P_2 V_2}{T_2} = RM$$

P = Absolute pressure (kPa)
V = Total volume (m³)
T = Absolute temperature
  (273° + °C = K)
R = Gas Constant (kJ/kg · K)
M = Mass (kg)

h.

$$TP = V_p + SP$$

i.

$$V_p = \frac{d}{2} \times V^2 = 0.602V^2$$

j.

$$V = V_m \left[\frac{d \ (\text{other than standard})}{1.204 \ (d = \text{Std. air})} \right]$$

k.

$$L/s = 1000 \times A \times V$$

l.

$$TP = C \times V_p$$

TP = Total pressure (Pa)
V_p = Velocity pressure (Pa)
SP = Static pressure (Pa)
V = Velocity (m/s)
V_m = Measured Velocity (m/s)
d = Density (kg/m^3)
A = Area of duct cross section (m^2)
C = Duct fitting loss coefficient

TABLE D.2   Fan Equations (Metric)

a.
$$\frac{L/s_2}{L/s_1} = \frac{m^3/s_2}{m^3/s_1} = \frac{rad/s_2}{rad/s_1} = \frac{rev/s_2}{rev/s_1}$$

L/s = Liters per second
m^3/s = Cubic meters per second
rad/s = Radians per second
rev/s = Revolutions per second
P = Static or total pressure (Pa)

b.
$$\frac{P_2}{P_1} = \left(\frac{rad/s_2}{rad/s_1} \right)^2$$

c.
$$\frac{kW_2}{kW_1} = \left(\frac{rad/s_2}{rad/s_1} \right)^3$$

kW = Kilowatts

d.
$$\frac{d_2}{d_1} = \left(\frac{rad/s_2}{rad/s_1} \right)^2$$

d = Density (kg/m^3)

e.
$$\frac{rad/s \ (\text{fan})}{rad/s \ (\text{motor})} = \frac{\text{Pitch diam. motor pulley}}{\text{Pitch diam. fan pulley}}$$

NOTE: m^3/h, cubic meters per hour, is used in lieu of m^3/s in some countries.

TABLE D.3   Pump Equations (Metric)

a.
$$\frac{L/s_2}{L/s_1} = \frac{m^3/s_2}{m^3/s_1} = \frac{rad/s_2}{rad/s_1} = \frac{rev/s_2}{rev/s_1}$$

L/s = Liters per second
m^3/s = Cubic meters per second
rad/s = Radians per second
rev/s = Revolutions per second
D = Impeller diameter

b.
$$\frac{m^3/s_2}{m^3/s_1} = \frac{D_2}{D_1}$$

c.
$$\frac{H_2}{H_1} = \left(\frac{rad/s_2}{rad/s_1} \right)^2$$

H = Head (kPa)

d.
$$\frac{H_2}{H_1} = \left(\frac{D_2}{D_1} \right)^2$$

e.
$$\frac{BP_2}{BP_1} = \left(\frac{rad/s_2}{rad/s_1} \right)^3$$

BP = Brake power

f.
$$\frac{BP_2}{BP_1} = \left(\frac{D_2}{D_1} \right)^3$$

NOTE: m^3/h, cubic meters per hour, is used in lieu of m^3/s in some countries.

TABLE D.4 Hydronic Equations (Metric)

a. $Q(\text{W}) = C_p \times d \times \text{L/s} \times \Delta t$

b. $Q(\text{kW}) = 4190 \times \text{m}^3/\text{s} \times \Delta t$

   $Q(\text{W}) = 4190 \times \text{L/s} \times \Delta t$

   $Q(\text{kW}) = 4.19 \times \text{L/s} \times \Delta t$

$\dfrac{\Delta P_2}{\Delta P_1} = \left(\dfrac{\text{m}^3/\text{s}_2}{\text{m}^3/\text{s}_1}\right)^2 = \left(\dfrac{\text{L/s}_2}{\text{L/s}_1}\right)^2$

d. $\Delta P = \left(\dfrac{\text{m}^3/\text{s}}{K_v}\right)^2 = \left(\dfrac{\text{L/s}}{K_v}\right)^2$

e. $WP \ (\text{kW}) = 9.81 \times \text{m}^3/\text{s} \times H \ (\text{m}) \times$ Sp.Gr.$/E_p$

f. $WP \ (\text{W}) = \dfrac{\text{L/s} \times H \ (\text{Pa}) \times \text{Sp.Gr.}}{1002 \times E_p}$

$E_p = \dfrac{WP \times 100}{BP} \ (\text{in percent})$

h. $\text{NPSHA} = P_a \pm P_s + \dfrac{V^2}{2g} - P_{vp}$

i. $h = f \times \dfrac{L_p}{D} \times \dfrac{V^2}{2g}$

Q = Heat flow (W or kW)

Δt = Temperature difference (°C)

C_p = Specific heat (4190 J/kg • °C for water)

d = Density (1 kg/L for water)

m^3/s = Cubic meters per second (used for large volumes)

L/s = Liters per second

ΔP = Pressure diff. (Pa or kPa)

K_v = Valve constant (dimensionless)

WP = Water power (kW or W)

m^3/s = Cubic meters per second

L/s = Liters per second

Sp.Gr. = Specific gravity (use 1.0 for water)

E_p = Efficiency of pump

H = Head (Pa or m)

BP = Brake power (kW)

NPSHA = Net positive suction head available

P_a = Atmospheric pressure (Pa) (Std. atm. press. = 101,325 Pa)

P_s = Pressure at pump centerline (Pa)

P_{vp} = Absolute vapor pressure (Pa)

g = Gravity acceleration (9.807 m/s²)

h = Head loss (m)

f = Friction factor (dimensionless)

L_p = Length of pipe (m)

D = Internal diameter (m)

V = Velocity (m/s)

TABLE D.5   Electric Equations (metric)

a. $kW = \dfrac{I \times E \times PF \times Eff}{1000}$ (Single phase)

b. $kW = \dfrac{I \times E \times PF \times Eff \times 1.73}{1000}$ (Three phase)

c. $E = IR$

d. $P = EI$

d. $\dfrac{\text{F.L. Amps*} \times \text{Voltage*}}{\text{Actual Voltage}} = $ Actual full-load amps

kW = Kilowatts

I = Amps (A)
E = Volts (V)
PF = Power factor
Eff = Efficiency
R = Ohms (Ω)
P = Watts (W)

*Nameplate ratings.

TABLE D.6   Air Density Correction Factors (metric units)

Altitude (m)	Sea level	250	500	750	1000	1250	1500	1750	2000	2500	3000
Barometer (kPa)	101.3	98.3	96.3	93.2	90.2	88.2	85.1	83.1	80.0	76.0	71.9
Air Temp (°C)											
0°	1.08	1.05	1.02	0.99	0.96	0.93	0.91	0.88	0.86	0.81	0.76
20°	1.00	0.97	0.95	0.92	0.89	0.87	0.84	0.82	0.79	0.75	0.71
50°	0.91	0.89	0.86	0.84	0.81	0.79	0.77	0.75	0.72	0.68	0.64
75°	0.85	0.82	0.80	0.78	0.75	0.73	0.71	0.69	0.67	0.63	0.60
100°	0.79	0.77	0.75	0.72	0.70	0.68	0.66	0.65	0.63	0.59	0.56
125°	0.74	0.72	0.70	0.68	0.66	0.64	0.62	0.60	0.59	0.55	0.52
150°	0.70	0.68	0.66	0.64	0.62	0.60	0.59	0.57	0.55	0.52	0.49
175°	0.66	0.64	0.62	0.62	0.59	0.57	0.55	0.54	0.52	0.44	0.46
200°	0.62	0.61	0.59	0.57	0.56	0.54	0.52	0.51	0.49	0.47	0.44
225°	0.59	0.58	0.56	0.54	0.53	0.51	0.50	0.48	0.47	0.44	0.42
250°	0.56	0.55	0.53	0.52	0.50	0.49	0.47	0.46	0.45	0.42	0.40
275°	0.54	0.52	0.51	0.49	0.48	0.47	0.45	0.44	0.43	0.40	0.38
300°	0.51	0.50	0.49	0.47	0.46	0.45	0.43	0.42	0.41	0.38	0.36
325°	0.49	0.48	0.47	0.45	0.44	0.43	0.41	0.40	0.39	0.37	0.35
350°	0.47	0.46	0.45	0.43	0.42	0.41	0.40	0.39	0.38	0.35	0.33
375°	0.46	0.44	0.43	0.42	0.41	0.39	0.38	0.37	0.36	0.34	0.32
400°	0.44	0.43	0.41	0.40	0.39	0.38	0.37	0.36	0.35	0.33	0.31
425°	0.42	0.41	0.40	0.39	0.38	0.37	0.35	0.34	0.33	0.32	0.30
450°	0.41	0.40	0.38	0.37	0.36	0.35	0.34	0.33	0.32	0.31	0.29
475°	0.39	0.38	0.37	0.36	0.35	0.34	0.33	0.32	0.31	0.29	0.28
500°	0.38	0.37	0.36	0.35	0.34	0.33	0.32	0.31	0.30	0.28	0.27
525°	0.37	0.36	0.35	0.34	0.33	0.32	0.31	0.30	0.29	0.27	0.26

Standard air density, sea level, 20 °C = 1.2041 kg/m³ at 101.325 kPa.

Metric Units and Equivalents

TABLE E.1 Metric Units (basic and derived)

Unit	Symbol	Quantity	Equivalent or relationship
ampere	A	Electric current	Same as U.S.
candela	cd	Luminous intensity	1 cd/m² = 0.292 ft lamberts
Celsius	°C	Temperature	°F = 1.8 °C + 32°
coulomb	C	Electric charge	Same as U.S.
farad	F	Electric capacitance	Same as U.S.
henry	H	Electric inductance	Same as U.S.
hertz	Hz	Frequency	Same as cycles per second
joule	J	Energy, work, heat	1J = 0.7376 ft-lb = 0.000948 Btu
kelvin	K	Thermodynamic temperature	$K = °C + 273.15°$ $= \dfrac{°F + 459.67}{1.8}$
kilogram	kg	Mass	1 kg = 2.2046 lb
liter	L	Liquid volume	1 L = 1.056 qt = 0.264 gal
lumen	lm	Luminous flux	1 lm/m² = 0.0929 ft candles
lux	lx	Illuminance	1 lx = 0.0929 ft candles
meter	m	Length	1 m = 3.281 ft
mole	mol	Amount of substance	—
newton	N	Force	1 N = kg · m/s² = 0.2248 lb (force)
ohm	Ω	Electrical resistance	Same as U.S.
pascal	Pa	Pressure, stress	1 Pa = N/m² = 0.000145 psi = 0.004022 in.w.g.
radian	rad	Plane angle	1 rad = 57.29°
second	s	Time	Same as U.S.
siemens	S	Electric conductance	—
steradian	sr	Solid angle	—
volt	V	Electric potential	Same as U.S.
watt	W	power, heat flow	1 W = J/s = 3.4122 Btu/hr 1 W = 0.000284 tons of refrig.

TABLE E.2 Metric Equivalents

Quantity	Symbol	Unit	U.S. relationship
Acceleration	m/s^2	meter per second squared	$1 \ m/s^2 = 3.281 \ ft/s^2$
Angular velocity	rad/s	radian per second	$1 \ rad/s = 9.549 \ rpm = 0.159 \ rps$
Area	m^2	square meter	$1 \ m^2 = 10.76 \ ft^2$
Atmospheric pressure	—	101.325 kPa	29.92 in. Hg = 14.696 psi
Density	kg/m^3	kilogram per cubic meter	$1 \ kg/m^3 = 0.0623 \ lb/ft^3$
Density, air	—	$1.2 \ kg/m^3$	$0.075 \ lb/ft^3$
Density, water	—	$1000 \ kg/m^3$	$62.4 \ lb/ft^3$
Duct friction loss	Pa/m	pascal per meter	$1 \ Pa/m = 0.1224$ in.w.g./100 ft
Enthalpy	kJ/kg	kilojoule per kilogram	$1 \ kJ/kg = 0.4299$ Btu/lb dry air
Gravity		$9.8067 \ m/s^2$	$32.2 \ ft/s^2$
Heat flow	W	watt	$1 \ W = 3.412 \ Btu/h$
Length (normal)	m	meter	$1 \ m = 3.281 \ ft = 39.37$ in.
Linear velocity	m/s	meter per second	$1 \ m/s = 196.9 \ fpm$
Mass flow rate	kg/s	kilogram per second	$1 \ kg/s = 7936.6 \ lb/h$
Moment of inertia	$kg \cdot m^2$	kilogram × square meter	$1 \ kg \cdot m^2 = 23.73 \ lb \cdot ft^2$
Power	W	watt	$1 \ W = 0.00134 \ hp$
Pressure	kPa Pa	kilopascal (1000 pascals) pascal	$1 \ kPa = 0.296$ in.Hg = 0.145 psi $1 \ Pa = 0.004015$ in.w.g.
Specific heat—air (C_p)		$1000 \ J/(kg \cdot K)$	$1000 \ J/(kg \cdot K)$ $= 1 \ kJ/(kg \cdot K)$ $= 0.2388$ Btu/lb $\cdot$ °F
Specific heat—air (C_v)		$717 \ J/(kg \cdot K)$	0.17 Btu/lb $\cdot$ °F
Specific heat—water		$4190 \ J/(kg \cdot K)$	1.0 Btu/lb $\cdot$ °F
Specific volume	m^3/kg	cubic meter per kilogram	$1 \ m^3/kg = 16.019 \ ft^3/lb$
Thermal conductivity	$(W \cdot mm)/(m^2 \cdot K)$	watt millimeter per square meter times K	$1 \ W \cdot mm/(m^2 \cdot K)$ $= 0.0069$ Btu/h $\cdot$ in./ft$^2 \cdot$ °F
Volume flow rate	m^3/s	cubic meter per second	$1 \ m^3/s = 2118.88 \ cfm$ (air).
	L/s	liter per second $1 \ m^3/s =$ 1000 L/s $1 \ mL =$ liter/1000	$1 \ L/s = 2.12 \ cfm$ (air) $1 \ m^3/s = 15,850 \ gpm$ (water) $1 \ mL/s = 1.05 \ gpm$ (water)
	m^3/h	cubic meter per hour	$1 \ m^3/h = 0.588 \ cfm$ (air) $1 \ m^3/h = 4.4 \ gpm$ (water)

TABLE E.3 Temperature Equivalents

To use table: Select temperature to be converted from bold column (°F or °C). Read Celsius equivalent to left and Fahrenheit equivalent to right.

1° TO 100°

°C	°F/°C°	°F	°C	°F/°C	°F	°C	°F/°C	°F	°C	°F/°C	°F
−17.2	1	33.8	−3.3	26	78.8	10.6	51	123.8	24.4	76	168.8
−16.7	2	35.6	−2.8	27	80.6	11.1	52	125.6	25.0	77	170.6
−16.1	3	37.4	−2.2	28	82.4	11.7	53	127.4	25.6	78	172.4
−15.6	4	39.2	−1.7	29	84.2	12.2	54	129.2	26.1	79	174.2
−15.0	5	41.0	−1.1	30	86.0	12.8	55	131.0	26.7	80	176.0
−14.4	6	42.8	−0.6	31	87.8	13.3	56	132.8	27.2	81	177.8
−13.9	7	44.6	0.0	32	89.6	13.9	57	134.6	27.8	82	179.6
−13.3	8	46.4	0.6	33	91.4	14.4	58	136.4	28.3	83	181.4
−12.8	9	48.2	1.1	34	93.2	15.0	59	138.2	28.9	84	183.2
−12.2	10	50.0	1.7	35	95.0	15.6	60	140.0	29.4	85	185.0
−11.7	11	51.8	2.2	36	96.8	16.1	61	141.8	30.0	86	186.8
−11.1	12	53.6	2.8	37	98.6	16.7	62	143.6	30.6	87	188.6
−10.6	13	55.4	3.3	38	100.4	17.2	63	145.4	31.1	88	190.4
−10.0	14	57.2	3.9	39	102.2	17.8	64	147.2	31.7	89	192.2
−9.4	15	59.0	4.4	40	104.0	18.3	65	149.0	32.2	90	194.0
−8.9	16	60.8	5.0	41	105.8	18.9	66	150.8	32.8	91	195.8
−8.3	17	62.6	5.6	42	107.6	19.4	67	152.6	33.3	92	197.6
−7.8	18	64.4	6.1	43	109.4	20.0	68	154.4	33.9	93	199.4
−7.2	19	66.2	6.7	44	111.2	20.6	69	156.2	34.4	94	201.2
−6.7	20	68.0	7.2	45	113.0	21.1	70	158.0	35.0	95	203.0
−6.1	21	69.8	7.8	46	114.8	21.7	71	159.8	35.6	96	204.8
−5.6	22	71.6	8.3	47	116.6	22.2	72	161.6	36.1	97	206.6
−5.0	23	73.4	8.9	48	118.4	22.8	73	163.4	36.7	98	208.4
−4.4	24	75.2	9.4	49	120.2	23.3	74	165.2	37.2	99	210.2
−3.9	25	77.0	10.0	50	122.0	23.9	75	167.0	37.8	100	212.0

100° TO 1000°

°C	°F/°C	°F	°C	°F/°C	°F	°C	°F/°C	°F	°C	°F/°C	°F
38	100	212	160	320	608	288	550	1022	416	780	1436
43	110	230	166	330	626	293	560	1040	421	790	1454
49	120	248	171	340	644	299	570	1058	427	800	1472
54	130	266	177	350	662	304	580	1076	432	810	1490
60	140	284	182	360	680	310	590	1094	428	820	1508
66	150	302	188	370	698	316	600	1112	443	830	1526
71	160	320	193	380	716	321	610	1130	449	840	1544
77	170	338	199	390	734	327	620	1148	454	850	1562
82	180	356	204	400	752	332	630	1166	460	860	1580
88	190	374	210	410	770	338	640	1184	466	870	1598
93	200	392	216	420	788	343	650	1202	471	880	1616
99	210	410	221	430	806	349	660	1220	477	890	1634
100	212	414	227	440	824	354	670	1238	482	900	1652
104	220	428	232	450	842	360	680	1256	488	910	1670
110	230	446	238	460	860	366	690	1274	493	920	1688
116	240	464	243	470	878	371	700	1292	499	930	1706
121	250	482	249	480	896	377	710	1310	504	940	1724
127	260	500	254	490	914	382	720	1328	510	950	1742
132	270	518	260	500	932	388	730	1346	516	960	1760
138	280	536	266	510	950	393	740	1364	521	970	1778
143	290	554	271	520	968	399	750	1382	527	980	1796
149	300	572	277	530	986	404	760	1400	532	990	1814
154	310	590	282	540	1004	410	770	1418	538	1000	1832

NOTE: Absolute temperature (K) = °C + 273.15° = °F + 459.67°/1.8.
Temperature difference: K = °C = °F/1.8.

Sound Design Equations

1. $L_w = L_p + 10 \log_{10} \left[\dfrac{1}{\dfrac{Q}{4\pi r^2} + \dfrac{4}{A}} \right] - 10.5 \text{ dB}$

L_p = Sound pressure level, dB re 0.0002 microbar
L_w = Sound power level, dB re 10^{-12} W
r = Distance from the sound source (ft)
A or $S\alpha$ = Total Sabins in the room
Q = Directivity factor

2. $R = \dfrac{S\alpha}{1 - \alpha}$ or $R = \dfrac{A}{1 - \bar{a}}$

R = Room constant
α = Absorption coefficient of the surface treatment
S = Surface area in square feet
$\bar{a}$ = Average Sabin* absorption coefficient for the room

3. $L_p = L_w - 10 \log_{10} \left[\dfrac{1}{\dfrac{Q}{4\pi r^2} + \dfrac{4}{A}} \right] + 10.5 \text{ dB}$

4. Strouhal number $(N_{str}) = \dfrac{fD}{V} \times 5$

D = Characteristic dimension (in.)
V = Velocity (fpm)
f = Octave band center frequency (Hz)

5. Octave band sound power level = $F + G + H$ in dB re 10^{-12} W

F = Function (use with charts)
G = Function (use with charts
H = Function (use with charts)

---

*__SABIN__—The unit of acoustic absorption. One "sabin" is equal to one square foot of "perfect" sound absorption material such as an "open" window.

6. $L_{\mathrm{WB}} = L_{\mathrm{WD}} - TL + 10 \log_{10} \dfrac{S}{A}$

L_{WB} = Sound power level breakout
L_{WD} = Sound power level in duct
TL = Transmission loss of duct wall
S = Radiating surface area of duct wall (ft^2)
A = Cross-sectional area of duct component (ft^2)

7. $L_{\mathrm{w}} = L_{\mathrm{p}} + 10 \log_{10} A - 10 \text{ dB}$

L_{w} = Sound power level that enters duct
L_{p} = Sound pressure level in source room
A = Opening in duct (ft^2)

8. $\lambda = \dfrac{c}{f}$

λ = Wavelength in feet (m)
c = Speed of sound, 1125 fps (343 m/s)
f = Cycles per second (Hz)

9. $L_{\mathrm{w}} = 10 \log \dfrac{W}{W_{\mathrm{ref}}}$

L_{w} = Sound power level (dB)
W = Acoustic power output of noise source
W_{ref} = 10^{-12} W

10. dB re 10^{-12} W = dB re 10^{-13} W $- 10$

11. $L_{\mathrm{p}} = 20 \log \dfrac{P}{P_{\mathrm{ref}}}$

L_{p} = Sound pressure level in dB
P = RMS sound pressure
P_{ref} = Ref. RMS sound pressure

12. $D = \dfrac{C_1}{T_1} + \dfrac{C_2}{T_2} + \cdots \dfrac{C_n}{T_n}$

C_n = Actual duration of exposure (h)
T_n = Noise exposure limit

13. $B_{\mathrm{f}} = \dfrac{\text{rpm} \times \text{no. of blades}}{60}$

B_{f} = Blade frequency

14. $L_{\mathrm{w}} = K_{\mathrm{w}} + 10 \log Q + 20 \log P$

L_{w} = Est. sound power level, dB re 10^{-12} W
K_{w} = Specific sound power level
Q = Volume flow rate (cfm)
P = Pressure (in.w.g.)

15. $NR = TL - 10 \log S + 10 \log A$

NR = Noise reduction (dB)
TL = The transmission loss of the partition separating the two spaces
S = The surface area of the partition (ft^2)
A = Total of the sound absorptive materials in the receiving room (Sabins, ft^2)

16. $TL = 20 \log M + 20 \log F - 33 \text{ dB}$

M = Mass of construction (lb/ft^2)
F = Frequency (Hz)

17. $D = 0.5 \sqrt{A}$

$D =$ Distance (in feet) from the noise source
$A =$ Total absorption in room (Sabins, ft^2)

18. $NR = 10 \log \dfrac{A_2}{A_1}$

$NR =$ Noise reduction in dB
$A_1 =$ Total absorption (Sabins, ft^2) in the room before adding the sound absorption
$A_2 =$ Total absorption in the room after adding sound absorption

19. $L_{P1} = L_{P2} - 20 \log D_1 + 20 \log D_2$

$L_{P1} =$ Sound pressure level at position 1
$L_{P2} =$ Sound pressure level at position 2
$D_1 =$ Distance (in feet) from noise source to position 1
$D_2 =$ Distance (in feet) from noise source to position 2

20. $L_p = L_w - 20 \log D - 0.5$ dB

$L_p =$ Sound pressure level in dB
$D =$ Distance (in feet) from the point source to the point where sound pressure is measured

21. $a = (\text{antilog } L_a/20)10^{-5}$

$a =$ Acceleration in m/s^2 (1 G = 9.8 m/s^2)
$L_a =$ Acceleration level in dB re 10^{-5} (m/s^2)

22. $v = (\text{antilog } L_v/20)10^{-8}$

$v =$ Velocity in m/s (1 G at 100 Hz = 0.015 m/s)
$L_v =$ Velocity level in dB re 10^{-8} (m/s)

23. $d = (\text{antilog } L_d/20)10^{-11}$

$d =$ Displacement (m)—(1 G at 100 Hz = 0.0249 mm)
$L_d =$ Displacement level in dB re 10^{-11} (m)

G

Logarithms

TABLE G.1 Decibel Equivalents of Numbers

N	$\dfrac{10 \log N}{\text{(dB)}}$	$\dfrac{20 \log N}{\text{(dB)}}$	N	$\dfrac{10 \log N}{\text{(dB)}}$	$\dfrac{20 \log N}{\text{(dB)}}$
1.0	0	0	400	26	52
1.25	1	2	500	27	54
1.6	2	4	630	28	56
2.0	3	6	800	29	58
2.5	4	8	1,000	30	60
3.2	5	10	1.250	31	62
4.0	6	12	1,600	32	64
5.0	7	14	2,000	33	66
6.3	8	16	2,500	34	68
8	9	18	3,200	35	70
10	10	20	4,000	36	72
12	11	22	5,000	37	74
16	12	24	6,300	38	76
20	13	26	8,000	39	78
25	14	28	10,000	40	80
32	15	30	12,500	41	82
40	16	32	16,000	42	84
50	17	34	20,000	43	86
63	18	36	25,000	44	88
80	19	38	32,000	45	90
100	20	40	40,000	46	92
125	21	42	50,000	47	94
160	22	44	63,000	48	96
200	23	46	80,000	49	98
250	24	48	100,000	50	100
320	25	50			

TABLE G.2   Five Place Logarithms

No.	Logarithm	No.	Logarithm	No.	Logarithm	No.	Logarithm
0.00005	−4.30103	26	1.41497	66	1.81954	250	2.39794
0.0001	−4.00000	27	1.43136	67	1.82607	275	2.43933
0.0003	−3.55287	28	1.44716	68	1.83251	300	2.47712
0.001	−3.00000	29	1.46240	69	1.83885	325	2.51188
0.003	−2.52287	30	1.47712	70	1.84510	350	2.54407
0.01	−2.00000	31	1.49136	71	1.85126	375	2.57403
0.1	−1.00000	32	1.50515	72	1.85733	400	2.60206
0.2	−0.69897	33	1.51851	73	1.86332	425	2.62839
0.3	−0.52287	34	1.53148	74	1.86923	450	2.65321
0.4	−0.39794	35	1.54407	75	1.87506	475	2.67669
0.5	−0.30103	36	1.55630	76	1.88081	500	2.69897
0.6	−0.22185	37	1.56820	77	1.88649	525	2.72016
0.7	−0.15490	38	1.57978	78	1.89209	550	2.74036
0.8	−0.09691	39	1.59106	79	1.89763	575	2.75967
0.9	−0.04575	40	1.60206	80	1.90309	600	2.77815
1	0.00000	41	1.61278	81	1.90849	625	2.79588
2	0.30103	42	1.62325	82	1.91381	650	2.81291
3	0.47712	43	1.63347	83	1.91908	675	2.82930
4	0.60206	44	1.64345	84	1.92428	700	2.84510
5	0.69897	45	1.65321	85	1.92942	725	2.86034
6	0.77815	46	1.66276	86	1.93450	750	2.87506
7	0.84510	47	1.67210	87	1.93952	800	2.90309
8	0.90309	48	1.68124	88	1.94448	850	2.92942
9	0.95424	49	1.69020	89	1.94939	900	2.95424
10	1.00000	50	1.69897	90	1.95424	950	2.97772
11	1.04139	51	1.70757	91	1.95904	1,000	3.00000
12	1.07918	52	1.71600	92	1.96379	2,000	3.30103
13	1.11394	53	1.72428	93	1.96848	3,000	3.47712
14	1.14613	54	1.73239	94	1.97313	4,000	3.60206
15	1.17609	55	1.74036	95	1.97772	5,000	3.69897
16	1.20412	56	1.74819	96	1.98227	6,000	3.77815
17	1.23045	57	1.75587	97	1.98677	7,000	3.84510
18	1.25527	58	1.76343	98	1.99123	8,000	3.90309
19	1.27875	59	1.77085	99	1.99564	9,000	3.95424
20	1.30103	60	1.77815	100	2.00000	10,000	4.00000
21	1.32222	61	1.78533	125	2.09691	20,000	4.30103
22	1.34242	62	1.79239	150	2.17609	30,000	4.47712
23	1.36173	63	1.79934	175	2.24304	40,000	4.60206
24	1.38021	64	1.80618	200	2.30103	50,000	4.69897
25	1.39794	65	1.81291	225	2.35218	100,000	5.00000

Index